Environmental Consequences and Management of Coastal Industries

Terms and Concepts

Estuarine and Coastal Sciences Series: Volume 3

Environmental Consequences and Management of Coastal Industries

Terms and Concepts

Edited by

Michael Elliott

International Estuarine & Coastal Specialists Ltd., Leven, United Kingdom

School of Environmental Sciences, University of Hull, Hull, United Kingdom

Andrew Wither

National Oceanography Centre, Liverpool, United Kingdom

Series Editors

Steve Mitchell and Michael Elliott

ELSEVIER

Elsevier
Radarweg 29, PO Box 211, 1000 AE Amsterdam, Netherlands
The Boulevard, Langford Lane, Kidlington, Oxford OX5 1GB, United Kingdom
50 Hampshire Street, 5th Floor, Cambridge, MA 02139, United States

Copyright © 2024 Elsevier Inc. All rights reserved.

No part of this publication may be reproduced or transmitted in any form or by any means, electronic or mechanical, including photocopying, recording, or any information storage and retrieval system, without permission in writing from the publisher. Details on how to seek permission, further information about the Publisher's permissions policies and our arrangements with organizations such as the Copyright Clearance Center and the Copyright Licensing Agency, can be found at our website: www.elsevier.com/permissions.

This book and the individual contributions contained in it are protected under copyright by the Publisher (other than as may be noted herein).

Notices

Knowledge and best practice in this field are constantly changing. As new research and experience broaden our understanding, changes in research methods, professional practices, or medical treatment may become necessary.

Practitioners and researchers must always rely on their own experience and knowledge in evaluating and using any information, methods, compounds, or experiments described herein. In using such information or methods they should be mindful of their own safety and the safety of others, including parties for whom they have a professional responsibility.

To the fullest extent of the law, neither the Publisher nor the authors, contributors, or editors, assume any liability for any injury and/or damage to persons or property as a matter of products liability, negligence or otherwise, or from any use or operation of any methods, products, instructions, or ideas contained in the material herein.

ISBN: 978-0-443-13752-5
ISSN: 2589-6970

For information on all Elsevier publications
visit our website at https://www.elsevier.com/books-and-journals

Publisher: Candice Janco
Acquisitions Editor: Maria Elekidou
Editorial Project Manager: Teddy Lewis
Production Project Manager: Sruthi Satheesh
Cover Designer: Mark Rogers

Typeset by STRAIVE, India

Dedication

This volume is dedicated to the late Dr. Roger Bamber who passed away far too young; Roger was a dedicated invertebrate taxonomist who had worked for the Natural History Museum, London, in which he described and named several new species. He also had a lifetime of working on applied marine and estuarine problems and in advising industries and regulatory bodies. This volume contains a fraction of Roger's expertise and ideas.

Contributing authors

Roger N. Bamber (1949–2015)[†]
Formerly of ARTOO Consultants, BM (Natural History) Museum, London, United Kingdom
The Fawley Marine Laboratory, CEGB, Fawley, United Kingdom

Stephen Colclough
SC^2 Aquatic Consultants, Kent, United Kingdom

Keith R. Dyer
Independent Consultant, Emeritus Professor, University of Plymouth UK, Somerset, United Kingdom

Michael Elliott
International Estuarine & Coastal Specialists Ltd., Leven, United Kingdom
Emeritus Professor, School of Environmental Sciences, University of Hull, Hull, United Kingdom

Henk A. Jenner
Consultant, Aquator, Ijsselstein, The Netherlands

Peter Holmes
Formerly Head of Marine Science, Scottish Environment Protection Agency, Glasgow, United Kingdom

Andrew I.L. Payne
Formerly Cefas, Lowestoft, United Kingdom
"Layshla," Cheshire, United Kingdom

Colin J.L. Taylor
Formerly Marine Environment Manager, New Nuclear Build, EDF Energy, Inverness, United Kingdom

Andrew W.H. Turnpenny
Independent Consultant, Southampton, United Kingdom

Andrew W. Wither
Formerly Principal Marine Scientist, Environment Agency. National Oceanography Centre, Liverpool, United Kingdom

[†] Deceased

Contents

Acknowledgments		xiii
Note from the editors and contributors		xv
Introduction		xvii
Section 1	**Science background**	**1**
1	Role and adequacy of science	3
2	Argument mapping	5
3	Systems analysis and conceptual models	8
4	Estuaries	11
5	Lagoons	16
6	Hydrodynamics of the system	19
7	Physical terms—Sea level and tides	21
8	Physical terms—Currents and waves	23
9	Numerical hydrodynamic modeling	25
10	Sediment terms	29
11	Geomorphological terms	31
12	Carrying capacity of systems	33
13	Micro-, meio-, macro-, and mega-fauna and flora	36
14	Epiflora, epiphytes, and sessile and mobile epifauna and infauna	38
15	Plankton	40
16	Ichthyoplankton	43
17	Non-indigenous, alien, invasive, and other non-native species (NIS, AIS)	45
18	Colonization by non-native organisms	49
19	Mechanical, thermal, and chemical stressors	51
20	Baseline and reference conditions	54
21	Causes of and solutions to estuarine, coastal, and marine degradation	56
22	Ecosystem restoration	58
23	Ecohydrology and ecoengineering	64
24	Ecosystem resilience, resistance, recovery	67
25	Climate change and its effects	71
26	Phenology	74

Section 2	**Fisheries terms**		**77**
	27	Precautionary principle and approach	78
	28	Spawning-stock biomass (SSB or B)	80
	29	Recruitment	82
	30	Fishing mortality (F)	84
	31	Catch per unit effort (cpue)	86
	32	Maximum sustainable yield (MSY)	88
	33	Total allowable catch (TAC)	90
	34	Reference points	91
	35	Fisheries management	93
Section 3	**Impacts and assessment**		**97**
	36	Environmental assessment	98
	37	Marine processes and human impacts	102
	38	Biological and ecosystem health	106
	39	Activity-, pressures-, effects- and management response-footprints	110
	40	Hazards and risk	113
	41	Interactions between the industrial plant and the marine system	118
	42	Water and substratum quality considerations	120
	43	Eutrophication and organic wastes	123
	44	Determination of a significant effect	127
	45	Standards, objectives, indicators	129
	46	Temporal and spatial physical scales	136
	47	Appropriate assessment—Habitats regulations assessment, habitat risk assessment	139
	48	Ecotoxicology assessment	141
	49	Underwater sound	145
Section 4	**Cooling water**		**149**
	50	Cooling water and direct cooling	150
	51	Water abstraction	153
	52	Cooling water discharge guidelines	156
Section 5	**Impingement and entrainment**		**159**
	53	Source and receiving waters	160
	54	Impingement of biota	162
	55	Stationary trawlers	165
	56	Equivalent adult value (EAV)	167
	57	Acoustic fish deterrent (AFD)	171
	58	Fish recovery and return (FRR)	174
	59	Disposal of impinged material	176
	60	Entrainment (biota)	178

Section 6	**Biofouling**		**181**
	61	Microbial and macrobial fouling	182
	62	Settlement of planktonic organisms	188
	63	Peak times of settlement by fouling organisms	190
	64	Biology of fouling organisms	192
	65	Antifouling measures	194
	66	Biofouling control by chlorination	198
	67	BioBullets	201
	68	Heat treatment	203
	69	Microbially influenced corrosion (MIC)	205
Section 7	**Chemicals**		**209**
	70	Biocides	210
	71	Chlorination chemistry	213
	72	Electro-chlorination plants (ECPs)	218
	73	Continuous and pulse dosing	223
	74	Non-oxidizing residuals	226
	75	Chemicals which may be prohibited for discharge	229
	76	Microbial pathogens—Chemical interactions	232
	77	Corrosion control: Oxygen scavengers	235
Section 8	**Discharge plumes**		**241**
	78	Plume characteristics and behavior	243
	79	Scouring of the seabed	247
	80	Impingement and entrainment (physical)	249
	81	Regulatory mixing zone	251
	82	Thermal tolerances of organisms	254
	83	Temperature thresholds which determine spawning times	257
	84	Thermal plume constituents (excluding heat)	259
	85	Salinity tolerances of organisms	261
Section 9	**Monitoring**		**263**
	86	Management framework monitoring types and definitions	265
	87	Survey, experimental and modeling approaches	269
	88	Coastal plant environmental monitoring	272
	89	Compliance monitoring	275
Section 10	**Management background**		**277**
	90	Sustainable environmental management	279
	91	Marine, coastal, and estuarine activities	283
	92	Endogenic managed pressures and exogenic unmanaged pressures	286

	93	Cause-consequence-response frameworks (DPSIR, DAPSI(W)R(M) approaches)	**290**
	94	Socioecological system—Ecosystem services and societal goods and benefits	**294**
Section 11		**Governance and management**	**301**
	95	Governance of the coastal and marine environment	**303**
	96	Integrated Coastal Zone Management (ICZM)	**308**
	97	Administrative and regulatory aspects	**309**
	98	Nature conservation designations	**311**
	99	Environmental and operational managers	**315**
	100	Nature conservation bodies	**318**
	101	Discharge consent, permit, license, authorizations	**321**
	102	Breaching regulations	**324**
	103	UN Conference on Environment and Development (UNCED); Convention for Biological Diversity (CBD)	**326**
	104	Mitigation, amelioration, enhancement, compensation	**329**
	105	Sustainable solutions	**332**
	106	Ecological, socioecological, and socioeconomic valuation	**334**
	107	Habitats and species legislation—Example of the EU Directive	**341**
	108	Integrated management of catchment, transitional and coastal waters	**343**
	109	Integrated marine management	**347**
	110	BAT, BATNEEC, best practice, and Integrated Pollution Prevention and Control (IPPC)	**354**
	111	Marine infrastructure environmental management	**357**
	112	Marine licenses and Maritime Spatial Planning	**359**
	113	Land-based infrastructural development—Planning process	**362**
Index			**365**

Acknowledgments

The contributors express their heartfelt thanks to Ms. Karen Nicholson, IECS Ltd., and Ms. Lisa Tabois, Cefas, Lowestoft, for their extensive support in helping format the large amount of information. The editors express their thanks to Stephen Roast, EDF Energy, for his support and encouragement; to the production staff at Elsevier, and to their many colleagues and friends who contributed knowingly or unknowingly to the entries. Any remaining errors are solely the responsibility of the contributors. The editors also thank their families and especially their partners for their patience during the preparation of this book.

Note from the editors and contributors

In order to ensure accessibility, each entry in this volume is designed to be complete in itself, although links to other entries are given where these will be helpful. This does mean that some degree of repetition and duplication is inevitable.

Although the intention is that this work will be of general applicability worldwide, legislation and practices vary significantly between different territories. It is clearly impractical to cover all alternatives and therefore examples, particularly in the sections on Management and Governance, are based largely on United Kingdom and European practice, but it is hoped that the entries will still provide a useful guide to the types of issues that will be encountered elsewhere worldwide.

The references given with each entry are not designed to be comprehensive but were selected to be of most value to enable the reader to progress further according to their wishes.

Unless indicated otherwise the figures and photographs are provided by the contributors.

Where proprietary equipment or technologies are referenced, these are for illustrative purposes only and do not imply any recommendation on the part of the authors.

While the entries in the volume have been generated by extensive work by the contributors with and for coastal industries and their regulators, we would be pleased to hear from users of the volume regarding entries which should be added to any future editions.

Introduction

This volume covers the engineering, natural, and social sciences aspects related to coastal industries and their operation and management. It gives the background and concepts to these features, problems, and solutions. This book comprises a set of sections that include those general concepts and principles, features, and biotic and nonbiotic components. Included are also the model types from conceptual to numerical, empirical (based on experience) to deterministic (based on underlying theory).

Sections of the book:
1. Science background
2. Fisheries terms
3. Impacts and assessment
4. Cooling water
5. Impingement and entrainment
6. Biofouling
7. Chemicals
8. Discharge plumes
9. Monitoring
10. Management background
11. Governance and management

Although many of the examples are from European, and especially United Kingdom, practices, the information presented here is relevant to coastal industries in all global situations and so allows practitioners to assess the environmental consequences of the industries and the means of addressing those challenges. Originally conceived to aid designers and operators in the power generation sector, the text has been expanded and aims to cover many coastal industries, including petrochemical plants, harbors and ports, desalination plants, and sewage treatment works also; there are also many entries with relevance to renewable energy plants. The volume therefore indicates, defines, and illustrates the terms and concepts used worldwide. This is important as engineers, scientists, and regulators often have an imperfect understanding of the requirements of the others, especially as similar (in some cases identical) terms may have very different meanings or are used in different circumstances.

The volume emphasizes that designing, managing, regulating, and studying coastal industries require a background appreciation of the nature and activities of the plant and industry; the threats and pressures they produce; and the physics, chemistry, and ecology of the receiving environment. The practitioners also need an appreciation of the prevailing management and governance principles and practices. Hence, designing the optimum and more sustainable system will involve multiple disciplines covering engineering and the physical, chemical, biological, and social sciences. It is unrealistic to expect that any one party will be fully familiar with all these aspects, and available literature may involve terms and concepts that are unfamiliar. Therefore this volume aims to provide an accessible and systematic route through the various factors and issues. Entries are intended to be self-contained but cross-references to other entries are given where appropriate.

The book is designed as a set of 11 sections that follow this general introductory chapter. Each section contains a general introduction and then contains annotated and illustrated explanations of terms with selected references for further reading. The structure of the book has been designed such that the various entries are clearly and logically structured. Cross-referencing and the index should allow the user to rapidly find the most appropriate entry even if this is not initially selected. The blocks of topics are designed to help the plant manager, engineer, and sustainability manager, in discussions with regulators, to know where the regulators are coming from, and what their demands are. It is hoped that the users will include all those with an interest in the environmental effects of coastal industries whether in teaching, research, advice, and practice in environmental science and management, and including those in environmental consultancies and regulation. Therefore the book will help prepare industries for new terms and even moving terms from industry to environment and vice versa. As an introduction and illustration of coastal features, Fig. 1 shows industrial and other features on the coast or in an estuary. The example is taken from the Humber Estuary, Eastern England, but shows features recognizable worldwide.

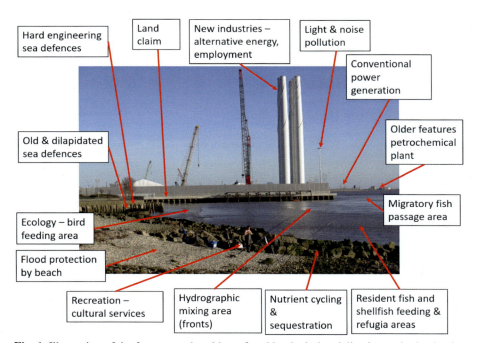

Fig. 1 Illustration of the features and problems faced by the industrialization and urbanization of estuaries and coastal areas.

The authors have built on many decades of experience working with and for regulators, industries, advisors, governments, etc.

The environmental framework

In the environmental and operational management challenges faced by coastal industries and their regulators there is only one big idea: *how to maintain and protect ecological structure and functioning while at the same time allowing the system to produce ecosystem services from which we derive societal benefits*. In other words: "*to look after the natural stuff and deliver the human stuff.*"

A coastal industry has a place within, and may be partly responsible for, what has been termed a "*Triple Whammy*" of current coastal problems that are associated with the need to cope with local, regional, national, international, and global problems. The three features are increasing industrialization and urbanization; an increased use of physical (space, energy, water, etc.) and biological (fish, shellfish) resources; and an impaired resistance and resilience to climate change and its attendant effects (temperature rise, ocean acidification, storminess, species distribution changes, the input of alien species, etc.).

The coastal manager will be subject to perceptions by the industry, the regulators, the public, etc. That manager will need to understand that, by definition, if something is of concern to a member or the public or regulator then by definition it is a problem to be dealt with, to be solved, or to be accommodated.

The coastal industry manager has the advantage of learning from many studies on the environment since the 1960s. These illustrate the importance of a knowledge of the natural system and especially ecohydrology, as the link between the physicochemical ecosystem features and its ecology within the human demands on that natural system. More recently, this has centered on the ability of the environment to produce ecosystem services that, after inputting human capital, can be turned into societal goods and benefits. The modern coastal industry manager and their regulators can learn from practice and experience worldwide such that there is no need to "reinvent the wheel" but rather to benefit from that experience elsewhere.

The effects of industries at the coasts, in estuaries, and the marine area also have to include the concepts of "unbounded boundaries" and "moving baselines." The former refers to problems or changes occurring anywhere in space and at any time; for example, a coastal manager should not just be mainly concerned with the biota in the immediate area, they then should be aware that this biota may rely on changes in the catchment (erroneously termed watershed in some areas) and further afield in the marine realm. Examples include the effects of coastal industries on wading birds and migratory fishes whose life cycle relies, respectively, on breeding sites in polar areas and intercontinental flyways or on the whole catchment and migrations routes to and from the sea.

"Moving baselines" refers to the problem that while a coastal industry environmental manager will be asked to judge the changes in the area in relation to their industrial activities, the features of the area will already be changing, notably through climate changes. Because of this, the coastal manager will be asked to be aware of the connectivity between areas in the ecology, hydro-physics, society, economy, etc., thereby compounding the complexity of fulfilling environmental objectives stipulated by the regulators. This connectivity and especially the connected nature of the oceans means, for example, that contaminants from a development in one area may be transported to and hence affect many other areas.

While the manager will be required to detect or predict change due to their activities, they should also be aware that inshore and estuarine communities are already adapted to a changing environment. As such, a naturally variable environment may be a "subsidy" to some previously acclimated organisms and allow them to thrive (as in estuaries) while being a stress to others that are poorly tolerant of variable environments. For example, a marine organism will be unable to tolerate freshwater from a discharge, whereas an estuarine organism will be able to tolerate this feature. This makes determining human-induced changes more difficult to monitor and manage and as such has been referred to as the so-called estuarine quality paradox.

The coastal industry manager will be faced with management and governance regimes and policies from the local to global—these include regional management schemes, diktats from both their own government and international blocs (such as the European Union), and regional seas conventions. Similarly, industries must deal with, and possibly satisfy, many bodies involved in environmental management: environmental protection agencies, nature conservation bodies, fisheries departments, other developers, municipal authorities, environmental health departments, port authorities, other industries, and nongovernmental organizations (NGOs).

An added complexity is that areas may also need to recover and cope with a historical legacy of previous activities although in many locations industries and activities are still endangering coastal and marine ecosystem functions. Hence, all countries need extensive legal and administrative frameworks in order to create a sustainable use of the environment that encompasses all sectors—industry, conventional and alternative energy generation, fishing, aquaculture, recreation, navigation, etc. Hence the national and international practices are aimed at increasing economic prosperity and the delivery of societal benefits while needing to protect human lives and livelihoods and safeguard the natural environment.

Industrial managers and their scientific advisors are confronted by a set of questions and challenges, some from environmental regulators regarding the effects of their industry on the environment, and others imposed by the market constraints of ensuring that an industry is sustainable. These constraints converge given that if an industry is not environmentally sustainable, then it is unlikely to obtain permission for its construction and operation. With regard to the industrial effect on the environment, both regulator and industry will be challenged to determine:

(i) the normal situation in their area and how it varies,
(ii) whether there has been or will there be a change in their area and if so, how big the change is in spatial and temporal terms, and then,
(iii) most importantly, what caused that change and whether (and should) something could be done about it, especially where this involves the actions of the industry.

Industry managers are therefore often charged by environmental regulators with determining the "signal-to-noise ratio," that is, what will be the change to the environment owing to their activities (the "signal") against a background of natural change or changes outside their control (the "noise").

The importance of understanding the natural features of coasts and estuaries is based on the axiom that if one does not understand the physics of the system and its physical structure, then there is little chance of understanding the ecological/biological behavior, components, and effects. As such, the primary duty of the industrial manager is to focus on protecting the health of natural and human systems within a successful and sustainable industry. This requires managers and their sustainability advisors to understand natural systems, especially the nature of the receiving environment including adjacent waters, sediments, and biota, that is, the "receptors" of their activities. This requires them to have a knowledge of their hinterland and catchment and its biology, physics, chemistry, engineering, governance, and administration. In essence, this focuses on the influence of the industry on the environment and the influence of the environment on the industry. Hence, at all stages, plant operators and their regulators should be trying to avoid problems before they occur, that is, by adopting the widely regarded precautionary approach.

Most industries need to understand the behavior of materials—physical, biological, and chemical and both solids and liquids—discharged by them into receiving waters and sediments, and hence the fate and effects of materials discharged. In turn, once those materials are discharged (either as particles or in solution), then there is the need to understand biological traits and behavior of organisms that may increase their exposure to toxicants and, once accumulated by organisms, the behavior of toxicants inside the organisms—whether they are stored (sequestered) or excreted. Such toxicants may be natural substances in unnatural amounts (e.g., copper) or unnatural substances in any amounts (e.g., PCBs). The effects then depend on the ability of an organism to tolerate, detoxify, store harmlessly, and/or excrete such materials. Accumulated contaminants may then be passed up the food chain with predation and so affect higher trophic levels. However, even "natural" materials may be regarded as toxic if in the wrong place or in the wrong amounts—for example, even freshwater may be a toxicant to marine organisms not adapted to freshwaters.

Roles and responsibility of industry

Industry has a set of obligations to help safeguard its environment and to monitor and assess change—while monitoring is not regarded as a management response measure, it is required to show whether management measures are required and whether any management measures have been successful. These aspects of monitoring and assessment include carrying out:

    (i) an Environmental Impact Assessment (EIA) (a process) linked to an outcome (an Environmental Statement as a precursor to being awarded planning permission);
   (ii) an Appropriate Assessment (for designated habitats and species);
  (iii) a Habitat Regulations Assessment, covering status and pressures;

- **(iv)** compliance monitoring (linked to license conditions) and a Cumulative Impact Assessment;
- **(v)** feeding data into a Strategic Environmental Assessment, linked to Maritime Spatial Planning;
- **(vi)** operational monitoring to determine plant performance;
- **(vii)** monitoring of their discharges to land, air, and waters, including for complex effluents, and linked to an Integrated Pollution Prevention and Control (IPPC) authorization (e.g., the H1 process of the UK Environmental Protection Agencies and the EpiSuite process of the US EPA counterparts); and
- **(viii)** a database toxicology assessment linked to discharge license creation, and showing the ability, fate, and effects of discharged materials to accumulate, be persistent, magnify, and/or be toxic.

The environmental manager of an industry should be aware of the assimilative and carrying capacities of the receiving environment. Assimilative capacity, as currently understood, is "the amount of an activity or activities allowed in a body of water before it adversely affects the quality." Similarly, carrying capacity can be regarded as "the ability of a body of water to support a given amount of activity or activities or ecological component."

The entries in this book show the importance of defining terms and having solid concepts, having environmental management decisions underpinned by good science, of having adequate training and resources, and increasingly having the ability to treat the industrial-environment system as a whole leading to systems analysis. This is what is now captured in the field of socio-ecological systems.

As emphasized here, ultimately, the environmental manager and their regulators are concerned with ensuring the health of the environment in which the industry is placed. In this, there are similarities between human (medical and well-being) and environmental health assessment, remediation of symptoms, and recovery (Box 1). As such, it is important to discuss environmental challenges in terms of the adverse signs and symptoms and change and even to refer to symptoms of "ecosystem pathology."

The environmental manager of a given industry will be concerned with each and every effect of their industrial activities from inception through construction and operation to decommissioning. The activities will each have a footprint (i.e., where and when the activities take place and where they are licensed), which then gives rise

---

Box 1 Equivalence between human and environmental health and well-being.

Medical/human well-being	=	Environmental
Diagnosis	=	Assessment
Prognosis	=	Prediction
Treatment	=	Remediation/creation
Recovery	=	Restoration
Prevention	=	Prevention

to perhaps larger pressures-footprints, as the mechanisms of effects, and then to perhaps even larger effects-footprints in space and time. However, the manager will also have to be concerned with the effects of other neighboring industries and indeed with changes inside an area due to factors outside the immediate area.

This complexity, of many activities, pressures, and effects all operating in an area, indicates the need to consider Cumulative Impacts Assessment (CEA). This has many challenges with the need to determine:

(i) ways of increasing our limited ability to measure the spatial and temporal effects-footprints of pressures from named activities;
(ii) the extent, duration, and frequency of the pressures from an activity, not just the activity itself in a given place at a given time (note that the term activity is not equivalent to the term pressure);
(iii) the relative effects of pressures from outside an area overlaying those caused by the industry itself;
(iv) the weighting given to the different effects-footprints in space and time, not just assuming they are added (synergistic) linearly and arithmetically (they could be antagonistic (cancelling) or even linked exponentially);
(v) knowing what is in an area, what activities are present, what receptors there are, and what is their relevance;
(vi) tackling the effects-footprints on the mobile receptors (mostly species, such as birds, fish, and marine mammals passing through an area) not just the sedentary receptors (habitats and species);
(vii) accepting the assumption that CEA relates to "all impacts of all activities" not just "all impacts of one activity/sector" (the latter is just an EIA carried out properly);
(viii) whether there is a tipping point or threshold to an undesirable state when all impacts are taken together and the effects-footprints overlap;
(ix) the process of the effect of an activity of an industry and its pressures increasing the chances of moving from an impact on the natural receptors to those on the human receptors, thus moving along the continuum from ecosystem structure and functioning, of ecosystem services, to societal goods and benefits; and
(x) finally, for larger water bodies, playing a role in tackling the conceptual difficulties in the continuum from an EIA to CEA to Strategic Environmental Assessment, Maritime Spatial Planning, and Integrated Coastal Zone Management.

The essence in management for an industry, whether environmental or operational, is undertaking risk assessment and risk management and, where possible, opportunity assessment and opportunity management. Risk assessment includes knowing:

(i) where the problems are and what changes they cause;
(ii) what their impact is on ecosystem structure and functioning;
(iii) what the repercussions are for ecosystem valuation based on economy-ecology interactions; and
(iv) what the future environmental changes and economic futures are for the industry.

Consequently, risk management includes knowing:

(i) what governance framework is there;
(ii) what stakeholders need, that is, what the priorities and concerns are of other stakeholders;
(iii) what successes and failures are;

(iv) what the industry, regulators, or society can do about the problems, hazards, and risks and how they could address them now and in the future; and
(v) how robust the decision-making is.

This requires knowing and using, as shown in the entries in this book, a hazard and risk typology showing what risks and hazards are created by industry and what are imposed on the industry by the natural and social environment. By definition, a hazard can occur in the environment, whether naturally or by human actions, and then it becomes a risk if it affects something that we value. Those risks can be exacerbated by human actions, for example, by removing natural coastal defenses thereby leading to greater coastal erosion.

Management response

As indicated before, the manner in which an industry influences its surroundings may be regarded as its footprint. There is foremost an activity-footprint that is that area and/or time, based on the duration, intensity, and frequency of an activity that ideally has been legally sanctioned by a regulator. This in turn gives rise to the pressures-footprint as the mechanism(s) of change resulting from a given activity or all the activities in an area once avoidance and mitigation measures have been employed (the endogenic managed pressures). These then further lead to the effects-footprint—the spatial, temporal, intensity, persistence, and frequency characteristics resulting from:

(i) a single pressure from a marine activity,
(ii) all the pressures from that activity,
(iii) all the pressures from all activities in an area, or
(iv) all pressures from all activities in an area or emanating from outside the management area; this includes the effects on both natural and societal systems (i.e., on the surrounding ecology and society).

Addressing these footprints then leads to the concept of a management response-footprint—the area and/or time covered by the management action and measures (or, in some legal instruments, the so-called Programme of Measures). Given the dynamic nature of the estuarine, coastal, and marine areas, this includes the areas covered by the distribution range of a species; this may cover the catchment of an estuary, areas at sea for migratory estuarine and diadromous fishes, or the high Arctic breeding grounds of estuarine and coastal birds.

As indicated here, the essence of the role of the industry manager and their regulators is in tackling cause-consequence-response chains and frameworks. In environmental management, this has been summarized as various acronyms, the most recent of which is DAPSI(W)R(M) (pronounced *dap-see-worm*). In this, the **D**rivers of human basic needs, for example, to get goods, enjoyment, or employment, requires **A**ctivities to be performed (fishing, industrial development, and tourism). In turn, these produce **P**ressures, as the mechanisms of change on the natural system (the State change) and on society, the Impacts (on human Welfare). Finally, these adverse changes require Responses (using management Measures) such as regulations and legislation carried out by environmental enforcement bodies.

The concept of IPPC (integrated pollution prevention and control) regulation will be familiar to coastal industry managers and is widely adopted. This is embedded in a set of principles that are adopted by countries worldwide. This means that the governance and management for the environment incorporates internationally recognized policies, politics, legislation, and administration, which involves horizontal and vertical integration of the management to accomplish ecosystem-based management (EBM) and hence the vision of the so-called ecosystem approach. Vertical integration of governance embraces all aspects from the local to global, whereas horizontal integration requires measures that cover all uses and users of the environment. The internationally adopted principles are as follows:

(i) ecologically sustainable development,
(ii) intergenerational equity,
(iii) the precautionary principle,
(iv) conservation of biological diversity and ecological integrity,
(v) ecological valuation and economic valuation of environmental factors,
(vi) the "damager debt" or "polluter pays" principle,
(vii) waste minimization, and
(viii) public participation—the role of individuals and ethics.

Given all of these aforementioned aspects, it is possible to emphasize the evidence needs by the regulators and the coastal industry managers and then to summarize the challengers faced by those actors (Box 2a and 2b).

Box 2a Evidence needs—Recipe leading to integrated marine management

- Need to understand how our activities lead to which pressures;
- Need to understand which pressures are within and outside our control;
- Need to understand ecological structure and functioning;
- Need to understand what state changes on the natural system occur from those pressures (effects-footprints);
- Lead to describing the impact on human welfare as effects on ecosystem services and societal goods and benefits;
- Lead to defining the appropriate responses as management measures;
- Require implementation of governance (defined as the summation of policies, politics, administration, and legislation);
- Within a multiuser system requiring resolution of conflicts among users;
- Communicated by working with stakeholders.

> **Box 2b A summary of the challenges and priorities for industry.**
>
> - Start off with SMART objectives (i.e., specific, measurable, achievable, realistic, and time-bounded);
> - Base management on good science;
> - Quantify the four footprints/cumulative impacts;
> - Emphasize that the natural system functions because of connectivity across all fields;
> - Collect data to use and use data collected;
> - Determine if management is working;
> - Have solid underpinning concepts;
> - Use ecological, socioecological, and socioeconomic valuation;
> - Harmonize the governance (policies, politics, administration, and legislation);
> - Focus on the global primary activity footprint for causes to climate change and the response activity footprint for the consequences.

Achieving solutions—Management actions and spatial scales

The management action of industries then includes many approaches (Box 3a) and at many scales (Box 3b).

The problems in the coastal environment imposed on and by industries can be regarded as of three types: materials put into the system, materials taken out of the system, and wider environmental changes such as climate perturbations. These create

> **Box 3a How are we managing industries?**
>
> - By management action
> - By developing programs of measures
> - By developing monitoring schemes
> - By linking monitoring to SMART indicators
> - By feedback to check if management is working
> - By implementing laws
> - By having many management bodies
> - By making industry get their house in order
> - By realizing the management footprint
> - By having visions, objectives, policies
> - By using good and fit for purpose science

> **Box 3b Where are we managing the environment?**
>
> - A small area (the activity-footprint)
> - A middle-sized area (pressures-footprints)
> - Middle to large areas (effects-footprints)
> - Whole estuaries
> - Whole catchments/river basins
> - Catchment-estuary-coastal areas
> - Seas and sea regions
> - Regional seas
> - Areas Beyond National Jurisdictions
> - The globe

pressures, the mechanisms of change, which occur within an area being managed or influenced by the industry in question (the endogenic managed pressures), and in which the causes and consequences of those pressures require to be managed. Alternatively, there may be exogenic unmanaged pressures in which the causes of those pressures emanate from outside the area being managed or being influenced by an industry but which create consequence for the area and the industry, that is, the management by an industry may have to be directed at the consequence rather than the cause. Today and in a given place, the industry will create endogenic pressures, but climate change will create exogenic pressures even though the industry has to deal with the consequences of both.

Industry has to comply with many different regulations and has many different regulators to satisfy. Most notably, industries will not receive a permit to operate unless they are prepared to tackle environmental aspects. Similarly, an industry will not operate sustainably unless it fulfills all aspects—hence the "environment of an industry" can be regarded as PESTLE—the political, economic, social, technical, and legal environment of the industry. This may be regarded as a shortened version of a set of ten features (the so-called 10-tenets Box 4) needed to fulfill sustainable and successful management measures and solutions. Such solutions provide a holistic framework covering the industry, its environment, and the local society.

Of greatest importance for any industry is its economic viability, but the underlying philosophy is that "what is good for the environment is also good for business and economy"—hence the mantra *"sustainable environment and sustainable business."* The industrial manager therefore is required to understand what problems are faced by the industry, whether from regulators, operational aspects, environmentalists, etc. In turn, the industrial manager will be required by or with the environmental regulator to define and implement solutions to problems actually or potentially created by the industry. It is of importance to note that only one of these ten-tenets relates to the natural system, its ecology, and that the majority of the tenets relate to societal, business, and governance aspects.

> **Box 4 The 10-tenets—Solutions to operational and environmental problems covering all aspects.**
>
> **Socially desirable/tolerable**—actions are desired (or at least understood and tolerated) by society;
> **Ecologically sustainable**—natural features and ecosystem services are safeguarded;
> **Economically viable**—costs are not prohibitive (using cost-benefit Analysis) and are sustainable;
> **Technologically feasible**—techniques and equipment for ecosystem and society/infrastructure protection are available;
> **Legally permissible**—the necessary regional, national, or international agreements and/or statutes are in place;
> **Administratively achievable**—statutory and nonstatutory bodies are in place and functioning;
> **Politically expedient**—approaches and philosophies are consistent with prevailing political climate and have support of political leaders;
> **Culturally inclusive**—local customs and practices are protected and respected;
> **Ethically defensible** (morally correct)—decision-making respects wishes and practices of individuals; and
> **Effectively communicable**—inclusive decision-making among stakeholders.

It is again emphasized here that determining the effects of industry on the environment and the effects of the environment on the industry focuses on cause, consequence, and response chains; a starting point for these is often to imagine the whole sequence as a conceptual model. This will cover the sequence from concept and planning (preconstruction) through construction and operation to decommissioning. At each stage, there is the need to carry out an Environmental Impact Assessment and produce an environmental statement in order to obtain an environmental permit. However, in addition to an EIA, industry will be confronted with other types of assessment—for example, a regulatory impact assessment, a health impact assessment, and an economic impact assessment.

It is axiomatic that the regulator does not have to demonstrate that an effect will occur but a developer or industry has to demonstrate that an unwanted effect will not occur. However, the planning process has the option to show that the effects are unavoidable and tolerated by society. The latter concept has led to the term IROPI—that an activity is allowed by Imperative Reasons of Overriding Public Interest irrespective of the environmental consequences. Despite this, it is very difficult to demonstrate the absence of an effect as there could always be an effect somewhere and at some time. In scientific terms, as the industry has to show no effect on biological health, whether of humans or other organisms, the effects could be at "any level of biological organization" from the cell, through individuals and populations, to biological communities and the ecosystem.

Similarly, those effects could be at spatial scales from the very local, through regional and national scales to global scales, for example, the results of fossil fuel combustion on climate change. Those effects may be short-lived or, conversely, last for decades or even centuries (as in the case of effluents containing radionuclides). Hence, in summary, industry has to be aware of and to cope with near, intermediate, and far-field effects and changes in short, medium, and long term and with short-, medium-, and long-term duration and frequency.

It is emphasized that industry is legally bound to follow regulations and statutory instruments although if an industry is subject only to guidelines, they have some discretion. Experience shows that if an industry is given a guideline, then it will discuss internally whether to follow it, whereas when presented with a piece of legislation that has legal repercussions if not followed, then the industry usually adheres to the legislation without question. However, based on wide analyses, practice shows that there is usually sufficient legislation within a country, but the challenge is often that it is not fully implemented or enforced.

The coastal industry manager is confronted by all environmental aspects of their industry and operations. Often this means that they are preoccupied by industrial waste and discharges to land, sea, and air, and so they need a good understanding of the behavior of those materials in relation to the behavior of the environment. That waste is regarded as contamination that may by definition then create pollution if the materials create harm to either the natural or social systems. Importantly, contaminants have been regarded as materials in the wrong place at the wrong time, a case of wasted resources that should be retained, reused, and/or recycled.

It is possible to take further the discussion about materials being added to the aquatic system and then causing changes with regard to infrastructure. A bridge or a harbor can be regarded as just a large particle put into the environment, each with its own behavior and which will change the local hydrodynamic patterns. In turn, these will change the sediments and the biotic communities. Similarly, any coastal industry, especially in the construction phase, may clear land of natural features, changing the physiography, and possibly the hydrographic patterns and water movements. This will also result in sediment pattern changes, with knock-on effects on the biota, both that biota inside a management area, and the biota passing through it and dependent on areas elsewhere.

Industry is regarded as having a set of obligations to monitor and to help to safeguard its environment—these include carrying out: EIA (a process) linked to outcome (an Environmental Statement) (as linked to planning permission); Appropriate Assessments (for habitats and species); a Habitat regulations Assessment; Status and pressures Monitoring and Cumulative Impact Assessment; feeding data into a Strategic Environmental Assessment, linked to Marine Spatial Planning; monitoring of their discharges to land, air, and waters linked to complex effluents and an IPPC authorization (e.g., the H1 process of the UK Environmental Protection Agencies and the EpiSuite of the US counterparts); and a Database toxicology assessment linked to discharge license creation (and showing the ability of discharged materials to accumulate, be persistent, magnify, and/or be toxic).

Furthermore, a coastal industry has to operate within the wishes and constraints imposed by a set of stakeholders, that is, those who have a stake in successful and sustainable industries and so are usually required to work together. Hence, it is important to know the categories of stakeholders and to know their roles and responsibilities. Such stakeholders have been grouped into several categories:

(i) those causing the problems (those putting materials, including infrastructure, into the environment and those removing materials, including resources and space, from the environment—the "*inputters*" and "*extractors*");
(ii) those regulating the environmental causes and consequences (the "*regulators*");
(iii) those affected by the problems and those benefitting for the actions (the "*affectees*" and the "*beneficiaries*");
(iv) those who have a role in influencing the outcomes—the policy-makers, nongovernmental organizations, advisors, researchers, and educators (the "*influencers*").

An integrated model

The earlier mentioned concepts and principles and those summarized in this book can be placed within an overall, integrated model (Elliott, 2023; Fig. 2). In this case, it is recognized that the ecosystem and its use by society has a central spine of a continuum going from the ecosystem having a suitable physicochemical structure and functioning to give an environment to be occupied by the biota which produces the ecological structure and biodiversity which in turn has its own functioning. The latter then creates a flow of ecosystem services which, after society inputs human capital, then produces provisions such as goods and benefits. That human capital is the ability to input time, money, energy, skills, knowledge, and the ability of being sentient, thereby obtaining goods and benefits.

The upper right-hand corner of the model indicates that drivers of basic human needs, such as the need for food, energy, shelter, satisfaction, and employment, are produced by societal activities, again the means of inputting human capital such as by creating coastal industries. The creation of societal goods and benefits then satisfies those basic human needs. However, our activities such as coastal industries have the potential to create pressures as adverse environmental effects which can lead to the degradation of both the natural and human systems. Those pressures can be created inside the management area (i.e., endogenic pressures, from inside the part of coast or estuary) or externally to a designated management area (i.e., the exogenic pressures created perhaps in a catchment or even globally). Those pressures will then lead to adverse environmental consequences such as deterioration of the natural system and loss of biodiversity or even reduction in the quality of human welfare (the upper half of the diagram which leads to showing the central spine to become gray indicating a degraded environment).

Such adverse environmental consequences then need to be prevented, addressed, mitigated, and reversed using responses by both the industry and its regulators, using

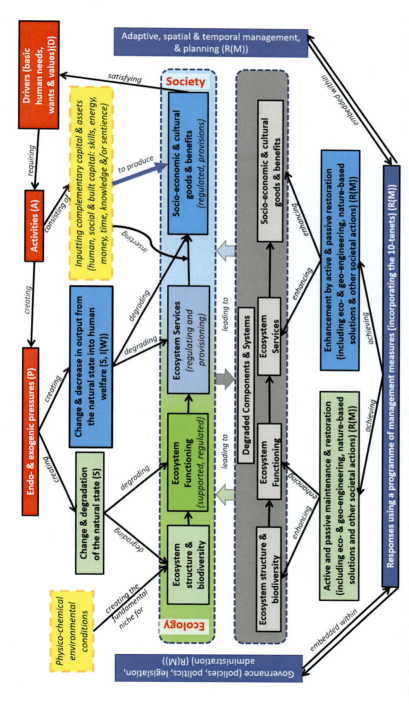

Fig. 2 The socio-ecological system showing the means of degrading and recovering the natural system, incorporating elements of the DAPSI(W)R(M) framework and the ecological structure and functioning to ecosystem services and societal goods and benefits continuum. Taken from Elliott, M., 2023. Marine ecosystem services and integrated management: "There's a crack in everything, that's how the light gets in"! Mar. Pollut. Bull. 193. https://doi.org/10.1016/j.marpolbul.2023.115177.

management measures. Of course, preventing environmental degradation in the first place is the most desirable but if this is not possible or successful then passive or active recovery and/or restoration techniques will be required. Passive responses may merely involve removing the cause of the pressures and allowing the degraded ecological system to recover. However, if this is not successful then active recovery and restoration will be required, by geo-engineering or ecoengineering, the latter now commonly termed nature-based solutions (as shown in the lower part of the diagram). Ecoengineering may be of two types—Type A in which the suitable habitat is created which then allows recruiting organisms to regain a population, or, if Type A is not successful, then Type B ecoengineering in which organisms are artificially introduced through restocking, replanting, or reseeding.

In turn, to have a successful and sustainable management then management measures and solutions implemented by the industry managers and their regulators will need to encompass the earlier described 10-tenets, given the governance, economic, and technological instruments and approaches at the disposal of the industry and its regulators. The governance should encompass the policies, politics, administration, and legislation under which the regulators operate. The coastal industry will then require adaptive management at the industry level but also it will be subject to adaptive management by the regulators.

In conclusion, it is hoped that this framework, together with the concepts and terms presented in this volume, will enable coastal industry managers to cope with the plethora of environmental management aspects and the demands of environmental regulators in order to achieve a sustainable and successful business but within a healthy, productive, diverse, and clean sustainable environment.

References

Elliott, M., 2023. Marine ecosystem services and integrated management: *"There's a crack, a crack in everything, that's how the light gets in"*! Mar. Pollut. Bull. 193. https://doi.org/10.1016/j.marpolbul.2023.115177.

Section 1

Science background

Introduction

This section covers the main philosophies relating to the role and adequacy of science in answering questions posed by new and existing coastal industries. It is argued that tackling any environmental problem is essentially a risk assessment and a risk management process in which a rigorous approach is needed toward defining and addressing the problem in hand, the challenges, and the solutions. As a means of achieving this, and to ensure that science covers the main elements required by management, there is the need to undertake argument mapping and derive conceptual models which summarize the main features of the system under study or requiring management. In this way the main issues are identified but within a framework suitable for natural and social scientists and managers alike. Increasingly, there is the need to quantify underlying natural, social, and engineering systems and in turn to produce empirical, deterministic, or stochastic numerical models as tools for understanding, exploring, and/or predicting the direction of change or the features of areas under defined scenarios; each of these features is given later as a set of relevant terms.

The industries will occupy space on the coast and in lagoons or estuaries and often intrude into wetlands and intertidal areas. For example, power generating stations with a requirement for ample cooling water are often designed to be in estuaries or lagoons or on open coastlines, whereas petrochemical industries require good transport access and even pipelines for feedstock delivery into the plant. Most industries produce waste, whether to the land, water, or atmosphere, and so there is the need to define and describe the way in which any structure and materials discharged will affect the overall environment, especially the hydrodynamics of these systems. This not only influences the behavior of the plant on the natural system but also the natural system on the plant. These ecosystems, subject to a set of entries in this review, have their own set of individual characteristics and their ability to maintain natural biological and physicochemical features, what may be termed their environmental carrying capacity. That carrying capacity relates to the functioning of the main ecological components, including the different groups of plants and especially the fish and animals most likely to be affected by the power plants.

Given the above, it is necessary to define and explain the differing types and sizes of organisms likely to be affected by the industrial activities. This includes the micro-, meio-, macro-, and mega-fauna and flora, and the fauna and flora defined according to their habitat; for example, the epifauna/attached flora and infauna. Furthermore, as shown by the other entries, the biodiversity of areas is constantly changing, not least because of the colonization by alien and invasive species (AIS) and non-indigenous species (NIS). This is especially the case where structures protrude into the water column or where water is taken in for cooling. In the case of direct cooled systems, this requires a knowledge of the changing environmental conditions—both inside and outside of the industrial plant—in relation to rock, concrete, or metal surfaces preferred for colonization and in relation to the temperature regimes created if there are cooling water discharges.

The set of entries in this section defines the natural environmental issues involved in quantifying and addressing the problems faced by coastal industries such as petrochemical and power plants. These include mechanical, thermal, and chemical stressors and so the individual and cumulative effects of these need to be addressed. In determining the causes and consequences of human actions, it is necessary to define the baseline against which changes are judged together with any reference condition, that is, the desirable status which might be the aim of management actions. This then calls into question the resilience, resistance, and recovery of marine and estuarine systems subject to natural and anthropogenic stressors—all features which are given entries here.

Superimposed on the changes to the natural and built environments caused by coastal and estuarine industries and the effects of those environments on the industries are the repercussions of climate change. This might be regarded as creating a moving baseline in which the occurrence and timing of natural events make more difficult the detection of industry-mediated effects. Hence the entries in this section cover phenology and the effects of climate change on the marine and estuarine system in relation to the changes brought about by the presence and operation of the industries.

Science background: 1. Role and adequacy of science

There are precise science requirements to address the questions "what are the effects of an industry on the marine and estuarine system" and "what are the effects of the estuarine and marine system on the operation of the coastal industry." The outputs of any study have to be scientifically rigorous and they must be scientifically, operationally, economically, and legally defendable as there are many consequences of having the right (and wrong) information. The evidence has to be defendable in legal proceedings (e.g., a public inquiry, compliance with permits, licenses, and authorizations) and will often influence the costs of building and operating the industrial plant. However, the evidence has to be proportionate and in context, that is, it has to be "fit-for-purpose" and thus give the required answers but with the required level of accuracy and precision. "Fit-for-purpose" science also includes the need to consider a set of questions (see Table 1 below).

While there are many aspects which academically could be studied, there is the need to emphasize that there is only certain information, data, and understanding which are specifically required; what may be termed the "nice-to-know" aspects which must be separated from the "need-to-know" aspects. Scientific curiosity often dictates that many aspects are studied but the science program should be designed to answer specific questions rather than as an academic exercise especially given that the costs of investigations need to be borne by the industry; this has long been termed the "polluter-pays principle" but perhaps it should now be referred to as the "damager debt principle." For example, an Environmental Impact Assessment could assess every part and characteristic of the receiving environment but should be more precise than this, that is., it should be: "what is the effect of this activity, being carried out in this way and at this place and time, with this level of mitigation and communicated in this way."

It is arguable that the main questions being tackled by the science are "so what?"—what are the repercussions of knowing certain information, of gathering certain data, and "what if?"—if an activity goes ahead or the system changes (or is changed) in a certain way then what will be the short-, medium-, and long-term effects. Increasingly,

Table 1 Fit-for-purpose science (criteria used by the to prioritize research).

1. When is the scientific information required—now, mid-term, long-term?
2. For how long will the information be obtained—short-term, medium-term, long-term?
3. For what area is it required—immediate area, near-field, mid-field, far-field?
4. What is the cost of obtaining the information—low, medium, high?
5. What is the overall benefit—low value, medium value, high value?
6. What is the importance of the information—its absence will only hamper our understanding, will be inconvenient, will delay development and/or operation, will lead to questions from regulators, will slow down the development or prevent operation, will be a "show-stopper"?

the latter includes considerations of climate change and so "long-term" may be related to the life of the industrial plant (from planning through construction and operation and to decommissioning) or longer to cover the global changes over decades or centuries.

References and further reading

Bamber, R.N., 1990. Environmental impact assessment: the example of marine biology and the UK power industry. Mar. Pollut. Bull. 21 (6) 270–274.

Glasson, J., Therival, R., Chadwick, A., 2011. Introduction to Environmental Impact Assessment. Routledge, p. 416.

Linked

Environmental assessment

Science background: 2. Argument mapping

Argument mapping is a method of producing a diagram of an argument, in the form of reasoning, inferences, debates, or cases. Pictures and diagrams (infographics) are thought to be a clearer way to illustrate and understand complex themes, Davies (2010) gives a summary of research and development into the visual representation of information, and Farnham Street (2019) focuses on mind mapping and conceptual mapping across the natural and social sciences. There are various systems of mapping, such as concept mapping, mind mapping, and argument mapping, which are sometimes used interchangeably but which Davies (2010) differentiates between; argument mapping allows the display of inferential connections and evaluation of the structure and basis of the argument.

Various software packages are available for computer-aided argument mapping (CAAM). Fig. 1 (after Davies, 2010) demonstrates the structure of an argument

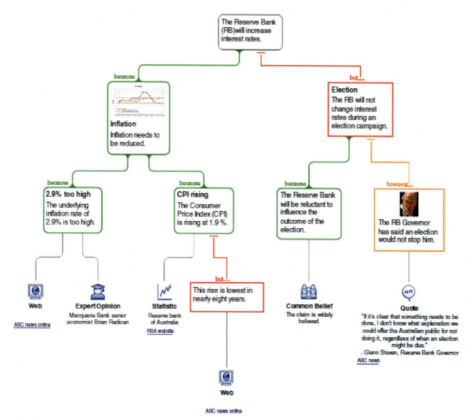

Fig. 1 Argument map using software rationale (Davies, 2010).

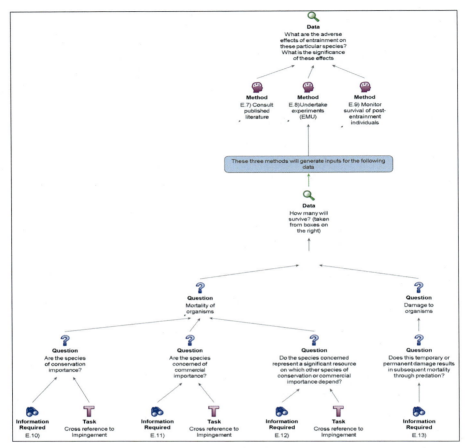

Fig. 2 bCisive™ argument map for consideration of entrainment impacts caused by cooling water abstraction.

map; at the top, there is a contention, followed by a supporting claim and an objection. Further claims, objections, and rebuttals follow these and the terminal boxes require evidence.

Fig. 2 gives an example of argument mapping using the software bCisive™ to keep track of arguments used in environmental case-making. It allows all the threads of a case to be tracked and provides an audit trail for any environmental position adopted in the power station planning process.

References and further reading

Davies, M., 2010. Concept mapping, mind mapping and argument mapping: what are the differences and do they matter? High. Educ. 62 (3) 279–301. https://doi.org/10.1007/s10734-010-9387-6.

Farnham Street, 2019. The Great Mental Models. vols. I–III Latticework Publishing Inc., Ottawa, ISBN: 978-1-9994490-0-1.

Twardy, C.R., 2003. Argument Maps Improve Critical Thinking. Argument. Citeseer.

Linked

Systems analysis and conceptual models

Science background: 3. Systems analysis and conceptual models

Coastal engineers and managers often need to structure the analysis of the ecosystem in relation to the cultural and social systems that drive governance and management processes to ultimately gain a better understanding of their interplay in managing coastal industries. This requires the use of the following definitions: a *system* is a set of things working together as parts of a mechanism or an interconnecting network; a set of principles or procedures according to which something is done; an organized scheme or method. A *process* is a series of actions or steps taken in order to achieve a particular end. *Management* is the process of dealing with or controlling things or people, and *governance* is the action or manner of governing through authority, decision-making, and accountability. *Governance* is the sum of the policies, politics, administration, and legislation systems under which an industry and the regulators have to operate.

Governing is having the authority to conduct the policy, actions, and affairs of a state, organization, or people. The term *framework* is used as a basic structure underlying a system, concept, or text and approaches can be separated into frameworks, theories, and models—hence coastal industry management is a framework that aims to identify the necessary components and the links between them prior to analyzing, interrogating, and then using the framework. This also allows the questions and hypotheses in sustainable resource management to be defined and addressed (see Elliott et al., 2020 for a discussion of systems analysis).

The integrated and comprehensive management of coastal and estuarine industries requires an understanding of their complex natural and social systems. It is argued that such a set of complex systems requires to be addressed via systems analysis. Systems analysis relies heavily on diagrams to show the complexity of the natural and human systems and its management; this could be taken further as there are possible mapping approaches of the system to represent system elements and connections, for example, *Actor maps*—covering which individuals and/or organizations are key players in the space and how they are connected; *Mind-maps*—which highlight various trends in the external environment that influence issues; *Issue maps*—covering the political, social, or economic issues affecting a given geography or constituency (often used by advocacy groups); and *Causal-loop diagrams*—interrogating the feedback loops (positive and negative) that lead to system behavior or functioning.

Mind mapping or conceptual models are regarded as an integral part of summarizing complex relationships within a system, what social scientists may call "wicked problems." They aim to present graphically (visually) the main components of a system, the pathways linking those components, and thus the nature of changes in one part of the system having repercussions for other parts of the system. While these are often qualitative diagrams, they form the basis of a quantitative assessment and thus future mathematical descriptions and predictive models; for example, conceptual models combined with Bayesian Belief Network Modeling can create descriptive and predictive systems. The boxes or nodes in a model may be regarded as the structural

Science background

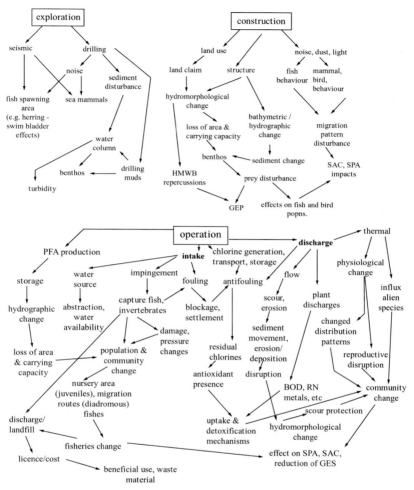

Fig. 1 Conceptual models relating to the repercussions of coastal power plants (exploration, construction, operation) (Elliott, 2012).

elements whereas the arrows (vectors) connecting these are functioning (rate processes) or cause-and-effect relationships (Fig. 1).

Owing to their inevitable complexity they have been referred to as "horrendograms" which may be regarded as being of greater value for the constructor than the reader. It is often more illustrative to present individual parts to the reader than to present the whole, for example to create a composite diagram showing all features and hence the complexity, but then deconstruct this in order to discuss, explain, and tackle the individual parts. As a related technique, the use of "Rich Pictures," a branch of Soft System Methodology (Avison et al., 1992), is increasingly used both as a tool for communicating complex ideas to stakeholders and the general public but also for determining, interrogating, and summarizing those ideas in semi-pictorial form. Similarly, there is an increasing amount of software to allow complex problem visualization thus solving, for example, bCisive (http://bcisive.austhink.com/).

References and further reading

Avison, D.E., Golder, P.A., Shah, H.U., 1992. Towards an SSM toolkit: rich picture diagramming. Eur. J. Inf. Syst. 1, 397–408.

Elliott, M., 2012. Preface—Setting the scene and the need for integrated science for the operational and environmental consequences of large industrial cooling water systems. In: Rajagopal, S., Jenner, H.A., Venugopalan, V.P. (Eds.), Operational and Environmental Consequences of Large Industrial Cooling Water Systems. Springer, Heidelberg/New York.

Elliott, M., Borja, A., Cormier, R., 2020. Managing marine resources sustainably: a proposed integrated systems analysis approach. Ocean Coast. Manag. 197, 105315. https://doi.org/10.1016/j.ocecoaman.2020.105315.

Gray, J.S., Elliott, M., 2009. Ecology of Marine Sediments: Science to Management. OUP, Oxford (260 pp.).

McLusky, D.S., Elliott, M., 2004. The Estuarine Ecosystem; Ecology, Threats and Management, third ed. OUP, Oxford, p. 216.

http://www.sydneycoastalcouncils.com.au/sites/default/files/systapproachfactsheet4.pdf.

Linked

Argument mapping; Cause-consequence-response frameworks

Science background: 4. Estuaries

An estuary is generally regarded as a water body transitional between the freshwater system and the coast and marine area, within which seawater is mixed with, and diluted by, the freshwater discharge from rivers and the land. An estuary generally forms the mouth of a river, or a riverine system, with a restricted width often forming a distinct boundary to the coastal area.

There has long been a debate regarding its definition of its limits but, as shown by the following references, there are three important components that need to be incorporated, namely coastal containment, the mixing of freshwater with seawater, and the presence of an estuarine biota. A general environmental definition can be stated as: "An estuary is a semi-enclosed coastal body of water which is connected to the sea either permanently or periodically, has a salinity that is different from that of the adjacent open ocean owing to freshwater inputs, and includes a characteristic biota" (Whitfield and Elliott, 2011). This definition encompasses those estuaries that close and are therefore not tidal, as well as those that are sometimes hypersaline, the latter being more common in arid and semi-arid areas.

The definition given before also includes coastal systems (e.g., certain estuarine lagoons and bays) that do not have inflowing rivers but where groundwater inputs provide conditions that support an estuarine biota. However, such definitions have been expanded to encompass water bodies worldwide which may be in arid areas. Hence, Potter et al. (2010) use the following definition to encompass the main characteristics of all estuaries: *An estuary is a partially enclosed coastal body of water that is either permanently or periodically open to the sea and which receives at least periodic discharge from a river(s), and thus, while its salinity is typically less than that of natural sea water and varies temporally and along its length, it can become hypersaline in regions when evaporative water loss is high and freshwater and tidal inputs are negligible.*

The mixing in estuaries is created by the action of tides in moving the water up and down the estuary and by the differential flow of the fresh and saltwater. Estuaries are thus distinguished by a gradient in salinity between the mouth and the river and a regular oscillation in its magnitude depending on the tidal phase. As a consequence, there is a distinctive ecology present that can tolerate these changes. There is a continuum of estuarine types depending on the relative magnitudes of the tidal flow and the river flow. With high relative tides the shape of the estuary becomes dependent on the tidal range when the coast is sedimentary, and extensive intertidal flats can be present, and the suspended sediment concentrations tend to be higher than in either the sea or the river. The highest suspended sediment concentrations occur at the turbidity maximum zone (TMZ) in the mid to upper region of an estuary. The latter and the consequent oxygen minimum (caused by the suspended sediment having an oxygen demand) affect the biological productivity and the chemical activity. With high river discharge the estuarine type depends upon the variability in the river flow and the effect of coastal sediment transport on the shape of the mouth. During low river flow seasons, the mouth may become closed, and a lagoon would form, a feature common in semi-arid areas in South Africa and Australia (see entry 5). Following these physical differences there are consequential biological differences. The parts of an estuary are illustrated as follows using the Humber Estuary, Eastern England, as an example:

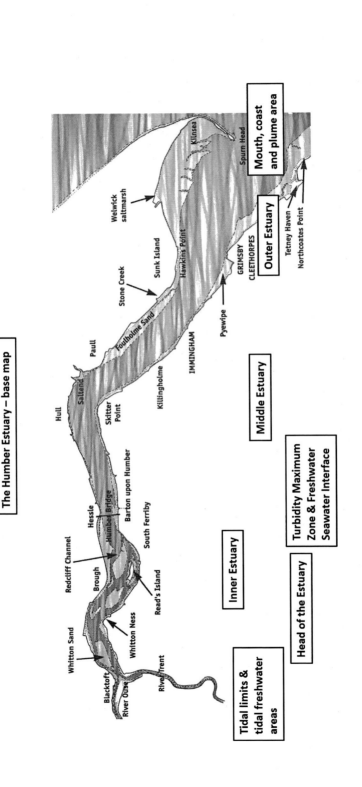

Under the European Union Water Framework Directive, estuaries are now regarded as a major type of transitional waters (e.g., Basset et al., 2012) and as such may be defined according to various disciplines: physical, chemical, biological, environmental quality, management, legal, and conservation (Elliott and McLusky, 2002). However, it is the physical characteristics that form the essential processes to which the other disciplines respond. It is emphasized that whereas all estuaries are transitional waters, the term transitional waters also includes fjords, fjards, rias, lagoons, deltas, etc., that is, not all transitional waters are estuaries (Table 1).

In order to summarize the current state of knowledge of estuarine systems, Elliott and Whitfield (2011) gave a set of paradigms based on estuaries worldwide (Table 2).

Table 1 The multi-disciplinary approach to defining estuaries.

Discipline	Character	Mechanism of definition
Physical	Physiography	Abrupt break between estuary and open coast
	Marine-based hydrographic processes	Penetration of tides (then subdivide by tidal regime)
		Penetration of marine waves
		Presence of density-driven currents
		Sea-derived sediment transport
Chemical	Salinity	Reduced from seawater but greater than freshwater
	Sensitivity	To nutrient enrichment
Biological	Community type	Penetration of marine plankton
		Presence of recognized sea fishes
		Migration route for diadromous fishes
		Presence of "estuarine" community
Environmental quality	Classification	Based on biology, aesthetics, chemistry, and with estuarine features
Management	Sea fisheries	Inland penetration of species "widely recognized as marine"
	Transition waters	Based on differences (by default—neither marine nor freshwaters)
Legal	With a recognized hinterland	Based on a catchment with a recognized coastline and transition area
	Area widely regarded as "estuarine"	Within a degree of precedence, an area regarded as such both by expert judgment and public perception
	Receiving a catchment	As the area where a river discharges
Conservation	Support of estuarine important biotopes and populations	As an area notable for its functioning (e.g., as for wading birds) or typical biotopes (e.g., saltmarshes, seagrass beds)

From Elliott, M., McLusky, D.S., 2002. The need for definitions in understanding estuaries. Estuar. Coast. Shelf Sci. 55(6), 815–827.

Table 2 Paradigms of estuarine features.

Paradigm 1: *An estuary is an ecosystem in its own right but cannot function indefinitely on its own in isolation and that it depends largely on other ecosystems, possibly more so than do other ecosystems.* Paradigm 2: *As ecosystems, estuaries are more influenced by scale than any other aquatic system; their essence is in the connectivity across the various scales and within the water body they are characterised by one or more ecotones.* Paradigm 3: *Hydromorphology is the key to understanding estuarine functioning, but these systems are always influenced by salinity (and the resulting density/buoyancy currents) as a primary environmental driver.* Paradigm 4: *Although estuaries behave as sources and sinks for nutrients and organic matter, in most systems allochthonous organic inputs dominate over autochthonous organic production.* Paradigm 5: *Estuaries are physico-chemically more variable than other aquatic systems but estuarine communities are less diverse taxonomically and the individuals are more physiologically adapted to environmental variability than equivalent organisms in other aquatic systems.* Paradigm 6: *Estuaries are systems with low diversity/high biomass/high abundance and their ecological components show a diversity minimum in the oligohaline region which can be explained by the stress-subsidy concept where tolerant organisms thrive but non-tolerant organisms are absent (cf. Estuarine Quality Paradox).* Paradigm 7: *Estuaries have more human-induced pressures than other systems and these include both exogenic unmanaged pressures and endogenic managed pressures. Consequently, their management has to not only accommodate the causes and consequences of pressures within the system but, more than other ecosystems, they need to respond to the consequences of external natural and anthropogenic influences.*

References and further reading

Basset, A., Barbone, E., Elliott, M., Li, B.-L., Jorgensen, S.E., Lucena-Moya, P., Pardo, I., Mouillot, D., 2012. A unifying approach to understanding transitional waters: fundamental properties emerging from ecotone ecosystems. Estuar. Coast. Shelf Sci. https://doi.org/10.1016/j.ecss.2012.04.012.

Dyer, K.R., 1997. Estuaries: A Physical Introduction, second ed. John Wiley, Chichester (195 p.).

Elliott, M., McLusky, D.S., 2002. The need for definitions in understanding estuaries. Estuar. Coast. Shelf Sci. 55 (6) 815–827.

Elliott, M., Whitfield, A., 2011. Challenging paradigms in estuarine ecology and management. Estuar. Coast. Shelf Sci. 94, 306–314.

McLusky, D., Elliott, M., 2004. The Estuarine Ecosystem; Ecology, Threats and Management, third ed. OUP, Oxford. (216 p).

McLusky, D.S., Elliott, M., 2007. Transitional waters: a new approach, semantics or just muddying the waters? Estuar. Coast. Shelf Sci. 71, 359–363.

Perillo, G.M.E. (Ed.), 1995. Geomorphology and Sedimentology of Estuaries. Elsevier, Amsterdam, p. 471p.

Potter, I.C., Chuwen, B.M., Hoeksema, S.D., Elliott, M., 2010. The concept of an estuary: a definition that incorporates systems which can become closed to the ocean and hypersaline. Estuar. Coast. Shelf Sci. 87, 497–500.

Whitfield, M., Elliott, M., 2011. Chapter 1.07: Ecosystem and biotic classifications of estuaries and coasts. In: Wolanski, E., McLusky, D.S. (Eds.), Treatise on Estuarine & Coastal Science. In: Simenstad, C., Yanagi, T. (Eds.), Classification of Estuarine and Nearshore Coastal Ecosystems, vol. 1. Elsevier, Amsterdam, pp. 99–124.

Whitfield, A.K., Elliott, M., Basset, A., Blaber, S.J.M., West, R.J., 2012. Paradigms in estuarine ecology—the Remane diagram with a suggested revised model for estuaries: a review. Estuar. Coast. Shelf Sci. 97, 78–90.

Linked

Integrated coastal zone management; Marine processes and human health

Science background: 5. Lagoons

Lagoons can have a variety of forms (Fig. 1). They can range in size from <1 ha to many km^2 in area, are relatively shallow, and are often separated from the sea by a barrier beach or barrier islands. There are two main groups: those that are connected to the sea allowing limited tidal exchange of water, and those that are isolated from the sea and depend upon rainfall or runoff to replace water lost by evaporation.

Group 1: Tidally connected lagoons with a permanent outlet to the sea. In this case water flows in during the flooding tide, but does not penetrate far because of the relatively low volume compared with the total lagoon volume. Consequently, there are likely to be strong horizontal gradients of salinity close to the inlet, and large areas of virtually freshwater elsewhere. Currents are strong at and around the inlet, but are low elsewhere.

Group 2: Isolated lagoons are those that depend upon seepage of water into underlying sedimentary rocks, or through a pervious barrier to the sea, to balance the input of freshwater from streams and rainfall. In hot climates, it is possible that evaporation may be exceeded by freshwater input at times, creating hypersaline conditions.

However, many lagoons exist in areas where connection to the sea may only be temporary, depending on seasonal variations in rainfall and river discharge. Additionally, the strength of the barrier preventing connection depends on the intensity of the littoral sediment transport. At times of strong winds and waves the barrier may become breached, especially when coinciding with precipitation. Such temporary closed features have been termed ICOLLs (intermittently closed and open lakes and lagoons) in Australia and Temporary Open/Closed Estuaries in South Africa. Although lagoons may be very biologically productive because of the trapping of nutrient and sediment inputs, their biodiversity is general poor and generally restricted to freshwater or brackish species. Their relative isolation makes them sensitive to alteration and anthropogenic disturbance.

Science background

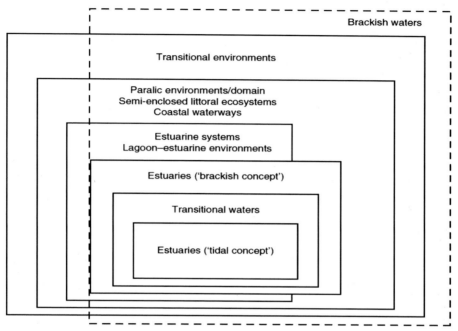

Fig. 1 Conceptual model linking different types of transitional water bodies.
From Tagliapietra, D., Sigovini, M., Ghirardini, A.V., 2009. A review of terms and definitions to categorise estuaries, lagoons and associated environments. Mar. Freshw. Res. 60, 497–509.

References and further reading

Basset, A., Barbone, E., Elliott, M., Li, B.-L., Jorgensen, S.E., Lucena-Moya, P., Pardo, I., Mouillot, D., 2012. A unifying approach to understanding transitional waters: fundamental properties emerging from ecotone ecosystems. Estuar. Coast. Shelf Sci. 132, 5–16. https://doi.org/10.1016/j.ecss.2012.04.012.

Kennish, M.J., Paerl, H.W., 2010. Coastal Lagoons: Critical Habitats of Environmental Change. CRC Press Taylor and Francis, Boca Raton, FL, p. 558.

McLusky, D.S., Elliott, M., 2007. Transitional waters: a new approach, semantics or just muddying the waters? Estuar. Coast. Shelf Sci. 71, 359–363.

Potter, I.C., Chuwen, B.M., Hoeksema, S.D., Elliott, M., 2010. The concept of an estuary: a definition that incorporates systems which can become closed to the ocean and hypersaline. Estuar. Coast. Shelf Sci. 87, 497–500.

Tagliapietra, D., Ghirardini, A.V., 2006. Notes on coastal lagoon typology in the light of the EU Water Framework Directive: Italy as a case study. Aquat. Conserv. Mar. Freshw. Ecosyst. 16 (5) 457–467.

Tagliapietra, D., Sigovini, M., Ghirardini, A.V., 2009. A review of terms and definitions to categorise estuaries, lagoons and associated environments. Mar. Freshw. Res. 60, 497–509.

Whitfield, A.K., Elliott, M., Basset, A., Blaber, S.J.M., West, R.J., 2012. Paradigms in estuarine ecology—the Remane diagram with a suggested revised model for estuaries: a review. Estuar. Coast. Shelf Sci. 97, 78–90.

Whitfield, M., Elliott, M., 2011. Chapter 1.07: Ecosystem and biotic classifications of estuaries and coasts. In: Wolanski, E., McLusky, D.S. (Eds.), Treatise on Estuarine & Coastal Science. In: Simenstad, C., Yanagi, T. (Eds.), Classification of Estuarine and Nearshore Coastal Ecosystems, vol. 1. Elsevier, Amsterdam, pp. 99–124.

Linked

Integrated coastal zone management; Marine processes and human health

Science background: 6. Hydrodynamics of the system

Hydrodynamics involves the description and quantification of the flow of water and the properties affecting its density, particularly salinity, temperature, and suspended sediment concentration, the distribution of which in turn affects the flow. All of these are likely to vary in three dimensions and at a variety of timescales. The hydrodynamics also involves the description of the bathymetry of the region under consideration at an appropriate resolution.

The flows are induced by both external and internal forces, and this will occur over a range of time and space scales. Thus the system will have a different scale depending on the time over which the hydrodynamics is examined. The forces and flows imposed by the surface water slopes are known as barotropic forces and currents. Those resulting from gradients of water density are known as baroclinic forces and currents.

The external barotropic forces are caused by (1) tides; (2) changes in atmospheric pressure, which alter the surface water slopes and cause surges; (3) wind stress on the sea surface, which also affects the surface water slopes; (4) waves and consequent wave setup; (5) river discharge imposed surface water slopes.

The internal baroclinic forces relate to the horizontal and vertical gradients in water density. These arise because of freshwater discharge and the resultant mixing, heating or cooling, or evaporation of the surface water by heat exchange with the atmosphere.

The flows resulting from the interaction of these forces are modified by friction at the seabed, and the differential flows of stratified layers which lead to mixing, in turn causing alteration of the horizontal and vertical density gradients. There is thus an interaction of the forces leading to a complex resultant balance.

The flows can be measured by following floats, the Lagrangian method, or by observations at a fixed location using stationary current meters or acoustic Doppler systems, the Eulerian method. These methods will only give the same results in steady uniform conditions, so that care has to be taken when comparing them. Taking averages of the velocities over the tidal period reveals steadier residual currents that are caused by longer term harmonics of the tide, by variations in the wind or atmospheric pressure, and by interactions with bathymetric changes. Temperature and salinity (density) gradients can be similarly measured by horizontal and vertical profiles allied to surveys or fixed station observations.

The hydrodynamics of the system, the evolution of currents, and the water density distribution as a result of the varying forces can be simulated by hydrodynamic numerical modeling. This simulation provides a means of examining the causes of observed measurements, and simulating the results of altering the imposed forces, and their effect on other factors, such as the ecology. It is emphasized that industries will need a good understanding of these processes in order to determine the fate and effects of any materials discharged into the surrounding waters.

References and further reading

Dronkers, J., 2005. Dynamics of Coastal Systems. Advanced Series on Ocean Engineering, vol. 25 World Scientific, New Jersey. 519 p.

Dyer, K.R., 1997. Estuaries: A Physical Introduction, second ed. John Wiley & Sons, Chichester. 195 p.

Svendsen, I.A., 2005. Introduction to Nearshore Hydrodynamics. Advanced Series on Ocean Engineering, vol. 24 World Scientific, New Jersey. 744 p.

Uncles, R.J., Mitchell, S.B. (Eds.), 2017. Estuarine and Coastal Hydrography and Sediment Transport. CUP, Cambridge, p. 351.

Linked

Numerical hydrodynamic modeling; Temporal and spatial scales

Science background: 7. Physical terms—Sea level and tides

Mean sea level can be calculated by averaging measurements of the sea level over at least a year to remove the main tidal variations. Mean sea level will vary due to eustatic changes, differences in the volume of the oceans due to temperature effects on density, or input of water from melting of land-based glaciers, and isostatic changes due to changes in the elevation of the land due to tectonic movements, or rebound from past glacial ice loading. At present, especially due to climate change effects and global warming, there is an annual eustatic increase in sea level of several millimeters which can be locally modified by other effects.

Tidal period is the time between subsequent high tides or low tides. The semi-diurnal tide has a period of about 12.5 h and is common around NW Europe and the Atlantic Ocean. There are also diurnal tides of about 24 h which are more common around the Pacific Ocean. Mixed tides can occur with semi-diurnal and diurnal tides dominating for different parts of the lunar cycle. In this section we will be emphasizing the semi-diurnal tide.

Tidal range is the difference in elevation between high and low tides, which occur at intervals of about 6.25 h during the semidiurnal tide in NW Europe. The tidal patterns show weekly, fortnightly, lunar monthly, seasonally, equinoctially, and annual cycles. The range varies between spring tides, when range is at a maximum, and neap tides, when there is a minimum range. The range at spring and neap tides varies during the year depending on the relative phases of the moon and the sun. Highest spring tides and smallest neap tides occur around the equinoxes. The tidal amplitude is half of the tidal range.

The highest tidal ranges occur in estuaries where the coastal configuration compresses the tidal energy and increases the high-water height. Estuaries can be classified according to the mean spring tidal range into hypertidal >6 m range, macrotidal <6 to >4 m range, mesotidal <4 to >2 m range, and microtidal <2 m range. The maximum tidal range has been measured in excess of 15 m. At the highest ranges the incoming flood tide may form a distinct series of waves as a tidal bore, especially in macrotidal estuaries where the narrowing and shallowing of the estuary with a progression inland causes the wave to increase in height; the tidal bore is especially prominent in the Severn Estuary, SW United Kingdom.

The tidal prism is the difference in volume of water in an estuary between high and low tides landwards of a particular cross section as far as the limit of tidal variation. The intertidal area is the area between high and low tide line. The area may be given for mean spring tides, maximum spring tides, or the highest astronomical tide. The tidal excursion is the distance traveling by a parcel of water from high to low water or vice versa. Hence the dispersion of effluent from any industry is dependent on all the features mentioned before.

References and further reading

Open University Course Team, 1999. Waves, Tides and Shallow-Water Processes. Butterworth-Heineman, Oxford.

Pugh, D., Woodworth, P., 2014. Sea-Level Science: Understanding Tides, Surges, Tsunamis and Mean Sea-Level. Cambridge University Press, p. 395.

Uncles, R.J., Mitchell, S.B. (Eds.), 2017. Estuarine and Coastal Hydrography and Sediment Transport. CUP, Cambridge, p. 351.

Linked

Estuaries

Science background: 8. Physical terms—Currents and waves

Residual or mean current is the magnitude of the current once the semidiurnal tidal variation has been removed by averaging over a complete tide. It generally shows the effects of longer term circulation due to weather, other tidal harmonics, and horizontal and vertical density gradients.

Dominant currents are the interaction of the semidiurnal tidal wave with shallow water and the coastline which then produces higher harmonics that are apparent by inequalities in the strength of the ebb or flood currents. Consequently, currents at a coastal or estuarine location may be ebb or flood dominant.

Surges are caused by variations of atmospheric pressure that can depress or raise the sea level. Rapid changes in the direction and magnitude of the sea surface winds also can force movement of the water as the wind changes. Their greatest effect occurs when the changes travel across the sea surface at about the same speed as the tidal wave. Surges can be exacerbated by the tidal state (especially spring tides), and the tide entering a funnel shaped, or a narrowing and/or shallowing area such as the North Sea, where the tidal wave, which travels in a north-south direction, can also be affected by strong north-south winds.

Swell waves are those waves created by winds at some distance from the point of observation. Once they travel outside their region of generation, they gradually increase in wavelength and decrease in height by a process of spreading and dissipation. Swell waves normally would have a period exceeding about 9 s. Because the longest period swell waves travel fastest, observed swell waves from an approaching storm would gradually become characterized by a shorter period.

Wind waves are those waves whose amplitudes are controlled by the local winds. However, there may be temporal and spatial delays in the response of the waves to the winds. Their height and period increase with fetch, the distance over which the wind blows, and with wind duration. The period of these waves is normally between about 2 and 9 s. In defining measured waves, the significant wave height (H_s), which is the mean of the highest one-third of the waves, and the zero-crossing period (T_z), the average period between instants when the surface rises above the mean water level, are terms used.

As the wave approaches the shallowing shore, the wave height increases eventually inducing a breaking wave. Wave breaking occurs in deep water when the generated waves achieve a crest angle sharper than about 120 degrees. When the water depth is less than about half of the wavelength of the wave, friction caused by the oscillatory motion of the near-bed water begins to slow down the wave progress, and it gains height, to eventually break when the wave height exceeds about 0.7 of the water depth.

When the wave approaches the shore obliquely, the slowing process will be greater in the area that is shallower water, and wave refraction occurs, with the crest swinging around to approach parallel to the shore.

The nearshore zone can be divided into a zone where refraction occurs. This comprises a surf zone between the break point and the shore, and the swash zone, which is defined by the highest point the breaking waves run up to on the beach, and the lowest point the water recedes to between waves. Because of the continual variation of the height of the breaking waves, there is also a wider breaker zone within which intermittent breaking occurs.

The breaking waves create a shoreward movement of the surface water, which is compensated to a degree by a seaward flow near the seabed. There is a consequent elevation of the mean water surface at the beach above the static level, termed wave setup. Wave setup is likely to be greatest on low gradient beaches where it can reach a third of the incoming wave height.

Within the surf zone, there are often circulation cells that discharge water created by longshore differences in the height of the breaking waves and by wave setup. These cells produce high velocity rip currents directed offshore at intervals along the beach. Rip currents can extend well outside the breaker zone. Longshore currents are produced within the surf zone by the combination of these circulation cells and the water movement caused by oblique approach of the waves to the beach.

As with all aspects of the bathymetry and hydrodynamics, the nature of the wave field in coastal areas near industries will affect the behavior of any materials discharged from industries. These features will affect the dispersion and dilution of materials, both in solution and particulate and covering physical, chemical, and biological materials.

References and further reading

Open University Course Team, 1999. Waves, Tides and Shallow-Water Processes. Butterworth-Heineman, Oxford.
Prandle, D., 1992. Dynamics and Exchanges in Estuaries and the Coastal Zone. Coastal and Estuarine Studies, vol. 40 American Geophysical Union. ISBN 0-87590-254-5.
Uncles, R.J., Mitchell, S.B. (Eds.), 2017. Estuarine and Coastal Hydrography and Sediment Transport. CUP, Cambridge, p. 351.
Zirker, J.B., 2014. The Science of Ocean Waves, Ripples, Tsunamis, and Stormy Seas. Johns Hopkins University Press. ISBN 978-1421410784.

Linked

Numerical hydrodynamic modeling

Science background: 9. Numerical hydrodynamic modeling

Knowledge of the hydrodynamics of the relevant estuarine and near coastal systems is fundamental for assessment of their effect in driving the environmental and ecological responses to change, whether natural or man-made. The prediction of the behavior of any discharged materials and of the natural and anthropogenic characteristics requires qualitative or especially quantitative numerical modeling. Numerical hydrodynamic modeling is the primary tool for carrying out this assessment, and the results have to be interpreted with a good scientific understanding of the processes involved.

Numerical hydrodynamic, or mathematical, modeling aims to simulate in time and in space the development of the flow and the properties of the water body as the boundary conditions change. It uses the Reynolds-averaged Navier-Stokes equations of mass, momentum, and energy conservation in iterative calculations that represent the effects of pressure fields created by the surface water slopes and the density differences due to salinity and temperature on the water flow and the mixing. This requires using assumptions about the rates of mass exchange and turbulent mixing that can be of different degrees of sophistication. The simplest technique is using coefficients of eddy diffusion and eddy viscosity that relate turbulent exchanges to the gradients of density and velocity, expressed similarly to coefficients of molecular diffusion or friction. The more common recent technique is to use the more sophisticated and realistic turbulent energy equation in so-called k-epsilon models.

Initially a choice has to be made of the number of dimensions used to represent the volume under consideration: 1 dimensional (1D), 2 dimensional (2D), and 3D models.

The 1D models average the conditions in a water volume cell defined over a width and depth between solid boundaries. The 2D models either average over depth (2DV) or over width (2DH). In both cases the assumption is made that the mixing of both mass and momentum can be effected by use of eddy coefficients. Thus the mixing coefficients will not have a precise physical definition, and the solutions to the equations can only be derived from comparisons with data, meaning that they are open to large errors. However, 2D regional scale models are often used to provide the boundary inputs for higher resolution localized three dimensional (3D) models, often termed "nested modeling."

Fully 3D models can represent the complicated bathymetry of estuarine and coastal areas and include detailed density and velocity gradients. In this case, the representation of the turbulent mixing processes is related to detailed quantification obtained from the turbulent energy equation, through theory, and from laboratory studies and measurements, and is therefore more robust. Another important consideration is how the vertical profiles of velocity caused by bed friction and density stratification are represented. Each of these models can then be developed further by increasing the time period under which the model operates.

A decision on the required resolution has to be made by the choice of the size of the elements of the grid and of the time steps within which the calculations are made.

Within each grid cell, the calculated values have to be assumed as constant so that the discrimination of gradients in calculated characteristics becomes a limit on the resolution of the calculations. In many cases, the grid size may be smaller near the coast or point of discharge of material and then larger at distance; this increases the resolution in areas where dilution and dispersion are greatest.

All models are set up using boundary conditions that require initial knowledge of the detailed bathymetry, and water level differences through space and time at the model limits. In addition, there is the need to add atmospheric pressure distribution and wind stress on the water surface, and the distribution of density at a number of points within the water body. Of course, the accuracy of the models depends upon the extent and accuracy of these initial observations. Parameterizations have to be assumed for the relationship between the near-bed water velocity and bed friction, and also for the efficiency of the turbulence in mixing the mass and momentum through the water column.

Using the defined boundary conditions and the set of comprehensive initial data, it is possible for the calculations to converge on a stable sequence of solutions representing the flow velocity and density distribution over the modeled area. These solutions can be stepped through time to represent the development through, for instance, a sequence of tidal periods. Validation can only be made by comparison with a separate observed calibration data set, and this can show large percentage errors. Testing enables these errors to be reduced by adjustment of the free parameters. Nevertheless, there are a number of other decisions that need to be made about the models used, including the computer resources available, which can limit the accuracy of numerical modeling. Vertical discretization depends on how the grid adapts to the changing water depths. Cartesian coordinates produce step-like changes from one cell to the next. Alternatively, use of sigma coordinates stretches the cells so that the sides are of unequal size, giving a smoother depth transition.

The specification of the turbulent mixing processes involved in the exchanges of mass and momentum can be examined by varying the definitions assumed for the turbulent mixing and the bed friction, and the results taken into account. There will be differences between the model and the calibration data that require decisions about the accuracy that is acceptable. Additionally, there are processes that may not always be included in the models, such as wind stress or waves. Consequently, the inherent errors in the results can be large and difficult to quantify. Despite these errors, it is still possible to use the models in scenario development and testing.

The simpler models use Cartesian grids with constant rectangular grid spacing, but recent developments have included curvilinear coordinates, variable grids, and alpha (logarithmic) vertical grids. As a result, modern computational schemes can cope with a higher amount of complexity, especially the large variation in temporal and spatial scales and with chemical and biological processes, and incorporate features such as effluent discharge plumes. Such models can also be used to predict pollutant concentration distribution and decay, and such parameters as suspended sediment concentration and movement, and sediment erosion and deposition, but all require thorough validation. Nevertheless, the additional complexity means that computational errors can accumulate, and these are difficult to define unless there is good validating data.

For use in estuarine and coastal situations, where there are many interacting hydrological driving forces, hydrodynamic models are best suited for testing and comparing hypotheses, examining sensitivity to variation in parameters, and in explaining observations, as well as testing hypothetical scenarios. However, modeling is the only technique that can be used to attempt prediction of the development of the system under changed conditions. For many applications, comparison of the results of several models can be enlightening and informative when the models include different processes, for example, with and without waves, or with and without inserted man-made structures. An example is shown in the accompanying figure of modeling a thermal discharge plume (Fig. 1). In this case the model has been run with and without the plume discharge and the calculated temperature differences (excess temperatures) presented.

There is a large and very specialized literature on the plethora of available mathematical models and their application in decision support systems for many purposes. Descriptions of some of the recent modeling schemes, often with free software, are accessible on the internet. They include Delft 3D (Delft 3D Suite), Mike 21 (DHI), Telemac (open TELEMAC), GETM (GETM—General Estuarine Transport Model), THREETOX (THREETOX model).

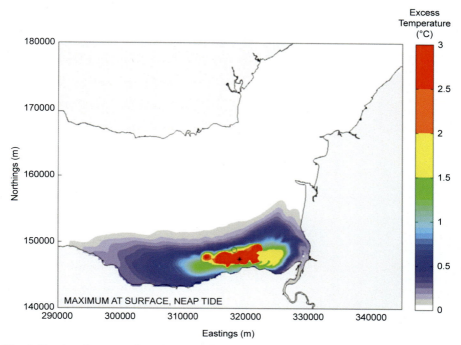

Fig. 1 Results of computed maximum surface excess temperatures for a discharge from Hinkley Point Power Station for neap tides in summer conditions.
From BEEMS technical report 085, Cefas, Lowestoft, United Kingdom.

References and further reading

Blumberg, A.F., Mellor, G.L., 1987. A description of a three-dimensional coastal ocean circulation model. In: Heaps, N.S. (Ed.), Three-Dimensional Coastal Ocean Models. Coastal and Estuarine Science, vol. 4. American Geophysical Union, pp. 1–19.

Hervuoet, J.-M., 2000. Telemac modelling system: an overview. Hydrol. Process. 14, 2209–2210.

Maderich, V., Heling, R., Bezhenar, R., Brovchenko, I., Jenner, H., Koshebutskyy, V., Kuschan, A., Terletska, K., 2008. Development and application of 3D numerical model THREETOX to the prediction of cooling water transport and mixing in the inland and coastal waters. Hydrol. Process. 22, 1000–1013.

Linked

Plume characteristics and behavior; Temporal and spatial scales

Science background: 10. Sediment terms

The type and behavior of sediment and the individual particles influence the behavior of materials discharged by industries. Natural and anthropogenic particulate materials therefore influence the water column and bed characteristics. Turbidity refers to colored and rather opaque water which is said to be turbid. This is caused by suspended particulate matter that may be both organic and/or inorganic, although color may also be conferred by the origin of waters, such as in peatlands discharging into estuarine catchments. Turbidity is quantified by measuring the attenuation of light through a given volume. When calibration is carried out using a semi-opaque standard liquid, relative magnitudes (turbidity units) result. When calibrated using simultaneously obtained filtered and weighed samples of the contained particles, concentration values ($mg\,L^{-1}$) can be interpreted. These will be of suspended matter concentration, or suspended sediment concentration, depending on how the samples are treated, and whether organic matter is measured separately.

Suspended sediment comprises those inorganic particles that are held within the water by turbulent exchanges of momentum that counteract the settling of the grains under gravity. The concentration of suspended sediment depends on the flow velocity and the availability of grains at the bed, the concentration contours moving up and down through the water column during the increase or decrease of tidal velocity. There is a threshold of suspension above which suspension occurs. Suspended particles are generally of fine sand, silt, or clay. Suspension concentrations of fine sand and silt are normally a few tens of $mg\,L^{-1}$ but can reach up to $10\,g\,L^{-1}$ especially in the turbidity maximum zone of highly turbid estuaries.

High concentration near-bed layers form by the settling of high concentrations of fine organic and inorganic particles. When the percentage of clay rises above about 5%–10%, cohesion between particles can become more important than gravity. Because of ionic charges on the particles (especially clay particles), and biological cohesion between them, individual particles can flocculate into low-density flocs that may attain diameters of centimeter scale with enhanced settling velocities. Settlement depends on the size, shape, and density of the particle and the current strength to allow settlement or to keep the particles in suspension (Stoke's Law). The inter-floc water is not easily squeezed out and the layers can remain above the bed where they may slowly dewater and consolidate, and become more resistant to reerosion. When suspended sediment concentrations reach the order of $g\,L^{-1}$ at peak flow, dense layers of a meter or more in thickness can occur at slack water and can persist over peak currents at neap tides. Occasionally these layers can be detected by high-frequency echo sounders and are then termed fluid mud.

Cohesive sediment refers to the threshold of erosion of sediment formed from settling flocs which is higher than that of sand and because of consolidation the sediment gains strength with time and with depth of burial. The organic content of cohesive sediment, generally called mud, is important in determining its physical and hydraulic properties as well as biogeochemical processes such as the sedimentary oxygen demand and the diagenetic processes, the post-depositional processes.

Bedload is formed of those particles moving along or close to the bed and driven by the shear stress of the water flow. The particles are generally of sand or coarser grains. The threshold of erosion at which movement commences is smaller than that for start of suspension.

Bedforms refer to the way in which sandy sediments are created by a sequence of bedforms generated with increasing shear stress above the threshold, with increasing wavelengths and amplitudes. They are termed ripples, dunes, and sand waves, and these bedforms increase the overall resistance of the bed to the flow. The wavelength of ripples is thought to be governed by the grain size, whereas dune size is related to water depth. However, the nomenclature and these size limits are a matter of intense debate. The larger features often show an asymmetry in their facing slopes which is used as an indication of their direction of transport. In areas where there is a deficit of sand, the moving sand will form into current-aligned narrow sand ribbons separated by bare rock. The ribbons often give way to patches of sand waves in the direction of sediment transport, thought to indicate slower transport rate. In accumulating (muddy) areas there are sometimes longitudinal features known as furrows, which are aligned with the direction of the maximum currents, with the direction of transport shown by a growth in size and a reduction in the number of furrows.

Sediment transport refers to the total mass of sediment moving through a unit cross section per unit time and hence giving the total mass transport rate. This may be separated into bedload and suspended load transport rates. It is generally considered that bedload transport is normally in the direction of the maximum current, and that suspended transport is in the direction of the residual current. In shallow water, these may not necessarily be the same direction. Near to beaches the longshore transport rate is generally integrated across the surf and swash zones. It can also be called the littoral drift. Although in all situations measurements may be made of the instantaneous transport rates at individual points, the longer term and tidally averaged rates are much more difficult to measure, and as a consequence their use in validation of numerically modeled results has to be carefully considered.

References and further reading

Dyer, K.R., 1986. Coastal and Estuarine Sediment Dynamics. John Wiley & Sons, Chichester. 342 p.

Komar, P., 1983/2018. Handbook of Coastal Processes and Erosion. CRC Press, Boca Raton, FL. 316 p.

Open University Course Team, 1999. Waves, Tides and Shallow-Water Processes. Butterworth-Heineman, Oxford.

Uncles, R.J., Mitchell, S.B. (Eds.), 2017. Estuarine and Coastal Hydrography and Sediment Transport. CUP, Cambridge, p. 351.

Whitehouse, R., Soulsby, R., Roberts, W., Mitchener, H., 2000. Dynamics of Estuarine Muds. Thomas Telford, London. 210 p.

Linked

Geomorphological terms; Hydrodynamics of the system; Water quality and substratum considerations

Science background: 11. Geomorphological terms

Banks: Offshore banks are generally composed of sand and occur where the sediment transport slows down and sand accumulates and hence may be areas of deposition of similarly sized particles, whether from natural or anthropogenic origins. Normally the sand circulates around and over the bank maintaining its position despite active sediment movement. Linear banks occur in areas of high tidal currents where there is convergence in the transport of sand from different directions. These banks generally occur between ebb and flood channels, channels on either side that have opposing predominant residual sediment transport directions. Shore-associated banks or Banner banks occur where the sand is held within the residual flow circulation cell in the lee of a headland (in some areas called a ness or point). A swale is a deeper saddle between two closely allied banks.

A Barrier beach is a long beach protecting low land behind. It is normally maintained by active longshore transport, so requires a sediment source at one end and a sink at the other. To reduce the overall transport and increase the width of the beach, groynes are often installed. A hydraulic groyne is a term that has been applied to the possible effect that an offshore structure, or disruption of the tidal water flow, may have on the longshore transport.

When the up-transport direction is constrained by a headland, there is a tendency for the beach to eventually develop a curve to face the approach direction of refracted waves, especially when there is another headland causing down-drift accumulation of sediment. The resulting outline is often called a logarithmic or zeta curve bay.

Beach profile is the profile of elevation taken in a normal direction down the beach from a known benchmark. These profiles will have step-like berms at a level that storm waves reach and at the spring tide height. Where the beach is not sufficiently high, storm waves may drive sediment across the top and deposit it on the back-beach through a process of overtopping, resulting in the beach intermittently transgressing back across the land behind. At a depth where the waves break there will be a breakpoint bar. This will be extensive in the longshore direction and will be the center of major longshore sand transport rates. Where two different significant wave heights occur at different times, perhaps due to the presence of an offshore bank, there may be two separate breakpoint bars at different depths. A common, though intermittent feature observed near high tide mark along beaches are beach cusps. They are regular crescentic forms of 10–100 m wavelength, often separating gravel horns from sandy bays. Their wavelength is thought to be governed by the wavelength of the incoming waves.

Tidal inlets occur where the drainage from the backing low-lying land breaks through a barrier beach to discharge into the sea. There may then be tidal water exchange through the inlet and the development of spits on either side resulting from converging longshore transport. The high tidal velocities carry some of the sediment from the longshore transport into the inlet on the flood tide, to be deposited as a flood delta, or out, creating an ebb delta. Nevertheless, it is possible that a proportion of the longshore transport may be exchanged across the mouth of the inlet through the ebb delta. When river discharge is small, the inlet occasionally may become blocked by sediment, and drainage may need to be maintained by the installation of a sluice to control water discharge at low water, or jetties, that allows water exchange throughout the tide.

Management terms:

> Sea level rise (SLR): the increase in the mean sea level caused by climate changes and other eustatic or isostatic processes.
> Coastal squeeze: the reduction in the width of the inter-tidal zone because of sea level rise, especially when the high tide is constrained by rising land or by coastal defenses.
> Set-back: the process of moving the coastal defense structures landwards at a rate equivalent to the rate of sea level rise to maintain the width of the intertidal zone.
> Managed retreat: the process of selectively setting back areas of low economic or societal value and protect areas of high value.

References and further reading

May, V.J., Hansom, J.D., 2003. Coastal Geomorphology of Great Britain. Geological Conservation Review Series, No. 28, Joint Nature Conservation Committee, Peterborough. 754 p. illustrations, A4 hardback, ISBN 1 86107 484 0.

Masselink, G., Hughes, M.G., 2003. Introduction to Coastal Processes and Geomorphology. Hodder-Arnold, London.

Pethick, J., 1984. An Introduction to Coastal Geomorphology. Edward Arnold, London.

Linked

Hydrodynamics of the system; Sediment terms

Science background: 12. Carrying capacity of systems

Industries both occupy areas in the environment and influence the surrounding areas with discharges and their wider activities. The environment may therefore be considered as having a capacity to accept those activities, especially without adverse effects being created. The term "carrying capacity" was originally used as an ecological concept, but is now used for both environmental and societal concerns, that is, what the natural system can accommodate, and what are the societal aspirations and requirements for a system. Baretta-Bekker et al. (1998) define it as "the maximum population size ... possible in an ecosystem, beyond which the density cannot increase because of environmental resistance." Similarly, the European Environment Information and Observation Network (EIONET) defines ecological carrying capacity as (i) the maximum number of species an area can support during the harshest part of the year or the maximum biomass that it can support indefinitely, (ii) the maximum number of grazing animals an area can support without deterioration (http://www.eionet.eu.int/gemet/concept).

Cohen (1997) similarly regards it as "the number of individuals in a population that the resource of a habitat can support," "the point at which the recruitment equals mortality," "the average size of a population that is neither increasing nor decreasing" or, as related to limiting conditions, "under steady state conditions, the population ... is constrained by whatever resource is in the shortest supply." In relation to commercial catchable stocks, Cohen (1997) gives five further definitions: population size at which the standing stock of animals is maximal, population size at which the steady yield of animals is maximal, animal population size being at that for maximal plants, the size of a harvested population that belongs to a sole owner, and the population size of an open access resource.

MacLeod and Cooper (2005) suggest carrying capacity is exceeded when population mortality exceeds recruitment because of environmental limitations (a stressor that a particular ecosystem can withstand before the ecological value is unacceptably affected)—a definition more widely adopted in fisheries science. However, they also acknowledge the difficulty of defining ecological value and unacceptable change—again implying a value judgment regarding what is acceptable change against a reference condition (see later).

Elliott et al. (2007) regarded it as "the maximal population (and/or community) that can be supported by the area's resources, principally space, food and reproductive partners." For temperate estuarine intertidal areas, a high carrying capacity can be their ability to support high numbers of over-wintering wading birds and/or juvenile fish. Hence, until recently, estuarine ecological carrying capacity was related to resources (principally food and space) available for use, a concept used more for wading birds than other organisms. Measures of both habitat quality and resource quantity are therefore needed to determine the population supported by an area, although, in the particular case of over-wintering bird populations, factors at their polar breeding sites away from temperate coasts will also have an influence.

The feature mentioned before is also true of certain species of fish such as sea bass *Dicentrarchus labrax*. Spawning takes place in coastal waters sometimes remote from the shallow intertidal waters which provide the optimal nursery habitats. Climate and fishing pressures at sea can have major impacts on the numbers of post-larvae appearing in estuaries in the summer months.

Where a resource such as food or space is limiting, it can be assumed that carrying capacity for birds is reached when one bird has to leave a site after the arrival of another. However, the idea of competitive interference between birds indicates that food resource competition alone cannot be used for determining carrying capacity as it underestimates the demands for space by birds, for example, the use of the concept for habitat compensation schemes. Modern studies of sea bass have demonstrated that the fish are very faithful to the same area of saltmarsh over their whole first summer. With over 80% of saltmarsh habitat lost over western Europe in the last 200 years, it is very likely that this lack of habitat is providing a constraint to further recruitment to the adult stock as this Lusitanian (warm water) species responds positively to climate change.

The term carrying capacity can also be used for societal aspects such as the ability of an area to support a given human activity or the sum of all the activities in an area. For example, a well-mixed, high energy area may have a high carrying capacity to absorb (and degrade, disperse, and assimilate) organic wastes without adverse effects being detected. This can also be described as the system assimilative capacity, a term often used to indicate the ability of an area to accommodate (again as in disperse, degrade, and assimilate) polluting discharges or industrial plant without damage. There is thus the need also to consider other definitions: Physical carrying capacity refers to space limitations, that is, the number of activities an area can withstand before there is some change to environmental quality, for example, number of berths in a marina. Social carrying capacity refers to the human population densities an area can sustain before numbers start to decline because of actual or perceptions of amenity decline, such as coastal tourism. Economic carrying capacity refers to the extent to which an area can become changed before the economic goods and services are adversely affected, for example excessive coastal development for tourism which reduces the desirability of the area.

Therefore, a composite definition is that:

Carrying capacity is the maximum number of users (population, community) that can be supported by the ecological or economic goods and services provided by an area. The aim of successful management and restoration therefore is to regain, maximize, or enhance the carrying capacity.

References and further reading

Baretta-Bekker, H.J., Duursma, E.K., Kuipers, B.R. (Eds.), 1998. Encyclopedia of Marine Sciences. Springer-Verlag, Berlin, Germany, p. 367.
Cohen, J.E., 1997. Population, economics, environment and culture: an introduction to human carrying capacity. J. Appl. Ecol. 34, 1325–1333.

Colclough, S., Fonseca, L., Astley, T., Thomas, K., Watts, W., 2005. Fish utilisation of managed realignments. Fish. Manag. Ecol. 12 (6) 351–360.
Elliott, M., Burdon, D., Hemingway, K.L., Apitz, S., 2007. Estuarine, coastal and marine ecosystem restoration: confusing management and science—a revision of concepts. Estuar. Coast. Shelf Sci. 74, 349–366.
Fonseca, L., Colclough, S., Hughes, R.G., 2011. Variations in the feeding of 0-group bass *Dicentrarchus labrax* (L.) in managed realignment areas and saltmarshes in SE England. Hydrobiologia 672 (1) 15–31.
Green, B.C., Smith, D.J., Grey, J., Underwood, G.J., 2012. High site fidelity and low site connectivity in temperate salt marsh fish populations: a stable isotope approach. Oecologia 168 (1) 245–255.
MacLeod, M., Cooper, J.A.G., 2005. Carrying capacity in coastal areas. In: Schwartz, M. (Ed.), Encyclopedia of Coastal Science. Springer, Heidelberg, p. 226.
McLusky, D.S., Elliott, M., 2004. The Estuarine Ecosystem: Ecology, Threats and Management, third ed. Oxford University Press, Oxford, p. 222.
Stillman, R.A., West, A.D., Goss-Custard, J.D., McGrorty, S., Frost, N.J., Morrisey, D.J., Kenny, A.J., Drewitt, A.L., 2005. Predicting site quality for shorebird communities: a case study on the Humber Estuary, UK. Mar. Ecol. Prog. Ser. 305, 203–217.
Van Cleve, F.B., Leschine, T., Klinger, T., Simenstad, C., 2006. An evaluation of the influence of natural science in regional-scale restoration projects. Environ. Manage. 37, 367–379.
Yozzo, D.J., Clark, R., Curwen, N., Graybill, M.R., Reid, P., Rogal, K., Scanes, S., Tilbrook, C., 2000. Managed retreat: assessing the role of the human community in habitat restoration projects in the United Kingdom. Ecol. Restor. 18, 234–244.

Linked

Biological and ecosystem health; Ecosystem services and societal benefits

Science background: 13. Micro-, meio-, macro-, and mega-fauna and flora

In determining the effects of coastal industries on the ecology and the effects of the ecology on coastal industries, it is necessary to distinguish the types and categories of organisms which may be affected, the so-called receptors or ecological components of environmental pressures created by activities in an area. Marine flora and fauna can be conveniently grouped into size classes for reference and as such are referred by these terms in licensing by statutory agencies. These size classes are not entirely arbitrary, but may relate to distinct taxa (biological groups) which exist within certain size ranges. The exception is the arbitrary distinction between macrofauna (from the Greek, meaning "long") and megafauna (from the Greek, meaning "large").

Microfauna: unicellular animals (and non-plant protists) are regarded as microfauna. Taxa regarded as microfauna are predominantly ciliates but also amoebae and smaller foraminiferans.

Microflora: unicellular algae and plants which may include diatoms and dinoflagellates; often reduced to the phytoplankton when in the water column or microphytobenthos (mpb) when on the bed or in the interstitial spaces; they may exude polymeric substances which can bind together the sediment grains.

Meiofauna: small multicellular animals, in practice those which predominantly pass through a 500-µm mesh, either by sieving bed sediment or the plankton. Taxa regarded as meiofauna are nematodes, foraminiferans (even though mostly unicellular), gastrotrichs, benthic copepods (mostly harpacticoids), rotifers, turbellarians, and certain families of polychaetes.

The separation based on mesh size is surprisingly not a totally arbitrary division, as those taxa accepted as meiofaunal (as listed before) generally all pass such a mesh, while those generally accepted as macrofaunal do not (at least as adults), although there are a few taxa which overlap this size boundary. Meiofaunal taxa are predominantly retained by a 30-µm mesh. In the deep sea, the size boundary between meiofauna and macrofauna is closer to 250 µm. Similarly, those taxa accepted as macrofaunal (as listed before) are generally all retained by such a mesh, while those generally accepted as meiofaunal are not, although there are a few taxa (such as the macrofaunal tanaidaceans) which overlap this size boundary.

Macrofauna: larger multicellular animals, which, at least as adults, are retained by a 500-µm mesh. Taxa regarded as macrofaunal are mollusks, crustaceans (other than copepods), almost all polychaetes, oligochaetes, sipunculans, cnidarians, and echinoderms, and also the very large unicellular foraminiferans with coatings of sand such as Astrorhiza.

Megafauna: a term of convenience for "particularly large" macrofaunal animals, such as edible crabs, lobsters, most fish, all marine mammals, colossondeid pycnogonids, and many jellyfish. There is no defined "mesh size" by which to represent their scale. Megafaunal species have large extents, may occur in low densities, and can be very difficult to sample quantitatively although remote scanning and photogrammetry techniques have allowed better quantification.

Macroflora: this includes the large, multicellular algae (the red, green, and brown algae such as kelps, wracks, and filamentous reds) and the saltmarsh, reedbed, seagrass plants; by definition the macroflora are large enough to be seen with the naked eye; the term macrophyte is often used as a composite term for these organisms.

Megaflora: while this is not a commonly used term in estuarine and coastal literature and has been used to describe the particularly large trees, it may be used to cover the mangrove trees and the willow and alder of upper estuarine wetlands.

References and further reading

Eleftheriou, A., McIntyre, A.D., 2005. Methods for the Study of Marine Benthos, second ed. Blackwells, Oxford.

Gray, J.S., Elliott, M., 2009. Ecology of Marine Sediments: Science to Management. OUP, Oxford, p. 260.

Levinton, J.S., 2021. Marine Biology: Function, Biodiversity, Ecology, sixth ed. OUP, Oxford.

What are Marine Macrofauna? https://www.utas.edu.au/about/news-and-stories/articles/2017/467-explainer-what-are-marine-macrofauna.

Linked

Biological and ecosystem health; Ecosystem services and societal benefits

Science background: 14. Epiflora, epiphytes, and sessile and mobile epifauna and infauna

Knowing where organisms live or can live given the appropriate conditions is important for determining the effects of materials discharged or structures created by coastal industries. Organisms may therefore be classified according to their biological characteristics, termed biological traits (Fig. 1). Hence, an ecological community can be classified by the taxonomic names or organisms or by those traits and groupings. Epiflora is a less commonly used term for plants and algae growing on the bed of the estuarine and coastal area, whereas epiphytes are plants and algae growing on another plant or alga. It is generally accepted that the plant or alga uses the substratum (the bed or other organism) for support and hence differentiates these from parasitic plants or algae.

Species of animal which live on another plant, animal, or the substratum are epifaunal. Serpulid polychaetes, barnacles, and sea mats (bryozoans) are epifaunal on algae, as well as on rock and man-made structures. Species such as these, which are attached to their substratum, are regarded as sessile epifauna. More motile species in such habitats, such as pycnogonids, nudibranchs, certain amphipods, or mites, are mobile epifauna. Species of the benthos which live on the surface of sediments, as opposed to within the sediment, are also epifaunal. Spider crabs are epifaunal on soft sediments, as are hermit crabs, some phoxichilidiid pycnogonids, sand gobies, and the common shrimp *Crangon crangon*.

Species of animal which live within a soft sediment or harder substratum are infaunal. Most species associated with muds and sands are infaunal, including those which live within the sediment surface layer and feed above it (sea pens, cumaceans, masked crabs, burrowing bivalve mollusks, red band fish, filter-feeding polychaetes, etc.). This also includes those organisms which burrow deeper and spend their entire lives away from the surface other than for releasing gametes (spatangoid urchins, paraonid polychaetes, caudofoveata mollusks, etc.). Species which live within the interstices of coarser substrata (clean sands and gravels) are also infaunal.

Most macroinfaunal species construct some sort of tube or burrow, often mucus lined, often temporary; infaunal species without a coelom do not and tend to be interstitial. Piddocks, gribbles, most species of the boring polychaete worm *Polydora*, and shipworms are also infaunal, in these cases in hard substrata (wood, clay, or rock).

Animals living inside other animals or the structures created or used by those animals are regarded as commensals, for example, certain polychaete worms which inhabit the same shell as hermit crabs. Those animals dependent on other animals and possibly causing harm to the organism are regarded as parasites. Some limnoriid isopods are infaunal within the blades of seagrass, although as they eat the seagrass, this too may be regarded as parasitism or merely a form of herbivory.

Sessile animals are those that remain fixed in one place, such as barnacles, whereas sedentary animals have some limited movement such as mussels and limpets; the terms sessile and sedentary are often used interchangeably but constitute the main

Science background

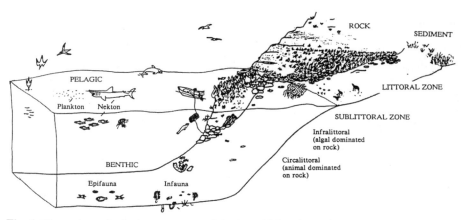

Fig. 1 The main ecological components of the intertidal and nearshore subtidal system (Hiscock, 1996).

groups of fouling organisms. In contrast mobile, motile, or errant organisms are usually unconstrained in their movement. Sedentary and sessile organisms usually have a motile or mobile dispersive stage.

References and further reading

Gray, J.S., Elliott, M., 2009. Ecology of Marine Sediments: Science to Management. OUP, Oxford, p. 260.
Hiscock, K. (Ed.), 1996. Marine Nature Conservation Review: Rationale and Methods, Coasts and Seas of the UK. Joint Nature Conservation Committee, Peterborough.
Levinton, J.S., 2021. Marine Biology: Function, Biodiversity, Ecology, sixth ed. OUP, Oxford.
Thrush, S.F., Hewitt, J.E., Pilditch, C.A., Norkko, A., 2021. Ecology of coastal marine sediments: form, function and change in the Anthropocene. OUP, Oxford.

Linked

Itcthyoplankton; Microfauna, meiofaunna, macrofauna and megafauna; Plankton

Science background: 15. Plankton

Given that many coastal industries discharge materials into the water column, there is possibly an adverse effect on the organisms in that water column. Similarly, materials discharged such as nutrients may act as a stimulant for the water column organisms. Furthermore, any changes to the water movements and the delivery of materials and dispersal stages of organisms will ultimately have ecological repercussions. Hence, it is important to have a knowledge of those organisms in the water column which constitute the plankton and nekton.

Planktonic organisms, though often capable of swimming, essentially float in the water column; their locomotion and dispersion are governed by the movement of the water body itself. They are commonly capable of vertical migration within the water column and can reposition themselves behaviorally to achieve directed migration by their passive floating.

Plankton predominantly comprises small (microscopic) organisms, including microalgae such as diatoms, protists such as dinoflagellates and ciliates, small crustaceans such as copepods, haustoriid amphipods, and branchiopods. In addition, this includes huge numbers and types of larvae of animal species which have non-planktonic adults, such as barnacles, bivalve and gastropod mollusks, bryozoans, and the eggs and larvae of some fish. There are also larger animals which spend their adult life in the plankton, notably medusae and siphonophores ("jellyfish"), mysids ("ghost shrimps"), and euphausiids ("krill"), as well as some polychaetes and gastropods.

Plant and algal species of the plankton are termed phytoplankton, whereas animal species are zooplankton with a further distinction according to size, hence picoplankton, microplankton, meioplankton, and macroplankton; the bacterial component of the plankton is termed bacterioplankton which are prokaryotes. Some of these are primary producers and others primary consumers in estuarine, coastal, and marine ecosystems. Notably, planktonic organisms are unable to cross significant density interfaces and so may be restricted by hydrodynamic fronts. So, for example, they would only be able to enter a thermal effluent plume as the plume water mixes with the ambient water, and thus the discharge becomes dilutes.

Holoplankton

Species of plant or animal which spend their entire life in the plankton are termed holoplankton. While a significant proportion of the organisms in the plankton is of larval or juvenile stages of animals and plants which are either benthic or pelagic as adults (meroplankton), the majority of species involved have planktonic juveniles and adults. The holoplankton includes microalgae such as diatoms, protists such as dinoflagellates and ciliates, small crustaceans such as copepods, haustoriid amphipods, and branchiopods, and larger animals such as siphonophores ("jellyfish"), ctenophores ("sea-gooseberries"), tunicates of the classes Thaliacea and Larvacea, mysids ("ghost shrimps"), and euphausiids ("krill"), as well as some polychaetes and gastropods. Some ctenophores can attain very large sizes, for example, the "Venus's girdle" (*Cestum*) reaching a length of over 1 m. The fish eggs, larvae, and post-larval stages in the plankton are termed ichthyoplankton.

The occurrence of holoplanktonic species is not seasonal, but their density may be. The holoplankton represents a major food resource, for pelagic feeders such as clupeid fish and baleen whales, for sessile (benthic) filter feeders such as barnacles and sponges, as well as for other members of the plankton.

Meroplankton

Species of plant, algae, or animal which spend only one stage of their life cycle in the plankton are called meroplankton. While the majority of planktonic organisms spend their entire life in the plankton (holoplankton), some species spend only part of their life cycle in the plankton. Normally it is egg and larval stages which are found in the meroplankton, such as barnacle nauplii (at times the most numerous members of the coastal plankton) and cypris larvae, decapod crustacean larvae (zoeae and megalopae of crabs and lobsters), bivalve and gastropod mollusk larvae (veligers and trochophores), some polychaete larvae, bryozoan cyphonautes larvae, echinoderm larvae (auricularia, pluteus, and bipinnaria), appendicularia larvae of some tunicates, and the eggs and larvae of some fish. Conversely, hydromedusae and scyphomedusae (jellyfish) are the planktonic adult stages of benthic hydrozoans ("hydroids," white weed) or scyphistomae, respectively.

Temporary planktonic life facilitates dispersion (pre-reproductive migration) by species which are otherwise sessile as adults or at least of restricted benthic locomotive ability. The occurrence of meroplanktonic species is mainly highly seasonal, particularly in temperate and polar waters. Large occurrences of the gelatinous forms, jellyfish and ctenophores, are often termed plagues and have been known to block cooling water screens in coastal power plants and other industries with seawater intakes, particularly if wind-driven currents push them shoreward. These planktonic stages will remain in the water column until metamorphosis enables them to locate a suitable area for settlement; that dispersal period may commonly last two to six weeks, during which time the larval stage could travel 40–60 km depending on the strength of local currents.

Coastal industries will need to know when organisms are reproducing, which stages are in the water column, and when planktonic stages of benthic species are likely to be settling on any surface, such as those inside cooling water intakes. The periods of planktonic and settlement stages of many temperate marine species are given in Rasmussen (1973).

References and further reading

Newell, G.C., Newell, R.C., 2006. Marine Plankton—A Practical Guide. ISBN-10 1-904690-41-6, Conservation Ltd (CD), Pisces.

Rasmussen, E., 1973. Systematics and ecology of the Isefjord marine fauna (Denmark). Ophelia 11, 1–507.
Raymont, J.E.G., 1980. Plankton & Productivity in the Oceans: Volume 1 Phytoplankton. Pergamon, Oxford.
Raymont, J.E.G., 1983. Plankton & Productivity in the Oceans: Volume 2 Zooplankton. Pergamon, Oxford.

Linked

Ichthyoplankton; Microfauna, meiofaunna, macrofauna and megafauna

Science background: 16. Ichthyoplankton

Ichthyoplankton are the eggs, larvae, and sometimes post-larvae of fish found mainly in the upper 200 m of the water column, also called the near-surface waters. The eggs are passive and drift along in the water body with the prevailing currents. Most fish larvae have almost no swimming ability initially; however, part way through their development they become active swimmers. Ichthyoplankton are a relatively small but vital component of total zooplankton. They feed on smaller plankton and are prey themselves for larger animals. Once they attain a swimming ability, they may undertake selective tidal stream transport (STST) in which they use the flood tides selectively to move into estuaries and then "rest" on the bed during the ebb tide to avoid being washed back out of the estuary. In this process they may travel the distance into or up an estuary equal to the tidal excursion (the distance moved by water from low to high tide).

In 1864 the marine biologist G.O. Sars investigating fisheries around the Norwegian coast found fish eggs, particularly cod eggs, drifting in the water. This established that fish eggs could be pelagic, living in the open water column like other plankton. Research interest increased when it emerged that, if ichthyoplankton was sampled quantitatively, the samples could indicate the relative size or abundance of spawning fish stocks. Determining the abundance of eggs and larvae in an area is usually less expensive than sampling the adults. For species such as sardine and anchovy, egg and larval counts are a good indicator of population size. Thus the egg and larval data can be used to monitor trends in population abundance of the adults and hence can be converted to "adult-equivalents." It is possible to tell when populations are declining, often more rapidly than if only the adults were monitored.

For species that are not captured by a fishery, monitoring their population trends by monitoring their eggs or larvae can provide an indication of a healthy or stressed ecosystem. It is unlikely that there would be estimates of the abundance, growth, or decrease of these species in any other way. However, it is emphasized that fish eggs and larvae, as with eggs and larvae of all organisms, have high mortality rates and so determining the mortalities during passage through cooling water systems is difficult due to the patchy and highly variable abundances of all plankton.

Christie et al. (2010) reported that fish larvae can drift on ocean currents and reseed fish stocks at a distant location. This work demonstrated what scientists have long suspected but have never proven that fish populations can be connected to distant populations through the process of larval drift.

References and further reading

Able, K.W., 2005. A re-examination of fish estuarine dependence: Evidence for connectivity between estuarine and ocean habitats. Estuar. Coast. Shelf Sci. 64 (1) 5–17. https://doi.org/10.1016/j.ecss.2005.02.002.

Christie, M.R., Tissot, B.N., Albins, M.A., Beets, J.P., Jia, Y., Ortiz, D.M., Thompson, S.E., Hixon, M.A., 2010. Larval connectivity in an effective network of marine protected areas. PLoS One 5 (12) e15715. https://doi.org/10.1371/journal.pone.0015715.

Kendall Jr., A.W., Ahlstrom, E.H., Moser, H.G., 1984. Early life history stages of fishes and their characters. In: Am. Soc. of Ichthyologists and Herpetologists Special publication 1: 11–22. South West Fisheries Science Center. http://swfsc.noaa.gov.

Newell, G.E., Newell, R.C., 1963. Marine plankton, a practical guide, fifth ed. Hutchinson of London, London, p. 244p.

Raymont, J.E.G., 1983. Plankton & Productivity in the Oceans: Volume 2 Zooplankton. Pergamon, Oxford.

Russell, F.S., 1976. The Eggs and Planktonic Stages of British Marine Fishes. Academic Press, London, p. 524p.

Todd, C.D., Laverack, M.S., 1991. Coastal Marine Zooplankton. A Practical Guide for Students. Cambridge University Press, Cambridge, p. 159p.

Whitfield, A.K., Able, K.W., Blaber, S.J.M., Elliott, M., 2022. Fish and Fisheries in Estuaries—A Global Perspective. vols. 1 and 2 John Wiley & Sons, Oxford, ISBN: 9781444336672, p. 1056.

Linked

Entrainment (biota); Plankton; Recruitment; Spawning-stock biomass

Science background: 17. Non-indigenous, alien, invasive, and other non-native species (NIS, AIS)

Coastal industries requiring to determine the actual or potential effects of their activities are required to have an understanding of which organisms are naturally (indigenous) in an area and which have been introduced. This may require an inventory of the local fauna and flora and, for organisms not local to the area, an indication of the vectors of transmission. This is particularly important for industries providing surfaces for settlement of these organisms and industries such as ports which may encourage the entry of such non-native species through ballast water discharges.

Plants and animals spend their lifetime within a certain habitat or environment, constrained by accessibility to essential resources (food, mates, shelter, etc.). Sessile organisms spend the greater part of their life (normally as adults) attached and immobile, but they also have dispersive larval or spore phases. Indeed, all animals show pre-reproductive migration as a result of evolutionary selection (Baker, 1978). The area within which populations of a species are consistently found is regarded as their "natural range" which is set by their ranges of tolerance and will include their optimal conditions. This range will be subject to environmental conditions, such as temperature, salinity, pressure, light, levels of dissolved oxygen, or pH, to which the species concerned are evolutionarily adapted (Solan and Whiteley, 2016). At the same time, such species are continually exploring the boundaries of this range, as it is evolutionarily advantageous to spread outside this range if conditions are or become favorable, as this enables exploitation of new resources while reducing intraspecific and interspecific competition.

Species which colonize a new area of habitat outside their natural range are considered to be non-native or non-indigenous in this new area. In practice, the baseline of a natural range is commonly arbitrary, and the extent of the range of a population (or species) fluctuates naturally in the longer term. However, where colonization occurs into new and favorable areas of habitat such that a new successful breeding population can establish itself, this range extension becomes established rather than short term, and the natural range itself is considered to have expanded. Equally, where such newly established populations become isolated from the original population, there would be a selection toward allopatric speciation through, at least, random genetic drift leading to a distinct species.

This is a continual and natural process. Further, it is enhanced by the natural variability of the marine environment in the longer term: as conditions in an existing range become deleterious (e.g., through warming, glaciation, over-predation, erosion), there is likely to be such range extensions, for example driving populations further toward the poles or further offshore.

Mobile species have some potential for continuous dispersion throughout their life—for example, pelagic fish. Sessile species commonly have a dispersive larval phase. Some taxa, such as isopods or pycnogonids, have limited dispersive ability, and their range extensions are slow, restricted, and more likely to lead to allopatric speciation.

Dispersion can, however, be assisted, such that distances involved are far greater than would be possible by the migration of larvae or even pelagic adults. Thus, sessile organisms are readily dispersed naturally by attachment to floating wood, turtle carapaces, or floating algae such as *Sargassum*, and by this process can be transported hundreds of kilometers, at the end of which they may colonize new habitats and establish new populations. Such new populations of non-native species have been referred to as "aliens" or, more correctly, "xenobionts."

This assisted distant recolonization can also be caused anthropogenically, for example via fouling on the hulls of ships or by transport of viable life stages in ship ballast waters. Cohen and Carlton (1997) list eight processes whereby human activities have been known to assist in the artificial introduction of xenobiont species into new areas outside their natural range.

Where these newly colonized populations establish and have a deleterious effect on the pre-existing ecology (e.g., by outcompetition, by over-predation, or by habitat destruction), they are termed "invasive" and are normally regarded as a problem. Examples are the Japanese seaweed *Sargassum muticum* ("Jap-weed"), the Antipodean barnacle *Austrominius modestus* (formerly *Elminius modestus*), and the Chinese mitten crab *Eriocheir sinensis* (a problem mainly in freshwater but with adults burrowing into the soft banks of estuaries), all of which cause problems for native species or habitat structure in their new range; conversely, the Indo-Pacific barnacle *Megabalanus tintinnabulum* (inter alia) and the bivalve *Mytilopsis leucophaeata* cause economic problems via fouling.

These previous examples are all of anthropogenically assisted invasion. *Sargassum muticum* was introduced anthropogenically into France (with oysters from Japan); its spread to UK waters is thought to be "natural," through floating plants or fragments. Species such as the white egret on the south coast of England are considered to be natural invasions, that is, range extensions achieved by the species own dispersive ability, although still deleterious to pre-existing species.

While it is difficult to determine such non-native species in any body of water, for example, a comprehensive list of marine non-native species in British waters up until 1997 (21 plant and 30 invertebrate species) was given by Eno et al. (1997); most of the species they list are not invasive. Under the Bern Convention, a European Strategy on Invasive Alien Species has been published (see later) which provides a framework for action, including encouraging the development of national strategies.

In practice, in any given marine or estuarine location, there is a continuous supply of organisms of xenobiont species, the very large majority of which find the habitat unfavorable or lack sufficient individuals to form a breeding population, such that they do not colonize or become established. The large majority of these go unnoticed, of course, but Eno et al. (1997) listed 18 marine xenobiont species recorded in British waters but which did not become established. As conditions in such a location change, for example by climatic warming, they may become more favorable to non-native species previously excluded (e.g., *Laminaria ochroleuca* in south-west England—Boalch, 1994), but of course at an equal rate they become more unfavorable to non-natives which may have colonized previously, with no predictable net increase in the presence of non-natives or invasive species.

European databases of non-indigenous species have been developed and methods for the control of these in marine waters have been developed (Olenin et al., 2011). However, while control in freshwaters and transitional waters (estuaries, lagoons, etc.) may be possible, control in open sea areas is more difficult if not impossible. Indeed, by the time such species are detected then this may imply the species has become established.

Non-indigenous species (synonyms: alien, exotic, non-native, allochthonous) are species or subspecies or lower taxa introduced outside of their natural range (past or present) and outside of their natural dispersal potential. This includes any part, gamete, or propagule of such species that might survive and subsequently reproduce. It also includes hybrids between an alien species and an indigenous species, fertile polyploid organisms, and artificially hybridized species irrespective of their natural range or dispersal potential. Their presence in the given region is due to intentional or unintentional introduction resulting from human activities, or they have arrived there without the help of people from an area in which they are alien. Increasingly, global warming will become a cause of species distribution change. Despite this, natural changes in distribution ranges (e.g., due to climate change or dispersal by ocean currents) do not qualify a species as being a non-indigenous one. However, the secondary spread of non-indigenous species from the area(s) of their first arrival could occur without further human involvement due to dispersal by natural means.

Invasive alien species (IAS) are a subset of established NIS which have spread, are spreading, or have demonstrated their potential to spread elsewhere and have an adverse effect on biological diversity, ecosystem functioning, socioeconomic values, and/or human health in invaded regions. Species of unknown origin which cannot be ascribed as being native or alien are termed cryptogenic species, some of these also can cause significant impacts. They may also demonstrate invasive characteristics and should be included in IAS assessments.

Often the impact of IAS has been interpreted as a decline in ecological quality resulting from changes in biological, chemical, and physical properties of an aquatic ecosystem. These changes include (but are not confined to) local elimination or extinction of sensitive and/or rare species, alteration of native communities, algal blooms or other outbreak formations and massive population expansions, modification of substratum conditions including shore zones, alteration of oxygen and nutrient concentration, pH and transparency of the water, and accumulation of synthetic pollutants.

The addition of any materials through human action to the aquatic system is regarded as contamination and then if the materials create adverse ecological consequences then it is termed pollution. This is the case with introduced organisms in that biological pollution is defined as the adverse impacts of invasive alien species at the level that disturb ecological quality by effects on one or more levels of biological organization: an individual (such as internal biological pollution by parasites or pathogens), a population (by genetic change, e.g., hybridization), a community (by a structural shift), a habitat (by modification of physical-chemical conditions), or/and an ecosystem (by alteration of energy and organic material flow) (Elliott, 2003; Olenin et al., 2010). The biological and ecological effects of biopollution may also cause adverse economic consequences.

References and further reading

Baker, R.R., 1978. The Evolutionary Ecology of Animal Migration. Hodder & Stoughton, London, p. 1012.

Boalch, G.T., 1994. The introduction of non-indigenous species to Europe: planktonic species. In: Boudouresque, C.F., Briand, F., Nolan, C. (Eds.), Introduced Species in European Coastal Waters, pp. 28–31. European Commission Ecosystems Research Report 8.

Cohen, A.N., Carlton, J.T., 1997. Transoceanic transport mechanisms: introduction of the Chinese mitten crab, *Eriocheir sinensis*, to California. Pac. Sci. 51, 1–11.

Elliott, M., 2003. Biological pollutants and biological pollution—an increasing cause for concern. Mar. Pollut. Bull. 46 (3) 275–280.

Eno, C., Clark, R., Sanderson, W., 1997. Non-Native Marine Species in British Waters: A Review and Directory. JNCC, Peterborough.

Olenin, S., Elliott, M., Bysveen, I., Culverhouse, P., Daunys, D., Dubelaar, G.B.J., Gollasch, S., Goulletquer, P., Jelmert, A., Kantor, Y., Mézeth, K.B., Minchin, D., Occhipinti-Ambrogi, A., Olenina, I., Vandekerkhove, J., 2011. Recommendations on methods for the detection and control of biological pollution in marine coastal waters. Mar. Pollut. Bull. 62 (12) 2598–2604.

Solan, M., Whiteley, N. (Eds.), 2016. Stressors in the Marine Environment: Physiological and Ecological Responses: Societal Implications. OUP, Oxford, ISBN: 9780198718826. Hardback.

European Strategy on Invasive Alien Species: https://wcd.coe.int/wcd/com.instranet.InstraServlet?command=com.instranet.CmdBlobGet&InstranetImage=1322677&SecMode=1&DocId=1440418&Usage=2.

Linked

Colonization by non-native organisms; Temporal and spatial scales

Science background: 18. Colonization by non-native organisms

In any given marine or estuarine location, there is a continuous supply of organisms of xenobiont species, arriving by both natural means (attached to floating wood, turtle carapaces, migrant seabirds or floating algae, or in oceanic eddies) or by anthropogenically assisted introduction (hull fouling, ship ballast water, by stepping in stages across large distances using oil rigs, etc.) (see Cohen and Carlton, 1997).

On arrival at the new location, their continued survival depends on a number of factors. Firstly, the environmental conditions (e.g., salinity, temperature regime, seasonality, current speed, substratum type, depth, light) must be appropriate. It is inevitably the case that most xenobiont species find the habitat unfavorable and so do not survive. In addition, in the case of temperate or polar waters, the conditions must remain appropriate throughout the year (e.g., the warm-water barnacle *Amphibalanus amphitrite* has often been found to settle on UK coasts, but has never survived normal winter conditions). In the case of species exhibiting commensalism, the appropriate host species must also be present.

Furthermore, sessile organisms rarely settle on clean substrata; rather, there is a requirement for an initial colonization by a biofilm of bacteria, such as *Thiobacilli*, yeasts and other fungi, or other microorganisms. Settlement selection by larvae is commonly triggered by the organic nature of the substratum. Secondly, the arriving organisms must be of an appropriate stage for settlement and so post-reproductive adults, for example, are usually not appropriate. Thirdly, given appropriate conditions, there must be a sufficient initial inoculation of individuals to allow persistence of the population. In the case of dioecious sexually reproducing organisms, specimens of both genders must be present. This is not necessarily a problem with organisms capable of asexual reproduction, and indeed the current UK population of the starlet sea anemone (*Nematostella vectensis*) may well derive from a single female introduced from the United States. Fourthly, the niche to which the arriving organisms are adapted must be available; otherwise, it may be outcompeted by pre-existing native species.

Power plant thermal discharges particularly create warmer conditions which may encourage survival of warm-water species once they arrive in an area. In this case, the presence of a population may mimic conditions in lower latitudes. The population of a non-native, warmer water species adjacent to a thermal discharge may be restricted by the temperature regime away from the thermal discharge. For example, the clam *Mercenaria mercenaria*, introduced from North America, colonized but is restricted to an area on the south coast of England adjacent to a power plant.

Thus, where such an organism is adapted to the conditions in the new habitat, colonization will occur if the arriving individuals are of an appropriate life stage and are not outcompeted by any existing organisms. Persistence of that colonization will depend upon the presence of sufficient organisms for successful reproduction and recruitment, as well as the environmental conditions remaining appropriate in the longer term.

Reference and further reading

Cohen, A.N., Carlton, J.T., 1997. Transoceanic transport mechanisms: introduction of the Chinese mitten crab, *Eriocheir sinensis*, to California. Pac. Sci. 51, 1–11.

Linked

Non-indigenous, alien, invasive and other non-native species (NIS, AIS)

Science background: 19. Mechanical, thermal, and chemical stressors

Coastal industries create many stressors, or pressures, as the mechanisms of change to the ecology or an area. Some industries create particular pressures, for example, live organisms entering a water intake undergo mechanical stresses, and in the case of directly cooled steam electric power stations, thermal and chemical stresses too. Filter screens are commonly used to prevent entry of fish and other debris into water conduits and heat exchangers. The larger items are swept onto the screens (impingement) and, on more modern installations, usually then pass through a fish recovery and return (FRR) system back to source water body (river, lake, or sea). Impingement effects on animals have been extensively studied and include exhaustion from prolonged swimming effort, exposure to biocides used to prevent fouling of the conduits, and being pinned on screens for extended periods where screens do not turn continuously (Turnpenny et al., 2010). The introduction of FRR systems since the 1970s has greatly improved survival rates for many species. However, screens that have not been specifically designed to be fish-friendly have resulted in large numbers of fish showing external or internal injuries.

Mechanical stressors to organisms exist owing to the high velocity of water passing through the power station (typically up to $3-4\,m\,s^{-1}$) generating hydraulic shear stresses, contact injuries with pump blades, and abrasion against solid surfaces. In an early paper considering the impacts on fish larvae and eggs of passing through a power plant (entrainment), Ulanowicz (1976) identified three major forces, often associated with high water velocity: pressure change, acceleration, and shear. Acceleration, the rate of change in velocity, can be assessed in three ranges. At the lower end of the scale is water speeding up in the intake and water slowing down as it leaves the outfall. At gravitational force level or below, this causes little harm. Intermediate accelerative forces are associated with turbulent eddies and can produce forces several times that of gravity, which could result in immediate or delayed damage to ichthyoplankton. At the highest end of the scale are instances where organisms collide with hard surfaces, which is probably fatal. Shear stress, or viscous force, occurs particularly where a fluid moves against a solid surface, exerting a viscous drag on the surface. Fig. 1 illustrates the potential effect of this force. Organisms can also be exposed to substantial hydrostatic pressure changes as a result of passage, for example, from deep underwater inlet tunnels to elevated condensers and back again to sea level.

Turnpenny et al. (2010) summarize the studies and technologies used in determining temperature, pressure, and chlorination stressors on entrained animals passing through the cooling water system. The combined effects of entrainment may be addressed by the use of an entrainment mimic unit or EMU which allows various stresses to be applied under controlled laboratory conditions to simulate passage through a cooling water system (Fig. 2).

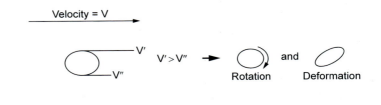

Fig. 1 The possible effects of shear stress resulting from water velocity on a fish egg. The egg velocity is subject to a fluid velocity on its outboard side (V^I) greater than that on its inboard side (V^{II}), inducing rotation and deformation.
From Ulanowicz, R.E., 1976. The mechanical effects of water flow on fish eggs and larvae. In: Saila, S.B. (Ed.), Fisheries and Energy Production: A Symposium. D.C. Heath, pp. 77–87.

Fig. 2 An Entrainment Mimic Unit (EMU2) at Cefas, Lowestoft, United Kingdom.
From Turnpenny, A.W.H., et al., 2010. Cooling Water Options for the New Generation of Nuclear Power Stations in the UK. Environment Agency Project no. SC070015/SR3, Bristol, UK, ISBN: 978-1-84911-192-8.

References and further reading

Turnpenny, A.W.H., et al., 2010. Cooling Water Options for the New Generation of Nuclear Power Stations in the UK. Environment Agency Project no. SC070015/SR3 ISBN: 978-1-84911-192-8, Bristol, UK.

Seaby, R., 2020. Nuclear power station cooling waters: protecting biota. Project no SC180004/R1. Environment Agency. ISBN 978-1-84911-448-6, Bristol, UK.

Ulanowicz, R.E., 1976. The mechanical effects of water flow on fish eggs and larvae. In: Saila, S.B. (Ed.), Fisheries and Energy Production: A Symposium. D.C. Heath, pp. 77–87.

Linked

Abstraction; Biocides; Biota impingement; Cooling water and direct cooling; Entrainment (biota)

Science background: 20. Baseline and reference conditions

A central feature of environmental assessment to be carried out by or for coastal industries is to determine whether that industry has changed or is likely to change the receiving environment, that is, the area where the industry is situated or where its source or discharged materials are from or moved to. This relies on a knowledge of what is an area like without the industry and how does that compare with the area after construction, operation, and/or decommissioning. The situation without the industry may be a termed a reference or baseline condition and so the impacts of a strategic action are the difference between the future environment without the strategic action and the future environment with the strategic action. The current environment and the future environment without the strategic action together are termed the environmental baseline. Determination of the baseline for a given site/development/action thus requires assessment of the current state of the environment and modeling the future state of that environment without the action, but allowing for evolution, whether by natural processes or by external factors such as climate change.

The various parameters determined for the baseline are thus the reference conditions against which any change as a result of future impact of a strategic action must be assessed. Determining such a change in time or space is termed a BACI-PS approach, for before-after-control-impact—paired series. However, marine and estuarine habitats are naturally highly variable and dynamic, both in response to larger environmental variability (year-to-year temperature decreases or increases, sunspot activity, climatic factors such as El Niño or the North Atlantic Oscillation, weather—as distinct from climate) and also stochastically. Detecting a required change, a signal, against the natural, inherent variability is defined as the *signal:noise ratio* such that the more inherently variable a habitat then the more difficult it is to detect such a signal or anthropogenic change. This is particularly the case in estuaries with highly variable natural conditions.

Unlike some terrestrial habitats, in which stable communities are achieved, the concept of a "climax community" in marine habitats is a misconception, and in these conditions the baseline itself is dynamic. Time-series studies can make the mistake of assuming the starting point to be some baseline, whereas in fact they are a snapshot within a changing pattern which may never return to that point. In particular, a background of climate change gives the concept of "moving baselines" in which the natural features, caused by external pressures, are changing. Despite this, true monitoring should be the detection of change against a known starting or required situation such that if the starting or required state is not known then detecting change is not possible and the efficacy of management measures cannot be determined.

Therefore, measurement of the baseline with respect to the animal and plant communities, and also, although to a lesser extent, the environmental parameters, such as salinity, temperature regime, must be undertaken over a long time span, at least 3 years (preferably 5 years although that period may differ in relation to the longevity of the ecological components, e.g., long lived species may require a greater time span for the reference conditions), and intensively, ideally monthly but also in relation to tidal or other cycles known to influence the biota. Should such measurement discover significant year-to-year variation it may become necessary to lengthen the measurement and monitoring time. Determination of sufficiency should be based not only on the perceived or actual variation in the measured parameters (deviation), but also whether the error margins in the modeling used for predicting the future baseline conditions are sufficiently narrow to allow testable prediction. In reality, long-term monitoring in both marine and estuarine habitats is now often rare, meaning establishment of any baseline in such dynamic environments is problematic.

References and further reading

CEQA Portal Topic Paper: Baseline and Environmental Setting: https://ceqaportal.org/tp/Baseline%20and%20Environmental%20Setting%20Topic%20Paper%2008-23-16.pdf.

Environment Agency (England): http://www.environment-agency.gov.uk/research/policy/32965.aspx.

Franco, A., Quintino, V., Elliott, M., 2015. Benthic monitoring and sampling design and effort to detect spatial changes: a case study using data from offshore wind farm sites. Ecol. Indic. 57, 298–304. https://doi.org/10.1016/j.ecolind.2015.04.040.

Gray, J.S., Elliott, M., 2009. Ecology of Marine Sediments: Science to Management. OUP, Oxford, p. 260.

Linked

Phenology; Temporal and spatial scales

Science background: 21. Causes of and solutions to estuarine, coastal, and marine degradation

Human activities and especially coastal industries have long caused the degradation of estuarine, marine, and coastal habitats and their physical, chemical, and ecological features including their species composition. These causes may include removing habitat or creating conditions no longer suitable for occupation by species and their communities. In some cases, the activities may create physical or water quality barriers which prevent connectivity between areas and so prevent colonization, recruitment, or recolonization by organisms.

Table 1 presents the dominant twelve categories of adverse changes to the estuarine, marine, and coastal system and their causes. These can be summarized as follows: enrichment by substances, loss of surface and habitat, biotic component loss

Table 1 Categories and causes of and solutions to the loss of estuarine and coastal degradation.

What?	Cause?	Reverse?
Land claim	Wetland removal/dyke construction	Restocking with vegetation, reconnection, resculpting
DO sag	Waste discharges	Reduction/treatment of inputs, reoxygenation, bubbling
Bivalve biogenic reef loss	Siltation, overharvesting	Adaptation, flushing, regulation, restocking
Eutrophication	Poor flushing, excess nutrients	Reconnection, regulation
Biota kills	Toxin input, WQ problems	Regulation, industry removal
Coral reef loss	Siltation, direct damage, bleaching	Runoff controls, re-creation, global rethinking
Loss of fish	Overharvesting, climate change, hydrodynamic barriers	Restocking, rethinking, adaptation, regulation
Salinity change	Upstream abstraction, impediments to flow	Removal, reconnection
Loss of seagrass	Smothering, nutrient excess, disease, hydrographic change	Reduction, removal, reconnection, replanting
Loss of flow	Diversion, abstraction, structures	Reconnection, reallocation
Seabed extraction	Aggregate removal, loss of sediment fraction	Reseeding, regulation, reallocation
Taxonomic changes	Non-indigenous species influx	Removal, eradication, prevention

From Lepage, M., Capderrey, C., Meire, P., Elliott, M., 2022. Chapter 8 Estuarine degradation and rehabilitation. In: Whitfield, A.K., Able, K.W., Blaber, S.J.M., Elliott, M. (Eds.), Fish and Fisheries in Estuaries—A Global Perspective. John Wiley & Sons, Oxford, pp. 458–552, ISBN 9781444336672.

or change, over-extraction of resources, and water and connectivity loss. Conveniently the categories of solutions can be summarized as follows: reversal, restocking, regulation, reconnection, resculpting, removal, revision, restoration, replanting, reduction, reallocation, reseeding, and reoxygenation.

Reference and further reading

Lepage, M., Capderrey, C., Meire, P., Elliott, M., 2022. Chapter 8. Estuarine degradation and rehabilitation. In: Whitfield, A.K., Able, K.W., Blaber, S.J.M., Elliott, M. (Eds.), Fish and Fisheries in Estuaries—A Global Perspective. John Wiley & Sons, Oxford, pp. 458–552, ISBN 9781444336672.

Linked

Ecosystem restoration; Sustainable environmental management

Science background: 22. Ecosystem restoration

Coastal industries or their forerunners may be or have been responsible for degrading estuarine and coastal areas. Restoration will then be required, perhaps as a license condition or to increase the presence of habitats (Figs. 1 and 2). For example, a degraded habitat or a habitat being degraded will require mitigation, re-creation, rehabilitation measures, or even compensation in which a new habitat is created. An industry wishing to expand by occupying wetland habitats will be required to create new wetland as a compensation mechanism. Successful and sustainable restoration will then require an understanding of structure and functioning which then informs what measures should be taken and where, how and why, especially working with rather than against nature.

Based on an understanding of the effects of human activities, links can then be made between the hydrophysical structure and processes and their effect on

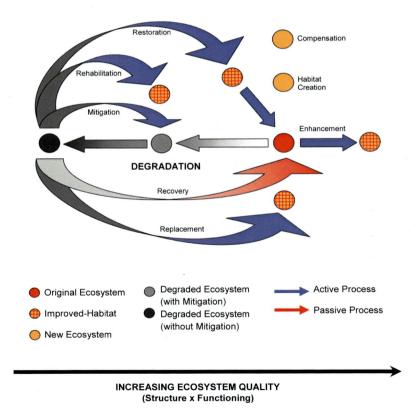

Fig. 1 Active and passive responses to ecosystem degradation.
From Elliott, M., Mander, L., Mazik, K., Simenstad, C., Valesini, F., Whitfield, A., Wolanski, E., 2016. Ecoengineering with ecohydrology: successes and failures in estuarine restoration. Estuar. Coast. Shelf Sci. 176, 12–35. https://doi.org/10.1016/j.ecss.2016.04.003.

Science background

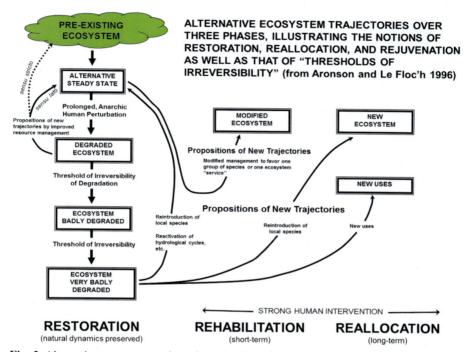

Fig. 2 Alternative ecosystem trajectories over three phases: restoration, reallocation, and rehabilitation plus "thresholds of irreversibility" to complete restoration to pre-existing conditions.
From Aronson, J., Le Floc'h, E., 1996. Vital landscape attributes: missing tools for restoration ecology. Restor. Ecol. 4, 377–387.

structuring ecosystems together with their goods and services. These links may be termed ecohydrology and so an understanding of this together with engineering and ecosystem modification mechanisms can then be used to improve or restore ecology and fulfill societal needs (Wolanski and Elliott, 2015). Table 1 lists measures to prevent a deterioration of an estuary or to remediate this after it has occurred, grouped according to hydrological, morphological, biological, chemical, and physical techniques/technologies. Those measures (in some legislation called a Programme of Measures) may be either sanctioned or required by governance (laws, policies, etc.) and should fulfill the 10-tenets of sustainability: ecologically sustainable, economically viable, technologically feasible, socially desirable/tolerable, legally permissible, administratively achievable, politically expedient, ethically defensible (morally correct), culturally inclusive, and effectively communicable (see entry).

Table 1 Ecohydrological measure categories, types, and examples to restore or recover habitats and areas.

Category	Ecohydrological measure type	Examples
Hydrology/ morphology	Measure to reduce tidal range, asymmetry, and pumping effects and/or dissipate wave energy	Breakwater, dyke and levee, dune, seawall, storm surge barrier, groyne, managed realignment, beach nourishment and/or slope development, oyster reef, coral reef, artificial reef, macrophyte planting (e.g., saltmarsh, seagrass, mangroves)
	Other measures for flood protection	Flood barrier, floodway/diversion canal, retention pond, dam/weir
	Other measures to stabilize coasts or improve morphological conditions	Groyne, gabion or other rock armor, coastal vegetation, coastal terracing, revetment
	Measure to decrease the need for dredging	Controlling sediment erosion from catchments, improved management of deep-draught vessel entry (e.g., tidal phase timing), more stringent justification of channel width/depth requirements
	Zoning measures	Managed realignment, set-back lines, building codes, multiple use zoning, special protection areas
	Measures to stop or reverse subsidence due to extraction of water and minerals	Aquifer recharge schemes, reinjection by oil and gas field production water
	Measure to restore longitudinal or lateral connectivity	Hydrological barrier removal (e.g., opening channels, removing weirs), wetland restoration, reducing water abstraction, increasing water release from dams for environmental flows
Physical/chemical quality	Measure to reduce nutrient loading (point and diffuse sources)	Removal/redesigning of land drainage systems (e.g., water sensitive urban design), changed agricultural and urban practices (e.g., fertilizer management), riparian buffer zones, set-back lines, improved land-use planning
	Measure to reduce persistent pollutant loading (point and diffuse sources)	Removal/relocation of discharge drains, improved waste treatment prior to disposal, changed land-use/industrial practices, riparian buffer zones

Table 1 Continued

Category	Ecohydrological measure type	Examples
	Measure to improve oxygen conditions	Oxygenation plants/bubblers, reducing nutrient loads (see examples under "measure to reduce nutrient loading"), increasing hydrological flows (see examples under "measure to restore longitudinal or lateral connectivity"), improving habitat conditions for bioturbators
	Measure to reduce physical loading (e.g., heat input by cooling water entries)	Diffusers, bafflers, heat sinks
	Measure to reduce sediment inputs and sediment loading	Shore stabilization, changed land-use practices
Biology/ecology	Measure to develop and/or protect specific habitats	Multiple use zoning, special protection areas, habitat restoration, managed realignment, maintaining environmental flows and connectivity (see examples under "measure to restore longitudinal or lateral connectivity")
	Measure to develop and/or protect specific species	Wildlife corridors, harvest quotas, minimum size at capture, restocking, habitat protection (see examples under "measure to develop and/or protect specific habitats")
	Measures to retain or restore natural gradients and processes, transition and connection	See examples under "measure to restore longitudinal or lateral connectivity" and "measure to develop and/or protect specific habitats"
	Measure to prevent introduction of or to eradicate/control against invasive species	Multiple use zoning, quarantine areas, vessel risk assessments, hull cleaning measures, biofouling, ballast water management, early detection pest monitoring programs, pest incursion response plans
	Measure for direct human benefit of ecological attributes	Multiple use zoning, special protection areas, habitat restoration, species protection

Continued

Table 1 Continued

Category	Ecohydrological measure type	Examples
Human safety	Measure for early warning/evacuation of natural disasters	Real-time flood forecasting, improved predictive systems, evacuation procedures, storm shelters and refuges
	Measure for improved resilience of housing and industry	Coastal set-back/roll-back, improved land-use planning, improved building design, see other examples under "measure to reduce tidal range, asymmetry, and pumping effects and/or dissipate wave energy," "Other measures for flood protection," and "Other measures to stabilize coasts or improve morphological conditions"

From Wolanski, E., Elliott, M., 2015. Estuarine Ecohydrology: An Introduction. Elsevier, Amsterdam, p 322, ISBN 978-0-444-63398-9; Elliott, M. Mander, L., Mazik, K., Simenstad, C., Valesini, F., Whitfield, A., Wolanski, E. 2016. Ecoengineering with ecohydrology: successes and failures in estuarine restoration. Estuar. Coast. Shelf Sci. 176, 12–35. https://doi.org/10.1016/j.ecss.2016.04.003.

References and further reading

Aronson, J., Le Floc'h, E., 1996. Vital landscape attributes: missing tools for restoration ecology. Restor. Ecol. 4, 377–387.

Elliott, M., Mander, L., Mazik, K., Simenstad, C., Valesini, F., Whitfield, A., Wolanski, E., 2016. Ecoengineering with ecohydrology: successes and failures in estuarine restoration. Estuar. Coast. Shelf Sci. 176, 12–35. https://doi.org/10.1016/j.ecss.2016.04.003.

Livingston, R.J., 2006. Restoration of Aquatic Systems. CRC Press, Taylor & Francis, Boca Raton, FL, p. 423.

Perrow, M.R., Davy, A.J. (Eds.), 2002. Handbook of Ecological Restoration—Volume 1—Principles of Restoration. Cambridge University Press, Cambridge, p. 444.

Perrow, M.R., Davy, A.J. (Eds.), 2002. Handbook of Ecological Restoration—Volume 2—Restoration in Practice. Cambridge University Press, Cambridge, p. 599.

Wolanski, E., Elliott, M., 2015. Estuarine Ecohydrology: An Introduction. Elsevier, Amsterdam, p. 322. ISBN 978-0-444-63398-9.

Linked

Causes of and solutions to estuarine, coastal and marine degradation; Sustainable environmental management

Science background: 23. Ecohydrology and ecoengineering

If estuarine, coastal, or marine industries damage the local ecology or as the result of expanding into wetlands are required to create new habitats which otherwise would be lost then they need a knowledge of habitat creation. Ecohydrology is the science and understanding of the links between the physical functioning and the means by which it creates the appropriate ecological functioning of an estuary (Wolanski and Elliott, 2015). It assumes that the ecology is primarily driven by the physics, which in turn affects the biological processes operating within a system. It includes changing the physiography and manipulating the freshwater flows from the catchment and it is also influenced by the anthropogenic users and uses of the estuary, some of which will have modified and impacted both the physics and the ecology. It is that knowledge which guides the management of the entire river basin from the headwaters down to the coastal zone, which ecohydrology views as an ecosystem.

The essence of ecohydrology is the role of the environment in influencing organisms and vice versa—of the organisms building, filling, and altering ecological niches. The hydrophysical regime creates the abiotic environment (the sediment and water fundamental niches) which is then colonized by biota and in turn the colonizing organisms interact to modify the system. Ecohydrology therefore has three interrelated and consecutive aspects required for an integrated water body (e.g., an estuary), river basin, or catchment management. Firstly, the hydrological processes at the management scale (either as an estuary, lagoon, water body, or catchment); secondly, the ecological structure and functioning at the relevant spatial and temporal scales and its relation to the ecological carrying capacity; thirdly, in turn, the socioecological system, the production of ecosystem services that can be used to deliver societal goods and benefits. This can then be phrased as testable hypotheses (from Wolanski and Elliott, 2015):

- H1: hydrological processes generally determine and initially regulate the structure of the biotic communities;
- H2: ecological functioning (rate processes) arises from the interactions between the elements of ecological structure (such as individuals, populations, and communities);
- H3: the biotic structure and functioning will induce feedback loops which can then help to structure the physicochemical environment (such as removing nutrients or changing water flows);
- H4: the previous three hypotheses can then be integrated with management measures (also termed a Programme of Measures), such as water control (e.g., compensation flows) or hydro-technical infrastructure (e.g., barriers), to achieve sustainable water management, the protection of ecosystem services, and the delivery of societal goods and benefits.

Successful ecological restoration is achieved by ecoengineering (often currently referred to as Nature-Based Solutions) which uses ecohydrology knowledge to modify and achieve the ecological aims for an area by one of two approaches. Firstly, by engineering the physics, including changing the physiography and manipulating the freshwater flows from the catchment. In turn these produce the ecological niches which then allows the ecology and habitats develop, especially if the colonizing species are ecological engineers, for example, seagrasses or clump-forming mussels. This is termed Type A ecoengineering. Type B ecoengineering is used where a required species has not or will not return unaided, especially if there are no connectivity pathways to allow colonization. In this case the ecology recovers by restocking (e.g., fishes, oyster beds) or replanting (reedbeds, saltmarshes, seagrasses), which then creates habitats or lets the ecological engineer species modify habitats, thus enhancing the physical-biological links.

Ecoengineering initiatives often aim to accelerate natural rehabilitation and sometimes harness dynamic variability. However, they often only achieve establishing a static system (the desired state) but this does not include all natural successional processes and stages. While ecohydrology aims to operate across the whole catchment-coast continuum, ecoengineering usually occurs at a smaller scale and will seldom recreate pristine estuaries given the intensive human populations living on their shores, although it aims to create ecosystems with at least some attributes of the original systems. It should be accompanied by regulating certain human activities and is more than just integrated river basin management. Primarily it aims to improve the ecology and provide benefits for the economy and the safety of society (i.e., so-called triple wins).

Ecoengineering aims to redress the balance after adverse historical changes, especially coastal and estuarine wetland removal, without unacceptable environmental trade-offs. Ideally, it provides relatively low-cost technologies for mitigating the impact on estuaries and coasts of human activities throughout the river basin (Table 1) (Elliott et al., 2016 gives worldwide examples). For example, the ecoengineering may use and enhance the natural capacity of the water bodies to absorb and process excess nutrients and contaminants and increase ecosystem resilience to accommodate global stressors such as climate change.

Table 1 Ecoengineering principles.

(1) Ecohydrological principles should be used to ensure a suitable and sustainable physicochemical system
(2) The design should encompass local features and so be site specific
(3) The design parameters and features should be kept simple in order to deliver the functioning required
(4) The design should use energy inside the system or coming from outside, such as flow conditions and working with nature, and that the system should be kept simple to minimize the information required for its execution, and lastly
(5) The ecoengineering design should aid the natural and social systems and so should have an ethical dimension; this may involve "over-engineering" the design in order to protect human safety and property

Modified from Bergen, S.D., Bolton, S.M., Fridley, J.L., 2001. Design principles for ecological engineering. Ecol. Eng. 18, 201–210.

In essence ecohydrology is the underlying process/abiotic drivers into which ecoengineering fits and by which ecoengineering is delivered, that is, ecohydrology provides the underlying science and ecoengineering is the mechanism for creating the ecology. If the ecoengineering is sustainable and successful then there are wins for the ecology (more habitat), human safety and welfare (flood and coastal protection), and economy (producing ecosystem services and societal benefits).

Early intertidal habitat creation designs in the period 1990–2010 were driven largely by engineering experiences drawn from the terrestrial and freshwater environments. Most of these sites underperformed for active fauna such as fish. The principles outlined before are now being integrated into future designs, following some early monitoring of the evolution and performance of a number of recent sites. Post-project appraisal is still scarce and is especially needed to further inform more efficient and more sustainable solutions.

References and further reading

Bergen, S.D., Bolton, S.M., Fridley, J.L., 2001. Design principles for ecological engineering. Ecol. Eng. 18, 201–210.

Burgess, H., Nelson, K., Colclough, S., Dale, J., 2019. The impact that geomorphological development of managed realignment sites has on fish habitat. In: Proc. Institute of Civil Engineers Conference, La Rochelle.

Elliott, M., Mander, L., Mazik, K., Simenstad, C., Valesini, F., Whitfield, A., Wolanski, E., 2016. Ecoengineering with ecohydrology: successes and failures in estuarine restoration. Estuar. Coast. Shelf Sci. 176, 12–35. https://doi.org/10.1016/j.ecss.2016.04.003.

Hudson, R., Kenworthy, J., Best, M. (Eds.), 2021. Saltmarsh Restoration Manual. UK & Ireland. Environment Agency, Bristol.

Wolanski, E., Elliott, M., 2015. Estuarine Ecohydrology: An Introduction. Elsevier, Amsterdam, p. 322. ISBN 978-0-444-63398-9,.

Linked

Biological and ecosystem health; Ecosystem resilience, resistance, recovery

Science background: 24. Ecosystem resilience, resistance, recovery

Ecosystem responses to human activities have been defined in conflicting ways. Resilience is most simply defined as "the ability of an ecosystem to return to its original state after being disturbed" or "how fast the variables return to equilibrium following perturbation," also termed as "robustness." Ecosystems may be regarded as being in stability states such that ecological resilience is the amount of disturbance that an ecosystem in one stability state can absorb before it is changed to another state, also referred to as "resistance." Tett et al. (2007) and Elliott et al. (2007) define resilience as the ability of the ecosystem to recover from disturbance, and state that an ecosystem shows resistance by initially reacting little to increases in pressure. Costanza (1992) defines resistance as "the degree to which a variable is changed following a perturbation." Holling (1986), however, calls this resilience, defined as "a system's ability to maintain structure and patterns of behaviour in the face of disturbance." In contrast, resistance is the ability of an area to withstand the effects of stressors.

Peterson (2000) defines ecological resilience as "the amount of change or disruption that will cause an ecosystem to switch from being maintained by one set of mutually reinforcing processes and structures to an alternative set of processes and structures." Hence, resistance and resilience are inherent properties of the ecosystem which indicate its ability to absorb change against a background of its complexity and/or variability. This can also be interpreted as redundancy in the system, for example if the system is sufficiently complex, it is unlikely that the loss of one or two species will cause a change in the system from having one set of characteristics, such as feeding (trophic) structure, to another. The latter, regarded as cascade effect (Kaiser et al., 2005), may occur under large-scale stressors such as fishing that selectively removes one group (e.g., demersal fish such as cod) to the benefit of another (e.g., pelagic species).

The structure and complexity of food webs center around connectivity (the number of links between species) and the length of food chains, among others. These properties of food webs change with scale, diversity, and complexity, and this is particularly the case with estuarine, coastal, and marine food webs which have large numbers of opportunist and generalist feeders. In particular, highly connected communities tend to be more robust (resilient) to species loss than low connected communities and so perhaps estuarine and marine communities have a greater structural robustness than other ecosystems. Ecosystem resilience can thus be exceeded when environmental and/or human-mediated stressors synergistically change the state. As such, Gunderson (2000) considers resilience as the time that a system takes to return to the stable state following a natural/human perturbation but also uses the term "adaptive capacity" as the processes that modify ecological resilience. Hence, while resilience may be measured as time, it depends on the amount of inherent complexity/variability of an ecosystem.

As an inherent property, all ecosystems are resilient but to differing degrees and a more specialized and less variable ecosystem may have a lower resilience than a naturally highly variable one. For example, a highly variable ecosystem such as an estuary is more likely to be able to withstand and/or absorb anthropogenic stress than a less variable one (so-called environmental homeostasis) (Elliott and Quintino, 2007). Similarly, the amount of resilience a system possesses relates to the degree of disturbance required to fundamentally disrupt the system causing a large-scale change to another state controlled by a different set of processes. In turn, reduced resilience increases the vulnerability of a system to smaller disturbances that could previously have been absorbed. However, even in the absence of disturbance, gradually changing conditions (e.g., nutrient loading, climate change, habitat fragmentation) may exceed threshold levels, resulting in an abrupt system response. Thus, it is suggested here that resilience and recoverability are synonymous so only the former is required.

The paths of decline and recovery of systems are regarded as trajectories or performance curves which although conceptually valid have not been defined quantitatively. Any attempt at restoration thus requires either an active or passive approach in which the habitat is made, respectively, to retrace or retraces without intervention the trajectory of decline. Aronson and Le Floc'h (1996) refer to three different options for recovery: restoration by reactivating (or allowing to be reactivated) natural processes including species reintroductions; rehabilitation, a short-term management measure to attain a specific ecosystem attribute, goods, or service; and reallocation where over the long term new trajectories produce new ecosystems and uses.

The conceptual model of Tett et al. (2007) takes this further to suggest resistance to change is the amount of (anthropogenic) stress (a pressure) that a system can accommodate before it deteriorates. Following the removal of the stress, the system will recover, though not necessarily always along the same trajectory of decline, the difference being termed hysteresis which differs with types of system and stressor. They then implied that a greater stress was needed to be removed to make the system recover, a feature they called resilience. Therefore, Elliott et al. (2007) revised their conceptual model to indicate that systems do not necessarily recover to their former state and also that their ability to recover is termed resilience (Fig. 1). For a given structural or functional parameter (which only defines one aspect of the multifaceted definition of ecosystem health, status, and function), resistance can be defined as the amount of a given pressure that can be applied without a deterioration in status (as defined by a specific measure). As a pressure is removed, Type 1 Hysteresis represents the lag in recovery; the status may not improve for some time after the pressure is removed. Given time, though, the status may recover, although it may not return to original levels. Resilience can thus be defined as the degree of recovery, based upon a given measure, compared to the original status—complete resilience results in a return to the original level, partial resilience is return to some lower (or higher) level, with Type II Hysteresis being the difference between the two. While the definition of resilience in Fig. 1 differs from those mentioned earlier, Elliott et al. (2007) showed that the terms resistance and resilience are used differently (and sometimes interchangeably).

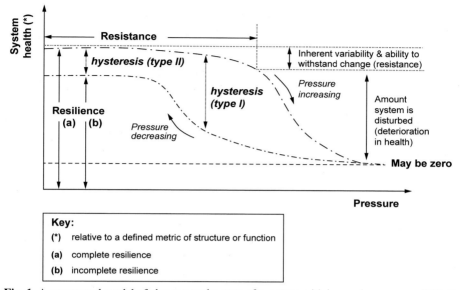

Fig. 1 A conceptual model of changes to the state of a system with increasing pressure (Elliott et al. 2007).
Adapted from Tett, P., Gowen, R., Mills, D., Fernandes, T., Gilpin, L., Huxham, M., Kennington, K., Read, P., Service, M., Wilkinson, M., Malcolm, S., 2007. Defining and detecting undesirable disturbance in the context of eutrophication. Mar. Pollut. Bull. 53, 282–297.

References and further reading

Aronson, J., Le Floc'h, E., 1996. Vital landscape attributes: missing tools for restoration ecology. Restor. Ecol. 4, 377–387.

Bengtsson, J., Engelhardt, K., Giller, P., Hobbie, S., Lawrence, D., Levine, J., Vila, M., Wolters, V., 2002. Slippin' and slidin' between the scales: the scaling components of biodiversity-ecosystem relations. In: Loreau, M., Naeem, S., Inchausti, P. (Eds.), Biodiversity and Ecosystem Functioning: Synthesis and Perspectives. Oxford University Press, Oxford, pp. 209–220.

Costanza, R., 1992. Towards an operational definition of health. In: Costanza, R., Norton, B., Haskell, B.D. (Eds.), Ecosystem Health: New Goals for Environmental Management. Island Press, Washington, DC, pp. 239–256.

Costanza, R., Norton, B., et al. (Eds.), 1992. Ecosystem Health: New Goals for Environmental Management. Island Press, Washington, DC, p. 279.

Dunne, J.A., Williams, R.J., Martinez, N.D., 2004. Network structure and robustness of marine food webs. Mar. Ecol. Prog. Ser. 273, 291–302.

Elliott, M., Quintino, V.M., 2007. The estuarine quality paradox, environmental homeostasis and the difficulty of detecting anthropogenic stress in naturally stressed areas. Mar. Pollut. Bull. 54, 640–645.

Elliott, M., Whitfield, A.K., 2011. Challenging paradigms in estuarine ecology and management. Estuar. Coast. Shelf Sci. 94, 306–314.

Elliott, M., Burdon, D., Hemingway, K.L., Apitz, S., 2007. Estuarine, coastal and marine ecosystem restoration: confusing management and science—a revision of concepts. Estuar. Coast. Shelf Sci. 74, 349–366.

Gunderson, L.H., 2000. Ecological resilience—in theory and application. Annu. Rev. Ecol. Syst. 31, 425–439.

Holling, C.S., 1986. The resilience of terrestrial ecosystems: local surprise and global change. In: Clark, W.C., Munn, R.E. (Eds.), Sustainable Development of the Biosphere. Cambridge University Press, Cambridge, pp. 292–317.

Kaiser, M.J., et al., 2005. Marine Ecology: Processes, Symptoms and Impacts. Oxford University Press, Oxford, p. 557.

Loreau, M., Naeem, S., Inchausti, P. (Eds.), 2002. Biodiversity and Ecosystem Functioning: Synthesis and Perspectives. Oxford University Press, Oxford, p. 294.

Peterson, G., 2000. Political ecology and ecological resilience: an integration of human and ecological dynamics. Ecol. Econ. 35, 323–336.

Pimm, S.L., 1984. The complexity and stability of ecosystems. Nature 307, 321–326.

Simenstad, C., Reed, D., Ford, M., 2006. When is restoration not? incorporating landscape-scale processes to restore self-sustaining ecosystems in coastal wetland restoration. Ecol. Eng. 26, 27–39.

Tett, P., Gowen, R., Mills, D., Fernandes, T., Gilpin, L., Huxham, M., Kennington, K., Read, P., Service, M., Wilkinson, M., Malcolm, S., 2007. Defining and detecting undesirable disturbance in the context of eutrophication. Mar. Pollut. Bull. 53, 282–297.

The Resilience Alliance, 2002. What is Resilience? http://www.resalliance.org/ev_en.php?ID=1004_201&ID2=DO_TOPIC.

Linked

Causes of and solutions to estuarine, coastal and marine degradation; Ecosystem restoration

Science background: 25. Climate change and its effects

Global climate change is regarded as an *exogenic unmanaged pressure* (see Fig. 1) in which a new industrial activity in an area has to respond to the consequences of climate change even if it is not alone in promoting such a change. The sequence and main features of concern may be divided into physical changes (hydromorphological repercussions) and the physiological responses by organisms, notably temperature change to which individuals, populations, and communities are exposed. In turn, these relate to the effects of the industry on the marine or estuarine system and the effects of the marine or estuarine system on the industry. Especially they may create a moving baseline against which changes due to the industrial presence and activities have to be judged. Conversely, and as a particular response, the measured impacts of industrial cooling water discharges may be used to simulate the possible impacts of future climate change.

These changes fit within the figure of the conceptual model given later and the main areas of concern which include the increased hazards and risk to the industry operation (see Tables 1 and 2) and so require long-term adaptation strategies. Such risks may include those indicated in the table given below but the adequacy of the

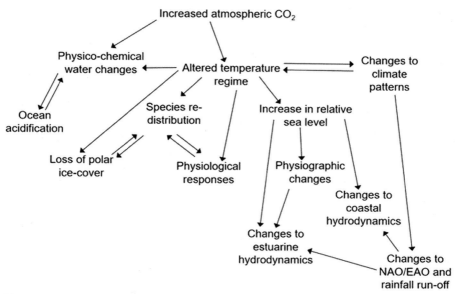

Fig. 1 Primary drivers and effects of climate change (NAO-EAO—North Atlantic Oscillation, Eastern Atlantic Oscillation).
Modified from Elliott, M., Borja, Á., McQuatters-Gollop, A., Mazik, K., Birchenough, S., Andersen, J.H., Painting, S., Peck, M., 2015. Force majeure: will climate change affect our ability to attain good environmental status for marine biodiversity? Mar. Pollut. Bull. 95, 7–27. https://doi.org/10.1016/j.marpolbul.2015.03.015.

Table 1 Changes to the physical environment through climate change.

- Repercussions for the ecology of increased energy (storminess) in the system (changes to the sedimentary environment and hence dependent communities);
- Repercussions for the ecology of loss or gain of habitat through relative sea level rise;
- Repercussions for the community-level ecology of temperature increases cumulatively between a thermal discharge (e.g., power plant) and climate change (changes to species distributions and community structure);
- Repercussions for the individual-level (physiological, reproductive, productivity) ecology of similar temperature increases (altered spawning cycles, growth rates, etc.).

Table 2 Examples of risk through climate changes.

- Risk of change to adjacent ecology through erosion, seawater incursion;
- Risk of beach adversely affected (erosion, size, etc.) as the result of hydromorphological changes;
- Risk of subtidal sandbanks moving and adversely affecting intakes and outfall;
- Risk of beach and defenses affected by changes in wave climate/variability/force/storm surges;
- Risk of beach and defenses affected by changes in wave climate/variability/force;
- Risk of overtopping of dykes by relative sea level rise, isostatic rebound;
- Risk of loss of intertidal carrying capacity due to SLR;
- Risk of loss of intertidal carrying capacity due to thermal plume (and acidity) compounding; climate-induced temperature change (and pH);
- Risk of introduction of alien and introduced species and spp. loss/gain;
- Risk of intertidal community change through changed reproduction patterns due to combined NNB and climate change thermal changes;
- Risk of estuary benthic community change due to salinity changes through runoff patterns (e.g., North Atlantic Oscillation, El Niño effects);
- Risk of water impingement and entrainment community changes due to changed ecological; structure and functioning due to runoff, temperature, salinity, pH;
- Risk of exceedance of water quality standards due to changed chemical behavior due to temperature, pH, and salinity changes from climate change.

evidence base, the level of uncertainty, and the errors of the predictions all require to be considered. Once the nature of the physical changes can be predicted or defined then the subsequent ecological repercussions can be determined.

References and further reading

Austin, D.E., 2009. Coastal exploitation, land loss, and hurricanes: a recipe for disaster. Am. Anthropol. 108 (4) 671–691.

Elliott, M., Borja, A., McQuatters-Gollop, A., Mazik, K., Birchenough, S., Andersen, J.H., Painting, S., Peck, M., 2015. Force majeure: will climate change affect our ability to attain

good environmental status for marine biodiversity? Mar. Pollut. Bull. 95, 7–27. https://doi.org/10.1016/j.marpolbul.2015.03.015.
Humphries, L., 2001. A review of relative sea level rise caused by mining-induced subsidence in the coastal zone: some implications for increased coastal recession. Clim. Res. 18, 147–156.
Klein, R.J.T., Nicholls, R.J., Ragoonaden, S., Capobianco, M., Aston, J., Buckley, E.N., 2001. Technological options for adaptation to climate change in coastal zones. J. Coast. Res. 17 (3) 531–543.
Pilkey, O.H., Young, R.S., 2005. Will hurricane Katrina impact shoreline management? here's why it should. J. Coast. Res. 21 (6) iii–x.
Smith, K., Petley, D.N., 2009. Environmental Hazards: Assessing Risk and Reducing Disaster, fifth ed. Routledge, Oxford.

Linked

Ecosystem resilience, resistance, recovery; Phenology

Science background: 26. Phenology

The spatial and temporal effects of industrial activities have to be determined over and above the natural seasonal cycles and spatial patterns. Phenology is the study of life cycles of plants and animals and the influence that seasonality and climate have over them. Phenology covers such topics as the timing of leaf and flower emergence, metamorphosis of butterflies, and the migration of birds and fishes and has received increased attention in helping understand the effects of climate change and potential mismatching of predator and prey abundances (e.g., phytoplankton and zooplankton in a temperate lake) (Winder and Schindler, 2004). Both day-length and temperature can influence the timing of these cycles, hence temperature changes such as those associated with the discharge of heated cooling water from power stations and other industries or cold water from desalination plants can affect the natural phenology.

The most important factor for bringing fish near to coastal developments and especially to a water intake structure is seasonal migration (Turnpenny and Coughlan, 2003). For example, the sea bass (*Dicentrarchus labrax*) spawns offshore and its fry are carried into estuaries by currents in the summer. They then return to the sea in autumn and return in successive years in shoals as increasingly larger individuals. A life history such as this makes the impacts of industrial water abstraction on bass quite predictable, with the young-of-the-year fish appearing in entrainment samples in summer, becoming impinged on filter screens as they head for deeper water in the autumn, and older bass taking advantage of intake currents for preying upon smaller fish throughout the summer. Power stations, in particular, have therefore proved to be useful tools in monitoring phenology as well as in considering juvenile abundance from year to year (Pickett and Pawson, 1994).

Cues for the seasonal behavioral patterns such as migration include temperature; Sims et al. (2005) showed that flounder (*Platichthys flesus*), which spend most of their time in estuarine waters, migrate earlier in colder years. Fish also respond to light levels; for example, the seasonal body condition of Norwegian, spring-spawning herring (*Clupea harengus*) is linked to the return of longer hours of daylight (Varpe and Fiksen, 2010). Much of the fish and weed catch at industrial water intakes can also be related to more minor cycles, such as the tidal state.

Power station cooling water is typically returned to its source at 8–12°C above the ambient temperature (Turnpenny and Coughlan, 2003). This "thermal plume" of discharged water may mimic summer seasonal changes, attracting species that prefer warmer waters. For example, the thermal discharge canal at Kingsnorth power station (United Kingdom) "fools" bass into remaining inshore all year round (Pickett and Pawson, 1994). The reverse effects may be associated with liquid petroleum gas (LPG) plants where thermal discharges are cooler than ambient water temperatures (Nigam et al., 2013).

References and further reading

Nigam, S., Padma, B., Rao, S., Srivastava, A., 2013. Effect of thermal discharge of cool water outfall from liquefied natural gas (LNG) plant into sea using CORMIX. J. Comput. Commun. 1, 1–5. https://doi.org/10.4236/jcc.2013.14001. Published Online October 2013 (http://www.scirp.org/journal/jcc) Available from: https://www.researchgate.net/publication/276494129. (Accessed 10 July 2022).

Pickett, G.D., Pawson, M.G., 1994. Sea Bass: Biology, Exploitation and Conservation. Chapman & Hall, London.

Sims, D.W., et al., 2005. Low-temperature-driven early spawning migration of a temperate marine fish. J. Anim. Ecol., 333–341.

Turnpenny, A.W.H., Coughlan, J., 2003. Using Water Well. Studies of Power Stations and the Aquatic Environment. Innogy plc. ISBN 095171726X.

Varpe, Ø., Fiksen, Ø., 2010. Seasonal plankton–fish interactions: light regime, prey phenology, and herring foraging. Ecology 91 (2) 311–318.

Whitfield, A.K., Able, K.W., Blaber, S.J.M., Elliott, M., 2022. Fish and Fisheries in Estuaries—A Global Perspective. vols. 1 and 2 John Wiley & Sons, Oxford, ISBN: 9781444336672, p. 1056.

Winder, M., Schindler, D.E., 2004. Climate change uncouples trophic interactions in an aquatic ecosystem. Ecology 85 (8) 2100–2106.

Linked

Climate change and its effects; Ecosystem resilience, resistance, recovery

Section 2

Fisheries terms

Introduction

Although not specifically related to the management issues and decisions relating to coastal industries, it is considered important that coastal industry managers have a knowledge of fisheries as they will often be required to discuss the effects of their activities on fish stocks, the local fisheries activities, and those carrying out and managing fisheries. Hence, some commonly used fisheries (management) terms need defining herein because temporal knowledge of the status of commercial and especially conservation species is required in order to know how such industries affect or are being affected by fish/shellfish and in certain cases their fisheries.

Fish and shellfish stocks are assessed and managed according to general fisheries principles, with the level of uncertainty of total and local stock size generally being associated with the accuracy and richness of the data available. The terms defined here therefore do not lend themselves to being written in the specific context of coastal industries so the definitions should be seen as background information which has been provided so that non-fisheries practitioners, including managers and decision-makers, can understand the necessary terms applied when managing fish stocks. It is also worth noting that the References cited in this section are a mix of specific citations mentioned in the text and general source references to the term.

Fisheries terms: 27. Precautionary principle and approach

Environmental management and the management of the activities, their pressures, and effects emanating from coastal industries have long been based around the precautionary principle. There are many definitions of the precautionary principle, but one of its primary foundations arose from the work of the Rio UNCED Conference of 1992, whose Principle 15 states: "In order to protect the environment, the precautionary approach shall be widely applied by States according to their capabilities. Where there are threats of serious or irreversible damage, lack of full scientific certainty shall not be used as a reason for postponing cost-effective measures to prevent environmental degradation."

In essence, therefore, the principle refers to the notion that something should be safeguarded when there is a likelihood that harm may result from (generally anthropogenic) actions. The term was coined with reference to management of the environment and ecosystems, and only subsequently applied to fisheries and marine resource management. Succinctly, though, if there is a belief based on appropriate and best science that likely irrevocable damage could be caused to an environment or ecosystem, then the approach or measures adopted under the principle of precaution need to be based on pragmatic assumptions regarding the uncertainties (Gray and Bewers, 1996).

In terms of fisheries and resource management, the principle if reasonably interpreted does offer an opportunity to progress from limited management intervention toward managing resources sustainably (Garcia, 1994). As an approach, for instance, precautionary total allowable catches (TACs; see entry 33 later) have been set for fisheries for which adequate (quantity and quality) data are available, sometimes based on averages of the previous few years of (low) catches as a means of keeping catches and effort low while information on the true status of stocks is achieved. A number of management measures/tools were defined by Garcia (1994) as a means of determining whether or not resources were being managed according to the precautionary principle. They include:

- adoption of the sustainable development principle;
- adoption of the principle of precautionary management;
- use of "best scientific evidence" available;
- adoption of a broad range of management benchmarks and reference points;
- criteria in place to be used in assessing present or potential impacts of development;
- a risk-averse management stand being taken;
- levels of impact and risk to a resource widely agreed as acceptable;
- resources being viewed holistically within their environment;
- management response time swift or at least speeding up;
- "non-fisheries users" included in management bodies;
- improving decision-making procedures;
- prior consultation procedures in place;
- strong monitoring, control, and surveillance, and heavy sanctions for non-compliance.

References and further reading

Cochrane, K.L., Garcia, S.M. (Eds.), 2009. A Fishery Manager's Guidebook, second ed. FAO, Rome. Print ISBN: 9781405170857; Online ISBN: 9781444316315.
Garcia, S.M., 1994. The precautionary principle: its implications in capture fisheries management. Ocean Coast. Manag. 22, 99–125.
Gray, J.S., Bewers, J.M., 1996. Towards a scientific definition of the precautionary principle. Mar. Pollut. Bull. 32, 768–771.

Linked

Baseline and reference conditions; Biological and ecosystem health

Fisheries terms: 28. Spawning-stock biomass (SSB or B)

It is important for coastal users and regulators to understand the nature and size of the resource being managed, whether for habitats or species. The spawning-stock biomass is defined as the total weight of the portion of a stock or population of fish or shellfish that is sexually mature and able to reproduce. Hence, it is the part of the stock/population that determines future levels of recruitment (see example Fig. 1 below, which is for Thames herring) (see entry 29) and hence future stock survivability, and as such is a crucial parameter in fisheries management (see entry 35).

For fisheries management and its actions to be effective, it is necessary that decision-makers and their advisors have available to them a reasonable estimate of SSB, and this is achieved in a number of ways. Direct surveys are generally expensive to conduct and usually yield only an index of abundance relative to previous years in a survey time series. Surveys carried out acoustically or on the basis of daily egg production (Lasker, 1985), however, do attempt to deliver an absolute estimate of SSB. Nevertheless, there are uncertainties associated with the results of such surveys because of the geographic coverage relative to the (sometimes uncertain) distribution of the whole population or stock, the seasonal timing of individual surveys, and even in some of the basic biological parameters required for the analysis (e.g., fish age at length and maturity). Therefore, even the best planned and conducted of these surveys can still yield at best an estimate of SSB. Also, mathematical models based on production/catch or population analysis are often used to generate an estimate of SSB, although those results are also subject to uncertainties, some similar to the ones associated with surveys (e.g., fish age at length) and others related to the appropriateness and accuracy of the commercial, recreational, or artisanal catch series on which they are based.

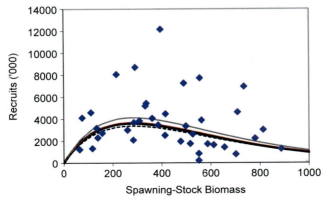

Fig. 1 The relationship between spawning-stock biomass and the ultimate number of recruits for Thames herring.

References and further reading

Cochrane, K.L., Garcia, S.M. (Eds.), 2009. A Fishery Manager's Guidebook, second ed. FAO, Rome. Print ISBN: 9781405170857; Online ISBN: 9781444316315.

Haddon, M., 2001. Modelling and Quantitative Methods in Fisheries. Chapman & Hall/CRC Press, Boca Raton, FL, p. 406.

Hilborn, R., Walters, C.J., 1992. Quantitative Fisheries Stock Assessment. Choice, Dynamics and Uncertainty. Kluwer, Boston, MA, p. 570.

Lasker, R. (Ed.), 1985. An egg production method for estimating spawning biomass of pelagic fish: application to the northern anchovy (*Engraulis mordax*), p. 99. US Department of Commerce, NOAA Technical Report, NMFS 36.

Quinn, T.J., Deriso, R.B., 1999. Quantitative Fish Dynamics. Oxford University Press, New York, p. 542.

Linked

Equivalent adult value (EAV)

Fisheries terms: 29. Recruitment

The size and nature of ecological populations in the vicinity of coastal industries are the result of many factors, both natural and anthropogenic. For example, changes in fish and shellfish population sizes are driven mainly by variability in recruitment, which is defined as the number of offspring surviving to enter a fishery or to a certain stage of the life history of that fish species, such as settlement or maturity. Recruitment is notoriously difficult to predict, being dependent on factors such as the size structure and numbers of the adult population (there is generally a relationship between recruitment and spawning-stock size—see Fig. 1 below—and age structure) and environmental suitability. Of these, the latter is highly influential at the early life history stage, being important *inter alia* in egg hatching, larval development and transport, and food availability.

The rate of mortality of spawning products (eggs, larvae, and juveniles) is much higher than that of larger adult fish, so a population of fast-growing fish should experience less natural mortality cumulatively over its life than a population of slow-growing ones. However, most commercially important species of fish and shellfish simply release their spawning products into the water, so the suitability of the environment during the usually short window of time in which a species spawns is crucial to the future health of each population and species. With environmental conditions varying dynamically seasonally and annually, the reason why recruitment varies becomes obvious.

In fisheries science, recruitment is usually predicted from analytical relationships between spawning-stock biomass and recruitment fitted over a data series covering many years (Hilborn and Walters, 1992; Quinn and Deriso, 1999). However, there is a problem with this concept in that the basic assumptions underlying the efficacy of such relationships are that the size of the spawning-stock biomass directly influences egg production and that all the eggs/larvae produced have an equal probability

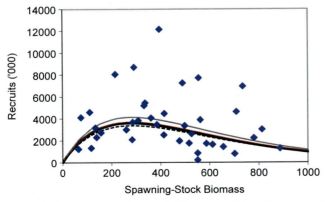

Fig. 1 The relationship between spawning-stock biomass and the ultimate number of recruits for Thames herring.

of survival. Given the statement before on how the environment on which early survival depends is both variable and very influential in recruitment, the flaw in the latter assumption is clear. The former assumption has also been openly questioned.

A concept related to recruitment also evident in fisheries management is that of so-called recruitment-overfishing (Hilborn and Walters, 1992), which results from fishing the adult (spawning-stock) biomass down to a level where it cannot produce sufficient young each year (the recruits) to replace the total annual loss to fishing (fishing mortality, F; see entry 30) and natural causes (natural mortality, M). Some currently clearly greatly depleted stocks have been exposed to such exploitation, others to growth overfishing (Hilborn and Walters, 1992), which is defined as fish being caught before they are able to make their optimum individual contribution through somatic growth to the biomass being exploited/caught.

References and further reading

Cochrane, K.L., Garcia, S.M., 2009. A Fishery Manager's Guidebook, second ed. FAO, Rome. Print ISBN: 9781405170857; Online ISBN: 9781444316315.

Haddon, M., 2001. Modelling and Quantitative Methods in Fisheries. Chapman & Hall/CRC Press, Boca Raton, FL, p. 406.

Hilborn, R., Walters, C.J., 1992. Quantitative Fisheries Stock Assessment. Choice, Dynamics and Uncertainty. Kluwer, Boston, MA, p. 570.

Quinn, T.J., Deriso, R.B., 1999. Quantitative Fish Dynamics. Oxford University Press, New York, p. 542.

Linked

Determination of significant effect; Phenology

Fisheries terms: 30. Fishing mortality (F)

The natural decline and inherent variability in the abundances of any ecological component have to be separated from the decline due to industrial activities, the signal: noise ratio. Fishing mortality is defined as the removal of fish/shellfish from a population by fishing, as opposed to the adverse effects of anthropogenic discharges causing mortalities. In fisheries science, fishing mortality is usually expressed as annual mortality or formally more commonly as the instantaneous rate F, the natural logarithm of the change in abundance caused by fishing per unit of time. It is actually the proportion of a fish population or stock that dies each year as a consequence of fishing. The value of F can range from 0 to values >1; high values are common for short-lived, heavily exploited pelagic species such as sardine and anchovy, but not for longer-living fish such as cod that only mature at several years of age.

In fisheries models, F and natural mortality (M), which is caused by predation, cannibalism, or such natural causes as old age and disease, are additive instantaneous rates that sum up to the total mortality (Z), which is known as the instantaneous total mortality coefficient, that is, $Z = M + F$ (Gulland, 1969).

Estimates of fishing mortality are generally included in mathematical yield models to predict yield levels obtained under various scenarios of exploitation. In the dome-shaped yield-per-recruit curve commonly applied in fisheries science (see the simplistic model shown later, in which fishing mortality is displayed in general terms as catch per unit effort), the point at which the fishing mortality is equivalent to or generates maximum sustainable yield, that is, F_{msy} (the top of the dome), is key, but because its enumeration is subject to great uncertainty, the more commonly used value is $F_{0.1}$, the estimated value on the ascending limb of the yield-per-recruit curve that lies where the slope is 10% of that at the origin of the curve or some other value lower than F_{msy} (Fig. 1).

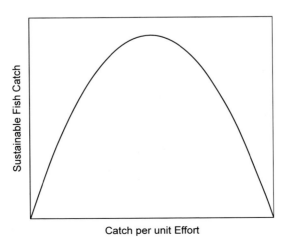

Fig. 1 Yield-per-recruit curve commonly applied in fisheries science (in which fishing mortality is displayed in general terms as catch per unit effort).

References and further reading

Cochrane, K.L., Garcia, S.M. (Eds.), 2009. A Fishery Manager's Guidebook, second ed. FAO, Rome. Print ISBN: 9781405170857; Online ISBN: 9781444316315.
Gulland, J.A., 1969. Manual of methods for fish stock assessment. 1. Fish population analysis. FAO Manuals in Fisheries Science, 4, p. 154.
Haddon, M., 2001. Modelling and Quantitative Methods in Fisheries. Chapman & Hall/CRC Press, Boca Raton, FL, p. 406.
Hilborn, R., Walters, C.J., 1992. Quantitative Fisheries Stock Assessment. Choice, Dynamics and Uncertainty. Kluwer, Boston, MA, p. 570.
Quinn, T.J., Deriso, R.B., 1999. Quantitative Fish Dynamics. Oxford University Press, New York, p. 542.

Linked

Determination of significant effect; Recruitment

Fisheries terms: 31. Catch per unit effort (cpue)

This is a term used widely in fisheries management, but also in conservation management, as a measure of abundance, that is, the availability of a target stock or species over time. It is calculated as the catch (usually weight, but it can be numbers) of a given species or species complex taken from the sea per unit of effort expended. For this purpose, catch and effort are as defined by Ricker (1958), and fishing being a finite activity in terms of resource, there is a relationship between cpue and yield (see generalized Fig. 1 below). Effort, which is taken to be the quantity of fishing gear in use for a specific period of time, can be measured in a variety of ways, depending on the gear deployed. For trawl fisheries, it is generally expressed as hours trawled; for seine fisheries, it can be the time, including search time, spent at sea; for line fisheries, the number of hooks and the soak time (the time the baited hook is left in the water); for trap fisheries, a combination of the number of traps and the time they are deployed; and for some net fisheries, the length or area of net combined with the duration the net is fished.

Because vessel size and hence its power, plus its access to technological advances in fish-finding equipment, and even fisher and skipper experience, vary across many individual fleets, attempts are often made by fisheries practitioners to standardize the unit of effort relative to a known unit level, despite such an exercise being difficult and subjective in nature. Notwithstanding, a cpue time series is by definition an indirect index of the relative abundance or availability of a target species, species complex, or stock over time in the area of study. A generally declining standardized cpue index over time can be indicative of either overexploitation or perhaps changing stock dynamics as resources move to other areas driven by factors such as climate change, prey availability, competing usage of the area by industries other than fishing, or even fishing pressure, whereas a stable index can be indicative of a stock being managed sustainably.

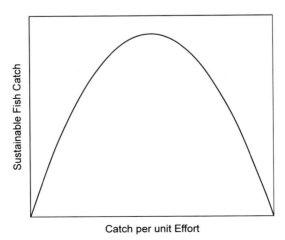

Fig. 1 Yield-per-recruit curve commonly applied in fisheries science (in which fishing mortality is displayed in general terms as catch per unit effort).

A cpue index can rarely be considered in isolation in management, because other dynamics come into play in its calculation, for instance the stability of the size and age spectrum of fish/shellfish being captured over time, with typically more large, older animals being taken early in the fishery. Consequently, the fishery gradually becomes dominated by smaller, younger fish/shellfish as the fishery matures (the latter state is often more productive than the former one). Another issue is that the "catch" in a cpue index refers to the quantity of fish actually taken from the sea, i.e. the landed fish plus the fish discarded before landing, which for the latter in many cases has to be estimated from direct observation or indirect extrapolation. To obviate this estimation problem, therefore, some fisheries are managed using an index of landings per unit effort, or lpue.

Although its calculation does require that fishers submit catch and effort information, the advantages of a cpue index include the fact that collection of the material does not interfere with routine fishing operations, and data are generally easily collected and analyzed, the latter even by non-scientists. Therefore, although not on a fleet-wide basis, the fisher is constantly reviewing their own cpue, for economic reasons. However, most good fishery management science these days combines fishery-dependent information (e.g., cpue) with fishery-independent information from (often expensive) research surveys. Rarely, however, are either or both of these sources of information used to define the absolute abundance of a species, stock, or stock complex.

Each type of commercial and recreational fishing and the scientific methods of fishing have its own estimation of catch efficiency (catchability) and gear efficiency which both influence cpue (see the description of methods and their use in Whitfield et al., 2022).

References and further reading

Cochrane, K.L., Garcia, S.M. (Eds.), 2009. A Fishery Manager's Guidebook, second ed. FAO, Rome. Print ISBN: 9781405170857; Online ISBN: 9781444316315.

Haddon, M., 2001. Modelling and Quantitative Methods in Fisheries. Chapman & Hall/CRC Press, Boca Raton, FL, p. 406.

Hilborn, R., Walters, C.J., 1992. Quantitative Fisheries Stock Assessment. Choice, Dynamics and Uncertainty. Kluwer, Boston, MA, p. 570.

Quinn, T.J., Deriso, R.B., 1999. Quantitative Fish Dynamics. Oxford University Press, New York, p. 542.

Ricker, W.E., 1958. Handbook of computations for biological statistics of fish populations. Bulletin of the Fisheries Research Board of Canada, 119, p. 300.

Whitfield, A.K., Able, K.W., Blaber, S.J.M., Elliott, M., 2022. Fish and Fisheries in Estuaries—A Global Perspective. Vols. 1 and 2 John Wiley & Sons, Oxford, UK, ISBN: 9781444336672, p. 1056.

Linked

Fishing mortality; Maximum sustainable yield; Total allowable catch

Fisheries terms: 32. Maximum sustainable yield (MSY)

In fisheries and ecological terms, the maximum sustainable yield or MSY is theoretically the largest yield (or catch) that can be taken from a stock of a species over an indefinite period. It has also occasionally been referred to as the maximum equilibrium catch (Ricker, 1975). The concept underpins the issue of sustainable management, and basically the term was coined to represent the largest annual catch that can be taken continuously under known, existing environmental conditions. The target of sustainable management would be to maintain a population size at its point of maximum growth by harvesting individuals that would normally be added to the population, allowing the population to continue to be productive indefinitely.

At small population size there is likely no constraint on the reproductive output of an individual but overall yield would be small because parent fish are few. At high population density, density-dependent factors such as competition for food and space can limit reproductive output to keep the population at its so-called carrying capacity and the yield would be small unless management wished to reduce the size of the stock. At intermediate levels of population density, however, fish can breed maximally provided that (environmental) conditions are appropriate, there will be a surplus of production available to harvest, and MSY can be achieved. Other concepts used in fisheries management related to but generally lower than MSY are optimum sustainable yield and maximum economic yield.

The concept of maximum sustainable yield is referred to extensively in fisheries management (Hilborn and Walters, 1992; Haddon, 2001), but its level varies with the stock, depending on the life history of the fish being harvested as well as the selectivity of the fishing gear being used. However, although MSY is often cited as a target for which to aim in single-species management, the generalized approach has been criticized widely because it overlooks some of the key factors involved in the dynamics of management, e.g. the size, age, and state of maturity and fecundity of the fish being harvested.

Some of the many fisheries collapses that have eventuated worldwide can no doubt be blamed on MSY being applied as a target for management, and there is also the issue that single-species fisheries management itself overlooks the necessity of maintaining ecosystem health and integrity through controlling bycatch and not impacting habitat unduly. Fisheries management covering several cooccurring species is even more complex and prone to additional uncertainties, but MSY remains in many cases the primary management target; it is essential therefore that the concept is not misused and misunderstood.

References and further reading

Cochrane, K.L., Garcia, S.M. (Eds.), 2009. A Fishery Manager's Guidebook, second ed. FAO, Rome. Print ISBN: 9781405170857; Online ISBN: 9781444316315.
Haddon, M., 2001. Modelling and Quantitative Methods in Fisheries. Chapman & Hall/CRC Press, Boca Raton, FL, p. 406.

Hilborn, R., Walters, C.J., 1992. Quantitative Fisheries Stock Assessment. Choice, Dynamics and Uncertainty. Kluwer, Boston, MA, p. 570.
Quinn, T.J., Deriso, R.B., 1999. Quantitative Fish Dynamics. Oxford University Press, New York, p. 542.
Ricker, W.E., 1975. Computation and interpretation of biological statistics of fish populations. Bulletin of the Fisheries research Board of Canada, 191, p. 382.

Linked

Recruitment; Spawning-stock biomass

Fisheries terms: 33. Total allowable catch (TAC)

The total allowable catch (TAC) is the total weight (or occasionally numbers) of fish/shellfish or even another marine resource that can be harvested from a stock or population in a specified period of time, generally per year or per season. Before taking effect formally, a TAC is usually subdivided by country or type of fishery and proportional allocations made often on the basis of past performance. It is the most common and easily understood form of fishery management in use worldwide today. For reasons of the precaution required because of the uncertainty in all fisheries evaluations, however, it does not equate to the maximum yield possible annually on the generalized yield vs cpue plot given elsewhere in these fisheries definitions, but is usually set slightly lower, on the ascending limb of the curve.

Issues of concern relating to TAC as a form of management, given that there is an overabundance worldwide of actual or potential fishing effort, relate to its effectiveness in terms of compliance and control, so in many cases management by TAC is accompanied by dedicated forms of effort or spatial management in terms of, for instance, caps on effort (e.g., days at sea), and completely or partially closed areas. Other controls, for example, minimum landing size, are also applied widely through *inter alia* mesh size limits in conjunction with TAC management. Given that fishing is on an "unseen" (by the fisher) resource, even with the use of modern electronic aids, there can always be incidental/accidental takes of marine resources over and above the TAC and the allocations within it. However, such catches are generally estimated or even recorded and taken up in the calculation of the scientifically calculated and advised initial TAC.

Another problem associated with management by TAC is that it is by definition applied to single stocks, populations, or species. Many species of fish are caught and harvested in combination with others, sometimes many others with occasionally their own TAC, and the establishment of a multispecies TAC is challenging. Furthermore, the control of dynamically varying proportions in catches is difficult for both individuals and countries. Therefore, the issue of discarding (of TAC and non-TAC species) allied to the necessary call for ecosystem rather than individual species/stock management is of crucial concern to modern-day fisheries management.

References and further reading

Cochrane, K.L., Garcia, S.M. (Eds.), 2009. A Fishery Manager's Guidebook, second ed. FAO, Rome. Print ISBN: 9781405170857; Online ISBN: 9781444316315.
Haddon, M., 2001. Modelling and Quantitative Methods in Fisheries. Chapman & Hall/CRC Press, Boca Raton, FL, p. 406.
Hilborn, R., Walters, C.J., 1992. Quantitative Fisheries Stock Assessment. Choice, Dynamics and Uncertainty. Kluwer, Boston, MA, p. 570.
Quinn, T.J., Deriso, R.B., 1999. Quantitative Fish Dynamics. Oxford University Press, New York, p. 542.

Linked

Fishing mortality; Maximum sustainable yield

Fisheries terms: 34. Reference points

As emphasized throughout this volume, detecting change in coastal areas, whether due to natural or anthropogenic causes, relies on a decision of against what baseline, threshold, trigger, or reference value that change is judged. It is also emphasized that whereas "surveillance" of environmental conditions is the detection of changes in spatial and/or temporal terms, perhaps with post hoc management actions, true monitoring requires setting predetermined thresholds and then monitoring against those on the basis that if they are surpassed then management action has previously been defined.

Limit reference points are biological or fishery management indicators that define the point at which action needs to be taken to safeguard a fish or shellfish stock (Caddy and Mahon, 1995). In order for stocks and the fisheries exploiting them to remain within safe biological limits (defined generally as the stock being above the minimum biologically acceptable limit, MBAL; see later), there needs to be a good probability that the spawning-stock biomass (B) is above the threshold where recruitment (new fish coming into the stock) might be impaired and that the fishing mortality (F) is below that which would likely drive the spawning stock to the biomass threshold, a condition that needs to be avoided. Hence, B_{lim} is the minimum spawning-stock biomass and F_{lim} the maximum acceptable fishing mortality to preclude the stock slipping into what would be regarded scientifically as a precarious position.

The certainty with which reference points can be identified varies with the quality of assessment data available. Therefore, however, it is generally deemed advisable also to stipulate precautionary reference points that specify higher biomass thresholds than B_{lim} and lower fishing mortality thresholds than F_{lim}, with a view to taking remedial action well before the limit reference points (one or both) are approached. These more cautious reference points are referred to as precautionary ones, so B_{pa} is the precautionary minimum spawning-stock biomass and F_{pa} the precautionary maximum fishing mortality.

Often, the value for B_{pa} will be the same as that identified as the minimum biologically acceptable limit, MBAL (defined as the estimated spawning-stock biomass below which a stock should not be allowed to fall for fear of imminent recruitment failure and stock collapse). However, in circumstances where the relationship between the exploited stock and the spawning stock is not that clear, as is the case with many deepwater fish currently being exploited, limit reference points are sometimes expressed with respect to the unexploited stock. Reference points are now widely used in fisheries management as the primary output of stock assessments to control different aspects of the fishery or population, and fishing regulations are generally set to meet these benchmarks. However, the benchmark in use for each stock depends on scientific advice and the management targets of the decision-maker and his/her regulating agency.

References and further reading

Caddy, J.F., Mahon, R., 1995. Reference points for fisheries management. FAO Fisheries Technical Paper, 347, p. 83.
Cochrane, K.L., Garcia, S.M. (Eds.), 2009. A Fishery Manager's Guidebook, second ed. FAO, Rome. Print ISBN: 9781405170857; Online ISBN: 9781444316315.

Linked

Baseline and reference conditions; Environmental assessment

Fisheries terms: 35. Fisheries management

It is emphasized here that coastal industry managers have to be aware of their influence on other activity sectors and of the influence of those other sectors on the coastal industries. Perhaps more than the management of any other sector, fisheries management has a longer history and a greater public and political profile than most other areas of coastal management. In particular, and with all other coastal activities, it has to engage the statutory bodies responsible, the role of statutory and non-statutory (also voluntary) management groups, the links between water quality and conservation management and fish and fisheries management, and also include estuaries and the adjacent freshwaters and the ocean.

Management of coastal, estuarine, or estuary-dependent stocks is conducted at several levels (international/European, federal/national, regional, state, and even local) and some management is well coordinated between and among agencies but there still can be jurisdictional issues or requirements to be met. The fisheries management framework (Fig. 1) that can be applied to coasts and estuaries includes many actions and measures.

These actions and management measures are as follows:

(a) environmental measures, which include spatial and temporal measures, ecosystem-based approaches, habitat measures, and assessment and monitoring;
(b) data and information handling measures, including stock assessment and stock structure as well as modeling approaches;
(c) technology and economic measures, including influences on and by markets, catch and effort limitations, and changes to the available technology, and
(d) socioeconomic measures involving macroeconomics and the wishes, desires, and behavior of society (from Whitfield et al., 2022, expanded greatly on the information in Link et al. 2020).

As shown in the figures, the previous fisheries terms given here all have a place by providing data against which the management measures can be judged.

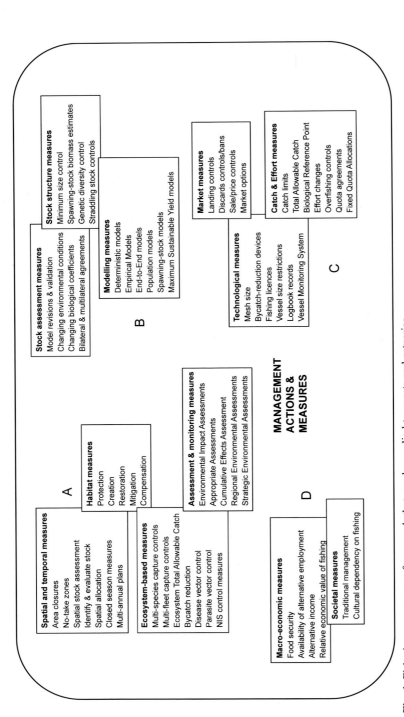

Fig. 1 Fisheries management framework that can be applied to coasts and estuaries. From Whitfield, A.K., Able, K.W., Blaber, S.J.M., Elliott, M., 2022. Fish and Fisheries in Estuaries—A Global Perspective, vols. 1 and 2. John Wiley & Sons, Oxford, pp. 1056, ISBN 9781444336672, expanded greatly on the information in Link, J.S., Huse, G, Gaichas, S., et al. 2020. Changing how we approach fisheries: a first attempt at an operational framework for ecosystem approaches to fisheries management. Fish Fish. https://doi.org/10.1111/faf.12438.

References and further reading

Jennings, S., Kaiser, M., Reynolds, J.D., 2001. Marine Fisheries Ecology. Blackwells, Oxford, p. 417.

Link, J.S., Huse, G., Gaichas, S., et al., 2020. Changing how we approach fisheries: a first attempt at an operational framework for ecosystem approaches to fisheries management. Fish Fish. https://doi.org/10.1111/faf.12438.

Whitfield, A.K., Able, K.W., Blaber, S.J.M., Elliott, M., 2022. Fish and Fisheries in Estuaries—A Global Perspective. vols. 1 and 2 John Wiley & Sons, Oxford, ISBN: 9781444336672, p. 1056.

Linked

Sustainable environmental management

Section 3

Impacts and assessment

Introduction

It is axiomatic that any development with the potential to adversely affect the natural or human-built environment needs permission from a regulatory or statutory authority. In particular, the law often requires an assessment of the expected or predicted environmental impacts of all industrial developments, especially large ones, from the exploration and construction phases, through operation and to decommissioning. Even in cases where an environmental assessment is not required legally, as with smaller developments or those for which minor environmental consequences are predicted, one is often performed in order to smooth the planning application process.

Most coastal industries have long been required to adopt formal environmental impact assessment (EIA) procedures, and it is emphasized that environmental impact assessments should be founded on good and defendable science. Being defendable ensures that the development or activity is legally and societally accepted and permitted. This may require developing a fundamental understanding of how an industry interacts with the environment, and vice versa, and the ability to provide a better understanding of risk. In turn, this allows potential issues to be predicted and mitigated and/or compensated. Most recently, this has also included an indication of how an industry will affect climate change mitigation measures but also how climate change will affect an industry or its activities.

The following topics outline the principles used in defining and assessing environmental impacts through formal Environmental Impact Assessment procedures, understanding concepts of ecosystem health, and assessing associated environmental risks. The most significant impacts on the marine or estuarine environment arise from the occupation of space by an industry, the use of any local resources such as minerals for construction or seawater for condenser cooling, and/or the discharge of waste materials. In particular, the need for cooling water and the environmental impacts associated with abstraction and discharge of cooling water are explained as notable examples of these features. However, it is emphasized that environmental assessments and procedures are similar to all industries.

Impacts and assessment: 36. Environmental assessment

Environmental assessments of one type or another are performed before a development project takes place, so that the environmental implications of the development are examined and evaluated prior to critical decisions being made. The results of such an assessment should then be used by regulators and stakeholders to inform the decision-making process to achieve the best environmental outcome from exploration and commencement of construction through operation and to decommissioning. Consultation with the public is an important feature in such assessments (Fig. 1).

Environmental impact assessment—projects which are likely to have significant environmental effects (covering both the natural and human-built environment) because of their nature, size, or location are subject to an EIA before permission to proceed is granted. For some large projects, such as nuclear power stations and other major developments (e.g., as listed in Annex I of the European EIA Directive), an EIA is mandatory. Smaller but still important projects listed in Annex II, such as land reclamation or wind farms, are subject to a *screening* process. This screening procedure estimates the degree of potential environmental issues that might arise as EIA are only required for cases where a significant environmental effect is predicted. If, after the screening, an EIA is deemed to be required, the process should start by *scoping* the potential impacts and then preparing an environmental statement giving details of the project and expected effects.

An EIA is a very precise procedure in determining the effect of the development "at this place, at this time, constructed, operated and decommissioned in a particular way, with this degree of mitigation and compensation and communicated widely." The EIA is a process which is then recorded or delivered as an Environmental Statement (ES) (also termed an EIA Report in some countries such as Scotland). The most recent update of the European EIA Directive requires that the effects of climate change on the development and the effect of the development on climate change mitigation should be considered in the EIA (Lonsdale et al., 2017).

The EIA and ES also give the opportunity to make improvements to the natural system and even to give mechanisms whereby the biodiversity and habitat structure and function can be improved, especially to compensate for past environmental degradation. For example, in England, the Environment Act 2021 has introduced, through changes to the planning legislation, the concept of Biodiversity Net Gain (BNG). This will be a mandatory requirement on land, in freshwaters, and to the low water mark in tidal waters. Marine environmental gain will apply as a complementary mandatory measure below the low water mark. Developments which are considered to be large in scope are required to provide an outcome that improves elements of our natural capital such as biodiversity, rather than simply protect against any negative impacts.

All developed countries have rigorous environmental assessment legislation; as an example in the case of Europe, several key directives are important in this regard,

Impacts and assessment

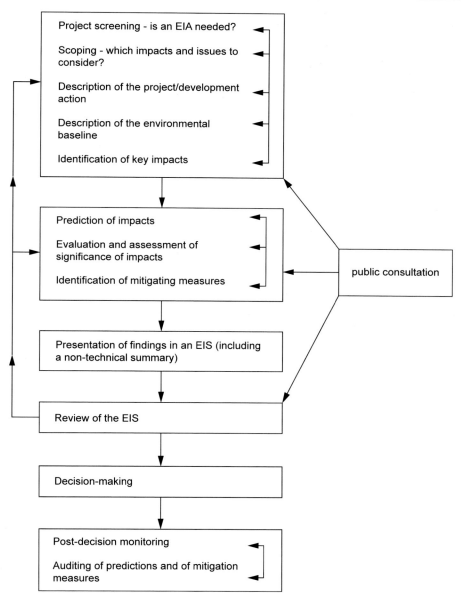

Fig. 1 The structure of an Environmental Impact Assessment leading to the production of an Environmental Statement (or Environmental Impact Statement, EIA Report).
From Glasson, J., Therivel, R., Chadwick, A., 2005. Introduction to Environmental Impact Assessment, third ed. Routledge, London, p. 448, 10: 0415338379.

namely the Strategic Environmental Assessment Directive (SEA) 2001/42/EC, the Habitats and Species Directive, and the Environmental Impact Assessment Directive (EIA) 85/337/EEC (updated to 2014/52/EU). The first relates to public plans and programs and the last to important large capital projects whereas the Habitats and Species Directive relates to the effects of a plan or project on a defined conservation feature or features in an area. Those conservation features may be priorities such as a species, a physiographic feature such as an estuary, or a habitat. In some cases, all assessments from these directives may be required for the same project. This will arise where a public body is involved in the planning leading to a project that having passed the SEA also becomes liable to an EIA and a habitats and species assessment (termed an Appropriate Assessment). In such a case, the SEA will have documented a lot of information that can help form the basis of the EIA and should be consulted and incorporated to ensure an integrated approach. It is often the case that a project or plan cofinanced by outside sources will have to comply with international standard for the assessment, often with ISO standards. Similarly, projects and programs cofinanced by the European Union have to comply with the EIA and SEA directives to receive approval for financial assistance, a measure which helps drive sustainable development.

Strategic environmental assessment—This directive is a systematic process for ensuring that environmental issues are considered at every stage in the preparation, implementation, monitoring, and reviewing of plans and programs. It also covers larger areas than just that covered by a development, as such incorporating the environmental assessments of many developments within a regional sea area. Such a framework will lead to a better integration of environmental considerations at the heart of public body's decision.

Appropriate assessment—This relates to the effect of a plan or project on the conservation objectives within or adjacent to a designated nature conservation area. It focuses on determining only the adverse effects on the conservation features, that is, those which were included when the nature conservation site was designated. It will include a Habitats Regulations Assessment (HRA) as laid down by legislation. This aspect has been criticized as omitting wider aspects, components, and features which may be important for the functioning of an area but which had not been included in the application for conservation feature designations.

Currently, there are many aspects requiring assessment (with, e.g., the relevant European or National legislation):

- Catchment quality (e.g., EU Water Framework Directive, US Clean Water Acts)
- Habitat and species conditions (e.g., EU Habitats Directive, US Conservation legislation)
- Marine regional quality (e.g., EU Marine Strategy Framework Directive, US Oceans Acts, Regional Seas Convention Quality Status Reports)
- Cumulative impacts assessment (e.g., EU CIA Directive)
- Strategic environmental assessment (e.g., EU SEA Directive)
- Environmental Impact Assessment (e.g., EIA legislation worldwide)
- Permit conditions for industry and marine activities

In turn, coastal industries have many obligations, through the polluter-pays or damager-debt principles:

- EIA (process) linked to outcome (Environmental Statement, EIA Report) (Directive, planning permission)
- Appropriate assessment (linked to EU Habitat and Species Directive HSD)
- Habitat regulations assessment (link to HSD)
- Status and pressures monitoring (linked to EU Water Framework Directive (WFD), Marine Strategy Framework Directive (MSFD))
- Cumulative impact assessment
- Strategic environmental assessment—linked to Maritime Spatial Planning (EU Maritime Spatial Planning Directive)
- H1/EpiSuite—linked to complex effluents, Integrated Pollution Prevention and Control authorization (IPPC Directive)
- Database toxicology assessment (linked to license creation, ability to accumulate, persistence, magnify, be toxic) (but there are limitations, cf. synergy/antagonism).

References and further reading

Glasson, J., Therivel, R., Chadwick, A., 2005. Introduction to Environmental Impact Assessment, third ed. Routledge, London, p. 448, 10: 0415338379.

Lonsdale, J., Weston, K., Elliott, M., Blake, S., Edwards, R., 2017. The Amended European Environmental Impact Assessment Directive: UK marine experience and recommendations. Ocean Coast. Manag. 148, 131–142.

http://ec.europa.eu/environment/eia/home.htm.

www.environment-agency.gov.uk/research/policy/33009.aspx.

http://ec.europa.eu/public_opinion/archives/ebs/ebs_372_en.pdf.

For a more generic background see:

http://www.eea.europa.eu/publications/92-826-5409-5/page042.html.

www.globalissues.org/issue/168/environmental-issues. https://consult.defra.gov.uk/defra-net-gain-consultation-team/consultation-on-the-principles-of-marine-net-gain/.

Linked

BAT, BATNEEC, best practice, and integrated pollution prevention and control (IPPC); Baseline and reference conditions; Integrated marine management

Impacts and assessment: 37. Marine processes and human impacts

As emphasized throughout this volume, the primary responsibility of a coastal industry environmental or sustainability manager is to determine the effects of the industry on the surrounding human and natural environment and the effect of the natural and human environment on the industry. This requires a good knowledge of the behavior of the industry and the features and behavior of the natural environment.

The natural marine and estuarine environment and its interaction with human systems can be considered to be a set of dominant critical and fundamental processes and these can be divided into physicochemical, ecological, and anthropogenic categories (see Fig. 1 and Table 1). They follow the general framework given in the Fig. 1—that the physicochemical system creates the habitat which is then colonized by organisms (this may be termed the environment-biology interactions) and so create the ecological structure (defined as the state at one time). The organisms then interact with each other at individual, population, and community level, therefore termed the biology-biology relationships, thereby creating the ecological functioning (defined as rate processes). Completing this cycle, the biology can then create a feedback mechanism and influence the physicochemical system (this is termed the biology to environment relationships). In essence, once the physical system has set up the conditions to be colonized by relevant organisms, those organisms then modify the system via feedback loops. Over these natural processes are superimposed the anthropogenic features and processes, including those from the coastal industries.

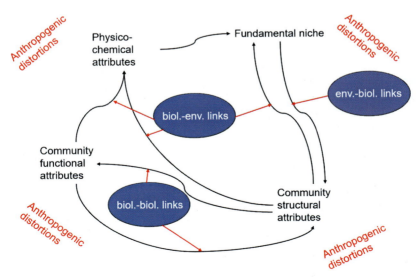

Fig. 1 Conceptual model of underlying ecological structure and functioning.

Table 1 Critical and fundamental processes and properties of ecological systems.

(a) Level 1: Physicochemical processes	
Process	**Specific regions to which the process applies**
Wind regime leading to the wave and swell regime creating the exposure regime	Primarily open coastal areas
Unrestricted wave energy to structure the shore exposure regimes	Primarily open coastal areas
Water flow/runoff creating the salinity regime/maintenance of salinity regime	Primarily estuarine areas
Appropriate temperature regime	All areas
Lack of mixing creating the stratification regime	Primarily restricted to exchange waters
Sediment to intertidal and subtidal areas	Primarily sedimenting areas inshore and in estuaries
Suitable (biological) substratum	All areas
Photic zone/light regime for primary producers	Primarily clear (less turbid) coastal areas
Production of favorable water quality for nektonic migration	Primarily estuarine migration routes
Supply of oxygenated water	Primarily waters of restricted exchange and/or organic enrichment
An uninterrupted flow creating the tidal regime	Primarily estuarine areas
Sediment capture and retention	Primarily estuarine areas
Energy dissipation coastal protection	Primarily open coastal areas
Habitat maintenance	All areas
Dispersal of propagules through water exchange/currents	All areas
Natural disturbance (scour, storm action, grazing, bioturbation)	All areas
(b) Level 2: Ecological processes—Biological inter-relationships and mediation	
Processes	**Ecosystems to which it applies**
Conditions for carbon fixation (autochthonous production)	Primary producing bed areas (saltmarshes, seagrass beds, rocky shores, etc.), water column
Nutrient sequestration in sediments	Primarily muddy intertidal sediments
Nutrient levels for primary production	Water column
Primary production of material for primary consumers	All producing areas
Net organic production/degradation creating oxygen regime	Within water column and bed sediments
Biomodification of sediments by fauna/flora (biostabilization, bioturbation, bioengineers)	Primarily intertidal muddy sediments

Continued

Table 1 Continued

(b) Level 2: Ecological processes—Biological inter-relationships and mediation	
Processes	**Ecosystems to which it applies**
Detrital processing and the delivery of detritus for decomposer food chain (allochthonous inputs)	Primarily estuarine and inshore areas
Delivery of recruiting organisms to an area	All areas
Net settlement patterns creating competition	All areas
Supply of food/nutrients to higher consumers	Primarily important feeding areas such as mudflats and subtidal sandbanks
Removal of waste products	All areas
Critical internal (within/between an organism or community) processes (reproductive ability, disease resistance, predator defense, damage repair, growth)	All areas

(c) Level 3: Anthropogenic influences on physicochemical-biological processes	
Process	**General examples**
Water quality creating barriers to migration	E.g., input of oxygen-demanding wastes such as sewage into estuaries
Physical barriers to migrations/dispersal	E.g., presence of weirs and amenity barrages in estuaries
Estuarine, coastal, and offshore structures creating hydrographic distortions	E.g., offshore wave, tidal and wind energy devices
Polluting inputs creating contamination/pollution responses	E.g., industrial discharges
Hydrographic and nutrient conditions creating eutrophication	E.g., diffuse runoff from agriculture and point source sewage discharges into waters of restricted exchange
Organic enrichment creating community and size-spectral response	E.g., paper mill waste, sewage discharges, and late-stage oil from spillages
Removal of size classes of biota and/or species creating community imbalance	E.g., fishing and shellfisheries targeting certain species and/or removing bycatch
Removal of a population or increase of cultured species reducing genetic diversity	E.g., by fishing and aquaculture escapees, respectively
Input of alien and introduced species	E.g., by transferal in ballast water or the creation of new niches for colonization such as hard bottoms in soft sediment areas
Loss of seabed/wetland reducing biological productivity	E.g., land claim in estuaries or building infrastructure on the seabed
Removal of prey populations and carrying capacity	E.g., loss of wetland and intertidal flats through land claim or hydrographical modifications due to barrage construction

The construction, operation, and decommissioning of coastal industries therefore has to consider the impact of these processes on the industrial plant and the impact of the plant on these processes. For example, the characteristics of the water column such as temperature, salinity, and oxygen concentration allow those organisms such as fish and plankton to colonize an area as long as the environmental tolerances of those organisms allow colonization (Solan and Whiteley, 2016). Once one set of organisms colonizes an area then interactions such as competition, recruitment cycles, and predator-prey interactions will modify the community and populations. Following this, the organisms may modify the environment, for example an excessive number of fish can reduce the water oxygen levels (Elliott and Hemingway, 2002) and certain benthic organisms will modify the bed sediment characteristics (Gray and Elliott, 2009). Construction and operation may then in turn either affect the water and sediment quality and quantity or the organisms themselves, and decommissioning will be required to allow the structure and functioning to return to pre-construction levels.

References and further reading

Elliott, M., Hemingway, K.L. (Eds.), 2002. Fishes in Estuaries. Blackwell Science, Oxford, p. 636.
Gray, J.S., Elliott, M., 2009. Ecology of Marine Sediments: Science to Management. OUP, Oxford, p. 260.
McLusky, D.S., Elliott, M., 2004. The Estuarine Ecosystem; Ecology, Threats and Management, third ed. OUP, Oxford, p. 216.
Rajagopal, S., Jenner, H.A., Venugopalan, V.P. (Eds.), 2012. Operational and Environmental Consequences of Large Industrial Cooling Water Systems. Springer, Heidelberg/New York.
Solan, M., Whiteley, N. (Eds.), 2016. Stressors in the Marine Environment: Physiological and Ecological Responses: Societal Implications. OUP, Oxford. Hardback ISBN 9780198718826.
Wolanski, E., McLusky, D.S. (Eds.), 2011. Treatise on Estuarine & Coastal Science, 12 Volumes. Academic Press, Waltham.

Linked

Ecohydrology and ecoengineering; Ecosystem resilience, resistance, recovery

Impacts and assessment: 38. Biological and ecosystem health

The environmental or sustainability manager of all coastal industries has a primary role in protecting the natural system, and indeed the adjacent human systems, from any adverse effects created by the industrial activities. In essence, this requires ensuring that the health of natural and human systems is not adversely affected by the industry, especially its occupation of space; use of materials; or its discharges to land, sea, or air.

Health is arguably a term more understood by the general public, who can relate it to their own well-being, and medical professionals than as an environmental or scientific term. Despite this, it has become the overriding concern in determining if an area has been degraded as the result of human activities. For example, in contrast to contamination, which is defined as the addition of materials to the system but without implying biological change, pollution is defined as the addition of materials through human activity but leading to biological harm and thus a deterioration in health of some part or all parts of the system under study (McLusky and Elliott, 2004).

That harm (due to human activities) can be defined as "the reduction in the fitness for survival" in one or more levels of biological organization—the latter levels are cell, tissue, individual, population, community, and ecosystem. Hence the approach to and determination of human health and environmental health may be analogous:

Medical		Environmental
Diagnosis	–	Assessment
Prognosis	–	Prediction
Treatment	–	Remediation/creation
Recovery	–	Restoration
Prevention	–	Prevention

An organism may be surviving environmental stress but still be impaired; for example its energy budget may be affected by higher temperatures from a heated effluent such that the organism is less able to put energy into reproduction, hence reducing the "fitness" of the population to survive. A healthy ecosystem can be regarded, therefore, as one that can maintain, support, and protect fundamental and final services while delivering benefits for society. Following the medical health analogy, symptoms of ecosystem pathology can be defined (see later) as adverse changes in natural features against which monitoring and management can be carried out. Hence any activity which adds or removes material from the aquatic system resulting in an adverse change to the ecological system can be regarded as potentially or actually impacting the system health.

The resulting changes to ecosystems through human activities are dependent on the ability of the component organisms to withstand the stressors and some organisms have greater tolerance to environmental variation than others; for example, estuarine organisms may be naturally stress tolerant owing to their occurrence in highly varying conditions. Hence such environmental variability may be a "stress" to some species but a "subsidy" to others, thereby allowing tolerant species to survive and thrive while removing, killing, or causing migration of more sensitive species.

Estuarine communities also naturally show features which in other systems may be regarded as symptoms of human-induced stress, for example, the presence of opportunistic species (the so-called estuarine quality paradox, see Elliott and Quintino, 2007, 2019). Hence the difficulty of detecting human-induced stress in estuaries against the natural inherent variability and characteristics. Each of these aspects requires the "normal" state and anthropogenic deviations from it to be defined, hence the very large effort expended in defining reference conditions, as an example, under various EU Directives (e.g., the definition of Good Ecological Status under the Water Framework Directive, Good Environmental Status under the Marine Strategy Framework Directive, and Favourable Conservation Status under the Habitats and Species Directive) (Borja et al., 2011 and references therein). Under those directives, and as an acknowledgment that few European habitats are pristine, the health of the system has to be "good," itself difficult to define, rather than pristine (see Mee et al., 2008). However, the health and condition of an ecosystem also relates to the uses of the ecosystem and so encompasses both natural and socioeconomic considerations (what has been termed "fit for purpose") (Tables 1 and 2).

Table 1 Concepts relevant to the assessment of ecosystem condition (Health).

Concepts	Definition	Example references
Ecosystem health	Relates to the normal state and successful functioning of the system	Costanza et al. (1992) and Costanza and Mageau (1999)
Ecosystem integrity	Relates to how pristine or undisturbed a system might be; the ability to maintain its organization	Kay (1991), Jørgensen (1997), Müller et al. (2000), and Campbell (2000)
Ecosystem quality	Condition of a particular ecosystem, measured in relation to each of its intended uses; it is usually assessed in relation to established guidelines, objectives, and indicators set by relevant agencies. Relates to chemical, physical, and biological aspects.	Harding (1992)
Ecosystem pathology	Relates to the symptoms which indicate departure from normality (healthy state?); reflects a decrease in fitness to survive	Rapport (1998) and Harding (1992)
Ecosystem distress	Relates to the state of less than good health and a decreasing fitness to survive; reflects a system reacting to change, functioning under stress or showing departure from normality	Rapport (1995, 1998)

Continued

Table 1 Continued

Concepts	Definition	Example references
Ecosystem dysfunction	Relates to the state of less than good health and poor fitness to survive; not functioning or functioning in a disturbed manner	Rapport (1995)
Ecosystem morbidity	Relates to a state of less than good health—showing signs of 'sickness' and low survival potential	Sherman (2000)

Modified from Elliott, M., Burdon, D., Hemingway, K.L., 2006. Marine ecosystem structure, functioning, health and management and potential approaches to marine ecosystem recovery: a synthesis of current understanding. Unpublished Report to Countryside Council of Wales, Bangor, Institute of Estuarine & Coastal Studies, University of Hull, Hull, UK (Accessed via https://www.iecs.ttd), with references therein.

Table 2 The seven aspects used for defining symptoms of ecosystem pathology.

(1) The fate and effects of nutrients, for example, the increase in concentration as the result of increased diffuse and point source discharges but also as the cause of eutrophication (an over-enhanced productive capacity)

(2) Primary production, for example, the organic production of a system which may be over-stimulated through increased sewage inputs adding nutrients to the system or depressed by the presence of turbid water which reduces the light regime required for photosynthesis

(3) Species diversity, for example, the removal of species which are intolerant of change under stressful conditions and the encouragement of stress-tolerant species; the introduction of alien or invasive species

(4) Community instability (biotic composition), for example, the increase in biological turnover due to the dynamics of stress-tolerant species

(5) Size and biomass spectrum, for example, the tendency toward smaller, r-strategist (usually small-bodied) organisms under stressed conditions

(6) Disease/anomaly prevalence, for example, the reduced tolerance of organisms to infection and the development of pathological anomalies such as ulcers when placed under stress

(7) Contaminant uptake and response, for example, the increased accumulation of conservative contaminants, such as trace metals, and perhaps the production of detoxification or excretion mechanisms after exposure in order to remove toxicants from the system

Modified from Harding, L.E., 1992. Measures of marine environmental quality. Mar. Pollut. Bull. 25, 23–27, and McLusky, D.S. Elliott, M., 2004. The Estuarine Ecosystem; Ecology, Threats and Management, third ed. OUP, Oxford, pp. 216.

References and further reading

Borja, A., Basset, A., Bricker, S., Dauvin, J.-C., Elliott, M., Harrison, T., Marques, J.-C., Weisberg, S.B., West, R., 2011. Chapter 1.08: Classifying Ecological Quality and Integrity of Estuaries. In: Simenstad, C., Yanagi, T. (Eds.), Treatise on Estuarine & Coastal Science. In: Wolanski, E., McLusky, D.S. (Eds.), Volume 1, Classification of Estuarine and Nearshore Coastal Ecosystems, Elsevier, Amsterdam, pp. 125–162.
Campbell, D.E., 2000. Using energy systems theory to define, measure, and interpret ecological integrity and ecosystem health. Ecosyst. Health 6 (3), 181–204.
Costanza, R., Mageau, M., 1999. What is a healthy ecosystem? Aquat. Ecol. 33 (1), 105–115.
Costanza, R., Norton, B., et al., 1992. Ecosystem Health: New Goals for Environmental Management. Island Press, Washington, DC, p. 279.
Elliott, M., 2011. Marine science and management means tackling exogenic unmanaged pressures and endogenic managed pressures—a numbered guide. Mar. Pollut. Bull. 62, 651–655.
Elliott, M., Quintino, V.M., 2007. The estuarine quality paradox, environmental Homeostasis and the difficulty of detecting anthropogenic stress in naturally stressed areas. Mar. Polluti. Bull. 54, 640–645.
Elliott, M., Quintino, V.M., 2019. The estuarine quality paradox concept. In: Encyclopaedia of Ecology, second ed. vol. 1. Elsevier, Amsterdam, ISBN: 978-0-444-63768-0, pp. 78–85, https://doi.org/10.1016/B978-0-12-409548-9.11054-1 (Editor-in-Chief B. Fath).
Harding, L.E., 1992. Measures of marine environmental quality. Mar. Pollut. Bull. 25, 23–27.
Jørgensen, S.E., 1997. Integration of Ecosystem Theories: A Pattern. Kluwer Academic, Dordrecht. 428 pp.
Kay, J.J., 1991. A non-equilibrium thermodynamic framework for discussing ecosystem integrity. Environ. Manag. 15 (4), 483–495.
Lackey, R.T., 2001. Values, policy, and ecosystem health. BioScience 51 (6) 437–443.
McLusky, D.S., Elliott, M., 2004. The Estuarine Ecosystem; Ecology, Threats and Management, third ed. OUP, Oxford, p. 216.
Mee, L.D., Jefferson, R.L., Laffoley, D.d'.A., Elliott, M., 2008. How good is good? Human values and Europe's proposed marine strategy directive. Mar. Pollut. Bull. 56, 187–204.
Müller, F., Hoffmann-Kroll, R., Wiggering, H., 2000. Indicating ecosystem integrity – theoretical concepts and environmental requirements. Ecol. Model. 130, 13–23.
Rapport, D.J., 1995. Ecosystem Health: Exploring the Territory. Ecosyst. Health 1, 5–13.
Rapport, D.J., 1998. Defining ecosystem health. In: Rapport, D.J., Costanza, R., Epstein, P.R., Gaudet, C., Levins, R. (Eds.), Ecosystem Health. Blackwell Science, Berlin, pp. 18–33.
Sherman, K., 2000. Why regional coastal monitoring for assessment of ecosystem health? Ecosyst. Health 6 (3), 205–216.

Linked

Biological and ecosystem health; Determination of significant effect

Impacts and assessment: 39. Activity-, pressures-, effects- and management response-footprints

The plethora of human activities and their pressures and impacts on estuaries, coasts, and seas require managing at local, national, regional, and international scales. This requires management responses in a programme of measures to determine (a) the time and area in which the human activities take place, (b) the time and area covered by the pressures generated by the activities on the prevailing habitats and species (in which pressures are defined as the mechanisms of change), and (c) the time and area over which any adverse effects (and even benefits) occur to both the natural and human systems.

The times and areas therefore can be regarded as footprints and hence the spatial and temporal scales of these lead to the concepts of *activity-, pressures-, effects-, and management responses-footprints* defined here. These footprints cover areas from tens of m^2 to millions of km^2, and, in the case of management responses, from a large number of local instruments to a few global instruments thereby giving rise to what is termed the management response-footprint pyramids. This may operate from either bottom-up or top-down directions, whether as the result of local societal demands for clean, healthy, productive, and diverse seas or by diktat from national, supranational, and global bodies such as the United Nations (see Cormier et al., 2022).

The coastal industry environment or sustainability manager will be charged with determining the footprint of the industry activity and the pressures and effects leading from that footprint. In turn, the regulators permitting that activity should understand the wide range of environmental control regulations, i.e. their footprint, both spatially and temporally (Table 1).

Table 1 Definitions for activity-, pressures-, and effects-footprints.

Footprint	Definition
Activity-footprint	The area and/or time, based on the duration, intensity, and frequency of an activity which ideally has been legally sanctioned by a regulator in an authorization, license, permit, or consent, and which should be clearly defined and mapped in order to be legally defendable; it should be both easily observed and monitored and attributable to the proponent of the activity
Pressures-footprint	The area and time covered by the mechanism(s) of change resulting from a given activity, or all the activities in an area, once avoidance and mitigation measures have been employed (the endogenic managed pressures). It does not necessarily coincide with the activity-footprint and may usually be larger but could be smaller. It also needs to include the influence and consequences of pressures emanating from outside the management area (the exogenic unmanaged pressures); given that these are caused by wide-scale events (and even global developments) then these are likely to have larger scale (spatial and temporal) consequences
Effects-footprint	The spatial (extent), temporal (duration), intensity, persistence, and frequency characteristics resulting from (a) a single pressure from a marine activity, (b) all the pressures from that activity, (c) all the pressures from all activities in an area, or (d) all pressures from all activities in an area or emanating from outside the management area. They include both the adverse and positive consequences on the natural ecosystem components and on the ecosystem services and societal goods and benefits. They need to include the near-field and far-field effects and near- and far-time effects because of the dynamics and characteristics of marine areas and the uses and users of the area. They may be larger in extent and more persistent than the causing activity-footprint and the resulting pressures-footprints. They also need to encompass the effects of both endogenic and exogenic pressures operating in that area
Response-footprints	The area and time covered by the governance methods and approaches of monitoring, assessing, and controlling the causes and consequences involved in the use of the marine environment through public policy-making, marine planning, and regulatory processes. The policies, marine plans, and technical measures produced by these processes indicate the means of determining if legal controls are satisfied and of providing information and data to national and supranational bodies. They focus on the area and/or time covered by the marine management actions and measures (e.g., programme of measures), including the distribution and range of a species

Based on Elliott, M., Borja, A., Cormier, R., 2020. Activity-footprints, pressures-footprints and effects-footprints—walking the pathway to determining and managing human impacts in the sea. Mar. Pollut. Bull. 155, 111201. https://doi.org/10.1016/j.marpolbul.2020.111201; Cormier, R., Elliott, M., Borja, Á., 2022. Managing marine resources sustainably—the 'management response-footprint pyramid' covering policy, plans and technical measures. Front. Mar. Sci. 9, 869992. https://doi.org/10.3389/fmars.2022.869992.

References and further reading

Andersen, J.H., Berzaghi, F., Christensen, T., Geertz-Hansen, O., Mosbech, A., Stock, A., Zinglersen, K.B., Wisz, M.S., 2017. Potential for cumulative effects of human stressors on fish, sea birds and marine mammals in Arctic waters. Estuar. Coast. Shelf Sci. 184, 202–206.

Borgwardt, F., Robinson, L.A., Trauner, D., Teixeira, H., Nogueira, A.J.A., Lillebø, A.I., Piet, G.J., Kuemmerlen, M., O'Higgins, T., McDonald, H., Arevalo-Torres, J., Barbosa, A.L., Iglesias-Campos, A., Hein, T., Culhane, F., 2019. Exploring variability in environmental impact risk from human activities across aquatic ecosystems. Sci. Total Environ. 652. https://doi.org/10.1016/J.SCITOTENV.2018.10.339.

Cormier, R., Elliott, M., Borja, Á., 2022. Managing marine resources sustainably—the 'management response-footprint pyramid' covering policy, plans and technical measures. Front. Mar. Sci. 9, 869992. https://doi.org/10.3389/fmars.2022.869992.

Elliott, M., Borja, A., Cormier, R., 2020. Activity-footprints, pressures-footprints and effects-footprints—walking the pathway to determining and managing human impacts in the sea. Mar. Pollut. Bull. 155, 111201. https://doi.org/10.1016/j.marpolbul.2020.111201.

Korpinen, S., Klančnik, K., Peterlin, M., Nurmi, M., Laamanen, L., Zupančič, G., Murray, C., Harvey, T., Andersen, J.H., Zenetos, A., Stein, U., Tunesi, L., Abhold, K., Piet, G., Kallenbach, E., Agnesi, S., Bolman, B., Vaughan, D., Reker, J., Royo Gelabert, E., 2019. Multiple pressures and their combined effects in Europe's seas. ETC/ICM Technical Report 4/2019: European Topic Centre on Inland, Coastal and Marine waters, Brussels, p. 164.

Verones, F., Moran, D., Stadler, K., Kanemoto, K., Wood, R., 2017. Resource footprints and their ecosystem consequences. Sci. Rep. 7, 40743.

Linked

Baseline and reference conditions; Ecosystem resilience, resistance, recovery

Impacts and assessment: 40. Hazards and risk

Coastal industries are subject to and may contribute to the many coastal hazards, each of which has causes and consequences. Hence, coastal environmental managers and operators need to be aware of the causes and consequences of hazards. In essence, hazards may occur either naturally or by human actions and they become risks when they adversely affect something valued by humans such as health, welfare, or property; in some cases, human responses to one hazard may make the consequences even more severe—for example, removing coastal protection due to mangroves may make the repercussions of cyclones and tsunamis even more severe. Hazard is the cause of an adverse effect compared to risk which in contrast is the probability of effect (i.e., the likely consequences) potentially leading to more severe consequences to humans. The severity of the hazard is measured by the number of people likely to be affected or the value of the assets likely to be affected. At its most severe, the concept of disaster then represents the interplay between the social and the natural systems.

Responses by society to hazards result from a perception of risk and the willingness to act depends on the perception and evidence for the consequences; for example, the placement of storm surge barriers occurred as a response to the 1953 storm surge in the North Sea and in anticipation of similar events in the future. Natural risk can be defined as the damage expected from an actual or hypothetical scenario triggered by phenomena or events following natural events (Smith and Petley, 2009).

Coastal hazards may be natural or anthropogenic (Table 1) and are divided into those over which individuals or communities have some control, for example, by agreeing not to inhabit vulnerable areas, and those where they have no control, for

Table 1 Typology of hazards in coastal and coastal wetland area.

Hazard	Natural or anthropogenic?	Examples
(A) Surface hydrological hazards	Natural but exacerbated by human activities	High tide flooding, spring tide, and equinoctial flooding; flash flooding, El Niño Southern Oscillation/North Atlantic Oscillation patterns; flow delivery repercussions of catchment modifications (land use increasing sediment loading, dams decreasing peak flows and sediment loadings, etc.)
(B) Surface physiographic removal by natural processes—chronic/long term	Natural but exacerbated by human activities	Gradual erosion of soft cliffs by slumping, estuary bank erosion by prevailing currents

Continued

Table 1 Continued

Hazard	Natural or anthropogenic?	Examples
(C) Surface physiographic removal by human actions—chronic/long term	Anthropogenic	Land claim, removal of wetlands for urban and agricultural area
(D) Surface physiographic removal—acute/short term	Natural	Cliff failure, undercutting of hard cliffs and intermittent erosion
(E) Climatological hazards—acute/short term	Natural but exacerbated by human activities	Storm surges, cyclones, tropical storms, hurricanes, offshore surges, fluvial and pluvial flooding
(F) Climatological hazards—chronic/long term	Natural but exacerbated by human activities or anthropogenic	Ocean acidification, sea level rise, storminess, ingress of seawater/saline intrusion
(G) Tectonic hazards—acute/short term	Natural	Tsunamis, seismic slippages, earthquakes
(H) Tectonic hazards—chronic/long term	Natural	Isostatic rebound, subsidence
(I) Anthropogenic microbial biohazards	Anthropogenic	Sewage pathogens
(J) Anthropogenic macrobial biohazards	Anthropogenic	Alien, introduced and invasive species, genetically modified organisms, bloom-forming species
(K) Anthropogenic introduced technological hazards	Anthropogenic	Failures or mismanagement of infrastructure, coastal defenses, catchment impedance structures (dams, weirs)
(L) Anthropogenic extractive technological hazards	Anthropogenic	Removal of space, removal of biological populations (fish, shellfish, etc.); seabed extraction and oil/gas/coal extraction leading to subsidence
(M) Anthropogenic acute chemical hazards	Anthropogenic	Pollution from one-off spillages, oil spills
(N) Anthropogenic chronic chemical hazards	Anthropogenic	Diffuse pollution, ocean acidification, litter/garbage, nutrients from land runoff, constant land-based discharges, aerial inputs

Table 1 Continued

Hazard	Natural or anthropogenic?	Examples
(O) Anthropogenic acute geopolitical hazards	Anthropogenic	Terrorism attacks leading to damage on infrastructure
(P) Anthropogenic chronic geopolitical hazards	Anthropogenic	Wars created by shortage of resources (e.g., land, water, minerals)

Modified from Elliott, M., Cutts, N.D., Trono, A., 2014. A typology of marine and estuarine hazards and risks as vectors of change: a review for vulnerable coasts and their management. Ocean Coast. Manag. 93, 88–99, in Elliott, M., Day, J. W., Ramachandran, R., Wolanski, E., 2019. Chapter 1—A synthesis: what future for coasts, estuaries, deltas, and other transitional habitats in 2050 and beyond? In: Wolanski, E., Day, J.W., Elliott, M., Ramachandran, R. (Eds.), Coasts and Estuaries: The Future. Elsevier, Amsterdam, pp. 1–28, ISBN 978-0-12-814003-1.

example, tectonic failure or extreme landslip (see Elliott et al., 2014, 2019). Such hazards then require to be tackled using technological, governance, and economic approaches, for example whether we have the capacity in methods, laws, and funding to modify and protect coastal landscapes against the influence of hazards or whether we need the capacity to mitigate the effects of hazards by financially supporting those affected by the hazards.

At the same time, coastal management and global agreements have to ensure that biodiversity is protected and the management actions ensure that the natural structure and functioning are sustainable in the long term. Developed countries have the financial and technological means to reduce the vulnerability and thus risk to the effects of change, for example with climate change, such as building defenses but this is an expensive option. Underlying this is the need and requirement to protect societal benefits such as infrastructure and urban areas, while at the same time protecting the natural system and the delivery of ecosystem services. Hence while we have the capacity to engineer the coastline to protect it from hazards, this would be at the risk of creating a non-natural system, thus contravening nature conservation agreements and laws. Coastal industries will be similarly faced with such decisions in protecting their assets while at the same time ensuring that their activities do not adversely affect natural systems.

Natural phenomena that can create risks can be divided into two main categories depending on the source of the causes: endogenous phenomena and exogenous phenomena in relation to their source within or on the Earth's surface. Endogenous causes, for example, include those which can release huge amounts of energy from seismic or tectonic events, and thus are seen as earthquakes and volcanic eruptions and the tsunamis, wrongly termed tidal waves, which emanate from these. Exogenous phenomena, such as landslides, floods and accelerated erosion of beaches and river beds, are often linked to extreme meteorological events which tend to modify the landscape. The nature of the landscape has to be such that it cannot accommodate such forces without resulting modification; for example flat landscapes will exhibit greater change by flooding than ones with greater elevation. Such phenomena are an expression of the internal and external geophysical dynamics and represent the natural evolutionary processes. However, by interacting with societal components (population, settlements, infrastructure, etc.) they frequently determine risk conditions. While natural systems have the capacity to adjust to such natural changes, it is only when the natural and societal aspects interact that we see hazards and risk, both terms being used in relation to human uses of the geographical space.

Floods, landslides, the instability of the coastline, abrupt subsidence, and substratum failure due to the presence of cavities in the subsoil are all either the cause or effect of natural events which are generally grouped together as hydrogeological phenomena. All of these are likely to adversely affect coastal industries. These changes are the result of interaction between meteorological events and the geological, morphological, and hydrological environment, in which humans either play an important role by making the landscape more susceptible to change or are significantly affected. Clearly, natural phenomena can cause disasters but more often human actions make them more severe. For example, in assessing the causes and consequences from the Katrina hurricane in August 2005, Austin (2009) showed that the situation was made worse in Southern Louisiana. These included its often-poor human population being less able to withstand the changes, a long history of coastal modification by natural and man-made levees, and other modifications through canal construction and the oil exploration and extraction industries. In essence, the loss of coastal wetlands removed a capacity to cope with natural events such as hurricanes, a poor and poorly prepared population were unable to cope with the aftereffects, and a large amount of unsuitable city infrastructure then increased the repercussions of the hurricane.

While determining the individual effects of each of these hazards and risks, the major challenge for the environmental and operational managers of coastal industries is to determine the effects of several or many hazards acting together. These interactions, termed cumulative impacts or effects assessment, may be synergistic (additive) or antagonistic (cancelling).

References and further reading

Austin, D.E., 2009. Coastal exploitation, land loss, and hurricanes: a recipe for disaster. Am. Anthropol. 108 (4) 671–691.

Barnett, J., Breakwell, G.M., 2001. Risk perception and experience: hazard personality profiles and individual differences. Risk Anal. 21 (1) 171–178.

Defeo, O., Elliott, M., 2021. Editorial—The 'triple whammy' of coasts under threat—why we should be worried! Mar. Pollut. Bull. 163, 111832. https://doi.org/10.1016/j.marpolbul.2020.111832.

Elliott, M., Cutts, N.D., Trono, A., 2014. A typology of marine and estuarine hazards and risks as vectors of change: a review for vulnerable coasts and their management. Ocean Coast. Manag. 93, 88–99.

Elliott, M., Day, J.W., Ramachandran, R., Wolanski, E., 2019. Chapter 1—A synthesis: what future for coasts, estuaries, deltas, and other transitional habitats in 2050 and beyond? In: Wolanski, E., Day, J.W., Elliott, M., Ramachandran, R. (Eds.), Coasts and Estuaries: The Future. Elsevier, Amsterdam, ISBN: 978-0-12-814003-1, pp. 1–28.

Mee, L.D., Jefferson, R.L., Laffoley, D.d.'.A., Elliott, M., 2008. How good is good? Human values and Europe's proposed marine strategy directive. Mar. Pollut. Bull. 56, 187–204.

Melchers, R.E., 2001. On the ALARP approach to risk management. Reliab. Eng. Syst. Saf. 71 (2) 201–208.

Pilkey, O.H., Young, R.S., 2005. Will hurricane Katrina impact shoreline management? Here's why it should. J. Coast. Res. 21 (6) iii–x.

Smith, K., Petley, D.N., 2009. Environmental Hazards: Assessing Risk and Reducing Disaster, fifth ed. Routledge, Oxford.

Linked

Ecosystem resilience, resistance, recovery; Marine processes and human impact

Impacts and assessment: 41. Interactions between the industrial plant and the marine system

The environmental consequences of human activities can be separated into three categories: what society puts into the sea (e.g., pollutants, hot water, litter), what we take out (e.g., cold water, fishes, aggregates), and the wider residual effects (e.g., climate change) (cf. exogenic unmanaged pressures and endogenic managed pressures) (see accompanying entry 92 and Elliott, 2011). Coastal industries thus have an impact by contributing to each of these during the construction, operation, and decommissioning of the plant in what may be summarized as "the effects of the industry on the marine system." Hence, the plant and its activities may be regarded as having a particular behavior in the marine system which then has to be accepted, mitigated for, or compensated for in order to prevent adverse and unwanted environmental effects.

Similarly, the marine system has an effect on the industry plant by impeding or affecting its operation; for example, cooling water intakes can lead to an excess of impingement of fishes and seaweed or entrainment of settling planktonic stages. These consequences then require management and operational decisions to prevent the efficiency of the plant being reduced. This may be summarized as "the effects of the marine system on the industry" (Table 1).

These various sets of potential problems on the main industrial and environmental components can be addressed and hopefully reduced either by the operators or the environmental regulators while still accepting, allowing for, and accommodating any cumulative, synergistic, or antagonistic effects. Each of these then has economic consequences for the plant operation and environmental consequences for the receiving environment. The former sets of changes can be accommodated under both an Environmental Impact Assessment (EIA) and a Habitats Regulations Assessment (HRA): with the EIA being "what is the effect of this development, at this time and place, carried out in this way, with this degree of mitigation and communication." The HRA is then more specific regarding the effect of this plan or project on the conservation features for which an area is designated.

Table 1 Examples of corresponding operational and environmental issues related to cooling water intakes.

Operational/production issues	Environmental issues
Blockage of cooling water intake screens by impingement of fish, seaweed, etc. to require removal and cooling efficiency loss Excessive settlement by planktonic larval stages in condensers to reduce cooling efficiency, requiring antifouling measures	Removal of biological material to affect natural populations; compromising natural population size Mortalities of entrained organisms compromising natural populations; liberation of antifouling residues with subsequent environmental consequences of pollutants

Table 1 Continued

Operational/production issues	Environmental issues
Land claim and occupation of wetlands and coastal areas to require mitigation measures, habitat creation	Removal of habitat and influences on conservation designations and objectives
Relative sea level rise, isostatic rebound, and changes to climatic conditions to require additional coastal protection and anti-erosion measures	Engineering interference with natural coastal processes, sediment supply, erosion and deposition cycles, shoreline configuration
Discharge of contaminants and polluting material to the receiving areas	Uptake of the contaminants by the sediments, suspended sediments, and biota leading to harm on various levels of biological organization in the near- and far-field receiving area

References and further reading

Elliott, M., 2011. Marine science and management means tackling exogenic unmanaged pressures and endogenic managed pressures—a numbered guide. Mar. Pollut. Bull. 62, 651–655.

Elliott, M., 2012. Preface—Setting the scene and the need for integrated science for the operational and environmental consequences of large industrial cooling water systems. In: Rajagopal, S., Jenner, H.A., Venugopalan, V.P. (Eds.), Operational and Environmental Consequences of Large Industrial Cooling Water Systems. Springer, Heidelberg/New York.

Elliott, M., Burdon, D., Atkins, J.P., Borja, A., Cormier, R., de Jonge, V.N., Turner, R.K., 2017. "*And DPSIR begat DAPSI(W)R(M)!*"—a unifying framework for marine environmental management. Mar. Pollut. Bull. 118 (1-2) 27–40. https://doi.org/10.1016/j.marpolbul.2017.03.049.

Kennish, M.J., Elliott, M. (Eds.), 2011. Treatise on Estuarine & Coastal Science. In: Wolanski, E., McLusky, D.S. (Eds.), Volume 8. Human-induced problems (uses and abuses) in Estuaries and Coasts, Elsevier, Amsterdam, p. 315.

Lepage, M., Capderrey, C., Meire, P., Elliott, M., 2022. Chapter 8 Estuarine degradation and rehabilitation. In: Whitfield, A.K., Able, K.W., Blaber, S.J.M., Elliott, M. (Eds.), Fish and Fisheries in Estuaries—A Global Perspective. John Wiley & Sons, Oxford, pp. 458–552. ISBN 9781444336672.

Linked

Activity-, pressures-, effects- and management response-footprints; Hazards and risk

Impacts and assessment: 42. Water and substratum quality considerations

The quality of waters and the bed (substratum) can be measured in terms of aesthetic, chemical, physical, and biological criteria to ensure it can support a healthy biodiversity and protect all other users of the environment, whether humans or other species. Water has to be suitable for human use, either for consumption or as process or cooling waters. With a few exceptions, water and substratum quality would be expected to be of highest quality when they are free from any anthropogenic pressures. For example, in a wider European context, the Water Framework Directive (WFD), Marine Strategy Framework Directive (MSFD), and Habitats and Species and Wild Birds Directives (HSD, WBD) follow this principle by assessing environmental quality against reference levels or benchmark sites. The desired quality is referred to, respectively, as Good Chemical and Ecological Status (WFD), Good Environmental Status (MSFD), and Favourable Conservation Status (HSD, WBD). Ecosystem quality for the ecological and environmental structure and function aspects are taken as the key measures. However, as these are difficult and/or expensive to quantify with precision, the approach is usually supported with discharge limits which, for example, for industrial discharges, may be chemical (annual mean or a peak concentration statistic) or physical (flow, salinity, temperature, etc.).

Almost any substance can cause disruption of natural systems if in sufficient quantity and it is instructive to think in terms of harmful concentrations rather than harmful substances. Toxic substances can either be natural substances in unnatural amounts (e.g., the metals copper and zinc which are essential metals for many organisms) or unnatural substances in any amounts (e.g., some mercury and arsenic compounds and many pesticides). It is useful to distinguish between contamination, as the addition of materials not normally in a place, and pollution per se, which implies an adverse biological response, defined as a reduction in fitness for survival (Rand, 1995).

It is also necessary to separate and distinguish a polluting oxygen-demanding discharge (e.g., sewage which can promote bacterial activity and deplete oxygen levels) from a poisonous (toxic) discharge (e.g., some metals and many chlorinated organic compounds) which either individually or in combination adversely affects the functioning of organisms. These do not necessarily have to kill the organism to be serious, as, for example, a substance that affects the breeding behavior or fertility of a species could prove just as serious for the continued survival of that species. While some trace metals, such as copper or zinc, may be essential and only toxic in high quantities, other metals such as mercury, cadmium, and arsenic are non-essential and so can eventually exceed the organismal capacity to detoxify them. Although an older reference, McIntyre and Pearce (1980) give an excellent introduction to pollution effects at levels of biological organization from the cell to the ecosystem.

Similarly, other substances, such as phosphates and nitrates, are required by biological systems but if the concentrations or proportions of these substances get substantially changed from natural concentrations (or ratios) then some species may grow in unnaturally high concentrations. This can lead to algal blooms including the potential for toxic species and other signs and symptoms of eutrophication (see the subsequent entry). More complex interactions can be caused by changes in pH, salinity, or temperature as these may have direct lethal effects and/or change the balance of species through behavioral responses as some species may prefer a different regime as well as altering other biological and physiological processes.

Microbiological contaminants as found in sewage discharges, can also be a problem with fecal organisms being of special concern for recreational bathing waters and shellfish destined for human consumption. Other biological problems can arise from the release of ballast water taken on by ships when visiting foreign waters where the discharge can introduce alien species which may outcompete local species and threaten the local ecosystem. As such, non-indigenous species may be regarded as macrobiological pollutants if they cause environmental problems (Olenin et al., 2011).

Water quality is affected by direct discharges, which are relatively easy to control and regulate, and diffuse sources, such as land runoff or rainfall, which can prove to be problematic in detection as the contaminants involved may not even be sourced from the country affected (i.e., in international rivers and catchments). Further sources of contamination can arise from accidents, equipment failures, or major incidents. All industries should therefore engage in risk assessment and risk management in one form or another. In some cases, it is possible to plan for failure, for example, by building bund walls around oil storage tanks, although in major emergencies any measure will be of secondary importance to protecting life and property.

Whereas the water column may disperse, degrade, and/or assimilate contaminant discharged from industrial plant, the sediment may be the ultimate sink for the materials which may get sequestered (buried) and stored. The form and behavior of materials discharged into the water column or settled on to the bed are influenced by the chemistry of the water and sediments. For example, the temperature and pH of the waters will affect the behavior of waterborne chemical contaminants and biogeochemical changes in the sediment will influence the behavior of the contaminants. For example, a metal which forms insoluble products in the absence of oxygen is likely to be sequestered as sulfides in anoxic sediments, i.e. where the creation of hydrogen sulfide has a major influence on both their chemical and biological nature (see Libes, 2009). Such changes in the sediments are termed diagenesis which may result in contaminants being sequestered at specific depths in the sediment. The materials will then only be released once the chemical environment changes such as sediment being moved by natural erosion or dredging.

References and further reading

Gray, J.S., Elliott, M., 2009. Ecology of Marine Sediments: Science to Management. OUP, Oxford, p. 260.
Lawrence, A.J., Hemingway, K.L. (Eds.), 2003. Effects of Pollution on Fish. Blackwell Science Ltd., Oxford.
Libes, S., 2009. Introduction to Marine Biogeochemistry, second ed. Academic Press, London, p. 90.

McIntyre, A.D., Pearce, J.B., 1980. Biological effects of marine pollution and the problems of monitoring. In: Rapports et Proces-Verbaux des Reunions no. 179, Conseil International pour l'Exploration de la Mer, Copenhagen.

Olenin, S., Elliott, M., Bysveen, I., Culverhouse, P., Daunys, D., Dubelaar, G.B.J., Gollasch, S., Goulletquer, P., Jelmert, A., Kantor, Y., Mézeth, K.B., Minchin, D., Occhipinti-Ambrogi, A., Olenina, I., Vandekerkhove, J., 2011. Recommendations on methods for the detection and control of biological pollution in marine coastal waters. Mar. Pollut. Bull. 62 (12) 2598–2604.

Rand, G.M., 1995. Fundamentals of Aquatic Toxicology: Effects, Environmental Fate and Risk Assessment, second ed. CRC Press.

Linked

Ecotoxicology assessment; Microbial pathogens—Chemical interactions; Non-indigenous, alien, invasive and other non-native species

Impacts and assessment: 43. Eutrophication and organic wastes

Many coastal and estuarine industries either influence or are influenced by excess nutrients and organic matter in the receiving waters. Discharges of these waste materials typically come from food processing industries, agriculture, urban discharges, and petrochemical plants and may be from point source discharges or diffuse runoff, the latter especially prevalent in the catchment of the estuaries.

While organic enrichment of water bodies is a natural phenomenon, excess organic matter and nutrients can lead to undesirable signs and symptoms, the suite of which is termed eutrophication. Hence industries will be confronted either by an operational problem, such as how to remove such wastes from their activities or the repercussions of these wastes being taken in cooling waters, or by an environmental problem in which their wastes are in danger of exceeding the assimilative capacity of the receiving water bodies thereby leading to what are termed "undesirable consequences" (Fig. 1).

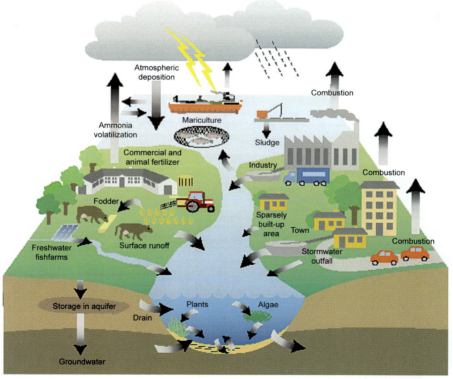

Fig. 1 The nitrogen cycle in the aquatic environment.
Danish Environmental Protection Agency, https://www2.dmu.dk/1_viden/2_miljoe-tilstand/3_vand/4_eutrophication/causes.asp.

Earlier definitions of eutrophication were restricted to an excess of nutrients, but more commonly it is the creation of adverse consequences (i.e., the signs and symptoms) that defines the term. Consequently, there are many legal instruments aimed at controlling such adverse signs and symptoms such as the European Urban Waste-water Treatment Directive (UWWTD), the Nitrates Directive and the *Programme of Measures* for the Water Framework Directive, and the US Clean Water Act. As an example, the UWWTD controls urban (sewage) discharges to a body of water, depending on its dispersing and degrading ability and according to whether "*an area is eutrophic or likely to become eutrophic.*" An area with a high assimilative capacity, such as an open coast, may be sanctioned to receive only primary or preliminary treated sewage whereas a more restricted estuary or lagoon will require more sophisticated (secondary or tertiary) treatment.

In order to make direct comparisons between all of these types of waste that are oxygen demanding, the organic component of the discharge can be converted to standard units indicating the likely scale of effect. For example, the EU UWWTD assumes that one person has a daily organic discharge of 60 g biochemical oxygen demand (BOD) and so, for example, food processing industries will have their organic discharges converted to population equivalents in order to determine their relative influence on receiving waters. As such, a highly organic waste from brewing or distilling can be equivalent to a town of 5000 inhabitants.

Fig. 2 shows the causes and consequences of eutrophication and emphasizes that organic matter and nutrients will lead to undesirable consequences if the prevailing

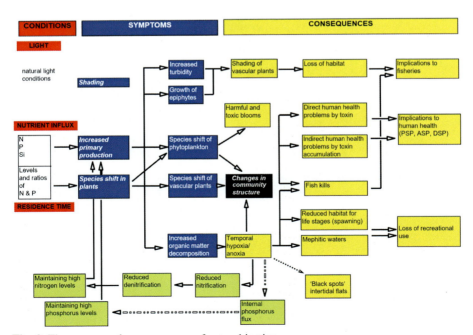

Fig. 2 The causes and consequences of eutrophication.
From de Jonge, V.N., Elliott, M., 2001. Eutrophication. In: Steele, J., Thorpe, S., Turekian, K. (Eds.), Encyclopedia of Ocean Sciences, vol. 2. Academic Press, London, pp. 852–870.

Table 1 Signs and symptoms of eutrophication—The causes, primary and secondary effects of eutrophication caused by increased nutrient inputs; high residence time/slow flushing rate.

Primary effects
• Occurrence of blooms of toxic or tainting phytoplankton forms • Increasing plant/algal biomass production • Occurrence of blooms of microalgae which may be a nuisance (and cause aesthetic pollution) through foaming (e.g., *Phaeocystis*, *Chaetoceros socialis*) • Decline or disappearance of certain perennial plants, often replaced by annual, fast growing opportunistic species such as foliose or filamentous green algae (e.g., *Ulva*, *Enteromorpha*) • Reduced diversity of the flora (and associated fauna) • Changes to photic regime through shading
Secondary effects
• Increased particulate and dissolved organic matter in seawater and sediments • Nuisance mat formation to hinder fishing and navigation • Nuisance mat formation producing anoxic conditions • Increase in microbial community and thus oxygen depletion, leading to hypoxic processes such as H_2S and CH_4 production • Development of opportunistic macrobenthic populations and thus changes along the Pearson-Rosenberg continuum • Poor water quality, especially water column oxygen depletion, thus affecting fishes and zooplankton • Mortalities of higher organisms through effects of neurotoxins • Hindrance to intertidal feeding by wading birds and ducks

environmental conditions are suitable. For example, harmful algal blooms will be created if the light regime and residence time (the inverse of the flushing rate) are sufficient to allow the nutrients to be assimilated leading to plant or algal growth. Similarly, high levels of organic matter, either directly discharged or as the result of the algal growth, can lead to depletion of dissolved oxygen creating barriers to fish movement, smothering of the benthic populations, etc. (Cabral et al., 2022). Table 1 presents the signs and symptoms of eutrophication as a suite of undesirable consequences.

References and further reading

Cabral, H.N., Borja, A., Fonseca, V.F., Harrison, T.D., Teichert, N., Lepage, M., Leal, M.C., 2022. Chapter 6. Fishes and environmental health. In: Whitfield, A.K., Able, K.W., Blaber, S.J.M., Elliott, M. (Eds.), Fish and Fisheries in Estuaries—A Global Perspective. John Wiley & Sons, Oxford, pp. 332–379.

de Jonge, V.N., Elliott, M., 2001. Eutrophication. In: Steele, J., Thorpe, S., Turekian, K. (Eds.), Encyclopedia of Ocean Sciences, vol. 2. Academic Press, London, pp. 852–870.

de Jonge, V.N., Elliott., 2002. Causes, historical development, effects and future challenges of a common environmental problem: eutrophication. Hydrobiologia 475/476, 1–19. https://doi.org/10.1007/978-94-017-2464-7_1.

Elliott, M., de Jonge, V.N., 2002. The management of nutrients and potential eutrophication in estuaries and other restricted water bodies. Hydrobiologia 475/476, 513–524.

McLusky, D.S., Elliott, M., 2004. The Estuarine Ecosystem; Ecology, Threats and Management, thirg ed. OUP, Oxford, p. 216.

Orive, E., Elliott, M., de Jonge, V.N. (Eds.), 2002. Nutrients and Eutrophication in Estuaries and Coastal Waters. Developments in Hydrobiology, vol. 164. Kluwer Academic Publishers, Dordrecht, p. 526.

Linked

Determination of significant effect; Eutrophic and organic wastes; Standards, objectives, indicators

Impacts and assessment: 44. Determination of a significant effect

The main purpose of the information relating to determining and defining the effects of industry on a receiving environment will be to detect the signal-to-noise ratio, that is, can an effect (i.e., the anthropogenic signal) be detected against a background of inherent variability (the "noise") in any measurements. Industries may be charged with determining whether they are having or are likely to have an effect and to demonstrate whether that effect is significant. However, such environmental significance can be expressed in statistical, ecological, or societal terms (see later).

Highly variable estuarine, coastal, and marine systems, in which there are changes over temporal scales from minutes to millennia and spatial scales from centimeters to global, behave as open dynamic systems, e.g., where the features at one area (such as sediment and hydrographic patterns, presence of organisms and species, or abundance of populations) may be dependent on events far away from the site and/or times previous to that being observed. In addition, the inherent variability in any monitoring parameters includes sampling, analytical, and statistical errors, each of which has the potential to introduce errors and lead to uncertainty in the conclusions being reached. Insufficient replication will increase sampling variability and poor AQC/QA (analytical quality control/quality assurance) in the field and laboratory methods will create inadequate data, whereas worker training and expertise will decrease that variability.

A BACI-PS (before-after-control-impact—paired series) (Schmidt and Osenberg, 1996) approach will allow the effects to be determined in time (comparisons of before and after an activity) and space (comparisons of a control area against the potentially impacted area). If sufficient samples are taken then there is statistical power to detect or predict a change against background variability; however, it is emphasized that the degree of variability is often very high such that the detection of change may be difficult. In addition, in order to use power analysis to determine the intensity of sampling required to detect an effect (i.e., the number of replicates), the magnitude of effect and the inherent variability in the determinants require to be defined at the outset. Determination of the number of replicates required to detect a given degree of change sanctioned by a regulator may indicate an unrealistic number of replicates and hence extremely high costs of carrying out the study and monitoring (Franco et al., 2015).

The significance of changes to the environmental system may be of three types: statistical (which is detectable at a given level of probability given sufficient data and a suitable survey design), ecological (i.e., changes may have an ecological significance even if they are not statistically detectable), or societal (if society believes there to be a problem, even if it cannot be shown ecologically or statistically, then by definition there is a problem). This sequence, as described here, is in increasing difficulty of detection. Hence, with sufficient data, a statistical significance test can be used, whereas the change to a single species in an area with many species may be ecologically significant but not statistically so. Furthermore, a change in an area may be highly regarded by society or perceived as such by society even if that change is neither statistical or ecological.

In defining and planning the science required to determine or predict the effects of a new coastal industry on the estuarine, coastal, or marine environment and the effects of the marine environment on the industry, the researchers have to be clear whether the outputs are site specific, that is, related to only one site in question, or whether they have a generic component and so the findings may have a wider value. For example, determinations of the habitats, populations, communities, and species at risk; the pathways of dispersion of pollutants under precise hydrographic patterns; and the climate and biogeochemical conditions affecting the fate and effects of the materials discharged will all have a lesser or greater site-specific aspect. Data relating to these may then be presented via graphical information systems, in overlaid maps or interactive .pdf in order to test scenarios. In contrast, for example, generic aspects include the response by organisms to temperature and contaminant discharge increase, the vulnerability of species as receptors to the construction and operational activities, and the influence of the activities on guilds and trophic groups of organisms.

References and further reading

Baker, J.M., Wolff, W.J. (Eds.), 1987. Biological Surveys of Estuaries and Coasts. Estuarine and Brackish Water Sciences Association Handbook. Cambridge University Press, Cambridge.

Bayne, B.L., Clarke, K.R., Gray, J.S. (Eds.), 1988. Biological effects of pollutants: results of a practical workshop. Mar. Ecol. Prog. Ser. 46 (1–3) 1–278.

Eleftheriou, A., McIntyre, A.D. (Eds.), 2005. Methods for the Study of Marine Benthos, third ed. Blackwell Science, Oxford. IBP Handbook No. 16.

Franco, A., Quintino, V., Elliott, M., 2015. Benthic monitoring and sampling design and effort to detect spatial changes: a case study using data from offshore wind farm sites. Ecol. Indic. 57, 298–304. https://doi.org/10.1016/j.ecolind.2015.04.040.

Gray, J.S., Elliott, M., 2009. Ecology of Marine Sediments: Science to Management. OUP, Oxford, p. 260.

Krebs, C.J., 1998. Ecological Methodology, second ed. Addison Wesley Longman, Harlow.

Schmidt, R.J., Osenberg, C.W. (Eds.), 1996. Detecting ecological impacts: concepts and applications in coastal habitats. Academic Press, San Diego, CA.

Southwood, T.R.E., Henderson, P.A., 2000. Ecological Methods, third ed. Blackwell, Oxford.

Linked

Ecosystem resilience, resistance, recovery; Temporal and spatial scales

Impacts and assessment: 45. Standards, objectives, indicators

As environmental management has become more focused on achieving results, governments have demanded, quite justifiably, that there should be procedures in place that demonstrate how well environmental policies are working. Indeed, this follows the axiom that "you cannot manage something unless you can measure it" and hence management needs to be accompanied by focused monitoring. Quantitative measurements using standards, indicators, and objectives are required to determine what management is required but also whether management has been successful. The term objective can be the environmental target that is aimed for which could be very specific or more tenuous and aspirational. However, if the targets are subjective, tenuous, or merely aspirational then it is difficult to determine when they have been achieved; for example, if the objective is merely to see an unquantified reduction in discharges to an area then one less liter of discharged effluent will achieve that aim even if there are no demonstrable environmental improvements.

To achieve these aims, standards may be adopted as the measure to reduce or stop the emission of a substance or alter a method of operation. An indicator would be the measure of some environmental factor that securely allows the tracking toward the objective, the overall aim for environmental management. The objective may be derived from a high-level vision of the coastal environment such as the UK government vision to achieve clean, healthy, productive, diverse, and resilient seas. As it may not be possible, because of high monitoring costs, to track all parameters leading to an objective, the indicator is usually taken as a representation of a group of these parameters.

Indicators can be of the nature of discharges, the state of the receiving waters or its ecological components, the number of activities and pressures in an area, the socioeconomic situation, the health of human or ecological populations, etc. These may then fit into a cause-consequence-response approach such as DPSIR or DAPSI(W)R(M) for analyzing environmental issues (Fig. 1 shows examples of indicators for each element in the latter framework) (see entry 93).

This approach is being used more and more in climate change reports and in marine studies. For example, the EU Marine Strategy Framework Directive (MSFD) and Water Framework Directive, the UK Marine Strategy, and the Regional Seas Conventions Quality Status Reports (such as OSPAR, the Oslo and Paris Convention) are all focusing attention on such a method. Of course, it is not always easy to select a reliable indicator and this is probably the most difficult part of the process. Improving standards do not necessarily have a linear effect on reduction of the indicator in the environment as many environmental indicators may have lag times of decades or even centuries before the pressure response is resolved. As an example, in a Marine Scotland workshop for the EU MSFD, one of its descriptors which has a requirement to reduce eutrophication was expanded to a number of possible targets; in more detail, it was suggested that that the "dissolved inorganic nitrate and phosphorous do not deviate more than 50% from a salinity related area specific background (Using OSPAR guidelines)." The indicator for this was the measurement of nutrients in the water

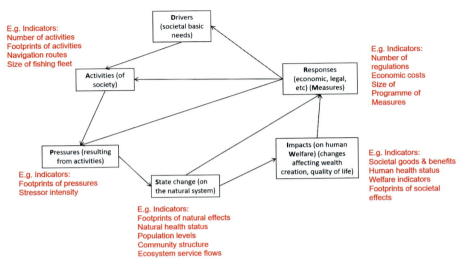

Fig. 1 Examples of indicators for each element in the cause-consequence-response chain.

column. To help achieve these, standards would be made more stringent to reduce nutrient loads, principally from sewage treatment works and from diffuse sources such as agriculture. However, staying with this simple example, the difficulties are considerable despite the fact this is one of the most studied relationship in marine science. Firstly, measurements are very varied spatially and temporally and as ship time is very expensive it is difficult to collect sufficient data to provide confident assessments. Secondly, the relationship between eutrophication and nutrients is not straightforward and has proved taxing for previous studies carried out as part of the OSPAR "Comprehensive procedure" and for Water Framework Directive assessments. Thirdly, although existing controls have already brought down the discharge of nutrients to coastal zones but this has hardly been reflected in recorded environmental levels.

A similarly complex relationship can be seen in most environmental indicators because of the complex inter-dependence between all the variables. Nevertheless, this approach is still the most appropriate and in times of economic pressure no government will be prepared to pour investment into environmental controls without seeing a meaningful return and determining which investment produces the greatest affect per unit cost. In the European example, a plethora of directives define monitoring requirements but as the cost rises governments are seeking to minimize their expenditure by combining monitoring programs or altering the frequency to increase their efficiency; hence this leads to the so-called *environmental assessment paradox* in which there are more demands on industry and regulators for monitoring but reduced resources for undertaking the monitoring.

Furthermore, in order for them to be effective in environmental management, indicators and monitoring should fulfill a set of criteria denoting properties necessary for environmental management. These can be reduced to the requirement to be SMART—specific, measurable, achievable, realistic, and time bounded but there are many other criteria within these (Table 1).

Table 1 The required properties of indicators and monitoring parameters for successful management with relevance to GES determination.

Property	Explanation and relevance
Anticipatory	Sufficient to allow the defense of the precautionary principle, as an early warning of change, capable of indicating deviation from that expected before irreversible damage occurs. Conducted in advance of construction, development, or implementation of projects that could modify marine structure and productivity
Biologically important	Focuses on species, biotopes, communities, etc. important in maintaining a fully functioning ecological community. Careful selection of key taxa and linkages is required to discern trends in ecosystem structure that might harm production in a healthy ecosystem; key ecosystem functions (such as primary production and organic matter delivery) also need to be included
Broadly applicable and integrative over space and time	Usable at many sites and over different time periods to give a holistic assessment which provides and summarizes information from many environmental and biotic aspects. These allow comparisons with previous data to estimate variability and to define trends and breaches with guidelines or standards; as being at the top of the food chain and often (comparatively) long lived, top predators in themselves integrate changes over the whole system
Concrete and results focused	Indicators are needed for directly observable and measurable properties rather than those which can only be estimated indirectly; concrete indicators are more readily interpretable by diverse stakeholders who contribute to management decision-making. Easily interpretable outcomes and trends that both managers and stakeholders can understand. Identifiable targets and thresholds for indicators upon which regulatory action would be taken
Continuity over time and space	Capable of being measured over appropriate ecological and human time and space scales to show recovery and restoration. Often annual surveys that provide indicators of abundance, age structure, recruitment, etc. Such time series are required for population dynamics and for recovery (or degradation) trends
Cost effective	Indicators and measurements should be cost effective (financially non-prohibitive) given limited monitoring resources, that is, with an ease/economy of monitoring. Monitoring should provide the greatest and quickest benefits to scientific understanding and interpretation, to society and sustainable development. This should

Continued

Table 1 Continued

Property	Explanation and relevance
	produce an optimum and defensible sampling strategy and the most information possible. Monitoring surveys are expensive. Monitoring survey designs that will reliably provide indicators useful for management decisions are required, if possible in real time
Grounded in theory, relevant and appropriate	Indicators should reflect features of ecosystems and human impacts that are relevant to achieving operational objectives; they should be scientifically sound and defensible and based on well-defined and validated theory. They should be relevant and appropriate to management initiatives and understood by managers. This is particularly relevant for components such as mammals, birds, fish and fisheries which are politically sensitive and understandable to many in society
Interpretable	Indicators should reflect the concerns of, and be understood by stakeholders. Their understanding should be easy and equate to their technical meanings, especially for non-scientists and other users; some should have a general applicability and be capable of distinguishing acceptable from unacceptable conditions in a scientifically and legally defensive way. Indicators of abundance and trends should be easily interpretable, reliable over years or, if not, changes must be easily explained. However, population trends have a high degree of variability as inherent "noise" in the data. Such changes have to be explained in relation to the "signal" of change being detected
Low redundancy	The indicators and monitoring should provide unique information compared to other measures. Often, indicators are used in combination but some may be duplicating information and so may need weighting to avoid double counting and redundancy
Measurable	Indicators should be easily measurable in practice using existing instruments, monitoring programs, and analytical tools available in the relevant areas, to the required accuracy and precision, and on the timescales needed to support management. They need to be capable of being updated regularly, being operationally defined and measured, with accepted methods and analytical/quality control/quality assurance and with defined detection limits. The capability to monitor and measure well-defined properties is essential to ensure that indicators can be faithfully provided in the long term. As mentioned before, the indicators should have

Table 1 Continued

Property	Explanation and relevance
	minimum or known bias (error), and the desired signal should be distinguishable from noise or at least the noise (inherent variability in the data) should be quantified and explained, that is, have a high signal-to-noise ratio
Non-destructive	Methods used should cause minimal and acceptable damage to the ecosystem and should be legally permissible. In situ, in the field measurements (e.g., from fish and fisheries surveys), or data can be obtained from commercial or recreational activities
Realistic/attainable (achievable)	Indicators should be realistic in their structure and measurement and should provide information on a "need-to-know" basis rather than a "nice-to-know" basis. They should be attainable (achievable) within the management framework. The measurements in estuarine surveys for fish and fisheries and the habitats in estuaries are highly relevant and easily achievable for estuarine management. Direct measures of the health of the individual fish and the stocks are relevant to management measures
Responsive feedback to management	Indicators should be responsive to effective management action and regulation and provide rapid and reliable feedback on the findings. Such feedback should be defined prior to using the indicator. Some information, such as for fishes, fisheries, and habitats, is rapidly obtained, and so can be disseminated almost in real time, it is valuable in feedback to management. In addition, it is also valuable to use monitoring to show a deleterious trend that is not immediately open to remedial action
Sensitive to a known stressor or stressors	The trends in the indicators should be sensitive to changes in the ecosystem properties or impacts, to a stressor or stressors which the indicator is intended to measure and also sensitive to a manageable human activity. They should be based on an underlying conceptual empirical or deterministic model, without an all-or-none response to extreme or natural variability, hence their potential for use in a diagnostic capacity. However, some monitoring is done in the absence of understanding or knowing factors that generate response, for example, fish recruitment. Despite this, it is valuable for managers to know the scale of variability and trends in recruitment to make decisions even though causes of recruitment variability or trends may be vague

Continued

Table 1 Continued

Property	Explanation and relevance
Socially relevant	Understandable to stakeholders and the wider society or at least predictive of, or a surrogate for, a change important to society. The high public and political perception of higher trophic levels makes the information understandable and highly relevant
Specific	Indicators should respond to the properties they are intended to measure rather than to other factors, and/or it should be possible to disentangle the effects of other factors from the observed response (hence having a high reliability/specificity of response and relevance to the endpoint). Such indicators cover the environmental factors relevant to communities, individuals, and populations such as DO, salinity, temperature, and the highly relevant biological determinants—condition factor, growth, yield, abundance, etc. Each of which can be used to create a SMART indicator
Time bounded	The date of attaining a threshold/standard should be indicated in advance. They are likely to be based on existing time-series data to help set objectives and also based on readily available data and those showing temporal trends. Once a baseline, reference, threshold, or trigger value is set then it should have a time boundary to determine if management measures have worked. For example, pollution controls implemented to remove a water quality barrier by a given date or the recovery of a population to a given size within a defined period
Timely	The indicators should be appropriate to management decisions relating to human activities and therefore they should be linked to that activity. This provides real-time information for feedback into management giving remedial action to prevent further deterioration and to indicate the results of or need for any change in strategy. For example, information on exceedance of water quality parameters to cause damage to the ecology is needed by managers at the time of licensing developments

Modified and expanded from Elliott, M., Houde, E.D., Lamberth, S.J., Lonsdale, J.-A., Tweedley, J.R., 2022. Chapter 12. Management of Fishes and Fisheries in Estuaries. In: Whitfield, A.K., Able, K.W., Blaber, S.J.M., Elliott, M. (Ed.), Fish and Fisheries in Estuaries—A Global Perspective. John Wiley & Sons, Oxford, pp. 706–797, ISBN 9781444336672.

References and further reading

Atkins, J.P., Burdon, D., Elliott, M., 2015. Chapter 5: Identification of a practicable set of indicators for coastal and marine ecosystem services. In: Turner, R.K., Schaafsma, M. (Eds.), Coastal Zones Ecosystem Services: From Science to Values and Decision Making. Springer Ecological Economic Series, Springer Internat. Publ, Switzerland, pp. 79–102. ISBN 978-3-319-17213-2.

Aubry, A., Elliott, M., 2006. The use of environmental integrative indicators to assess seabed disturbance in estuaries and coasts: application to the Humber Estuary, UK. Mar. Pollut. Bull. 53 (1-4) 175–185.

Cormier, R., Elliott, M., 2017. SMART marine goals, targets and management—is SDG 14 operational or aspirational, is 'Life Below Water' sinking or swimming? Mar. Pollut. Bull. 123, 28–33. https://doi.org/10.1016/j.marpolbul.2017.07.060.

Ducrotoy, J.P., Mazik, K., Elliott, M., 2011. Bio-Sedimentary Indicators for Estuaries: A Critical Review. Union des océanographes de France, Paris, ISBN: 978-2-9510625-2-8, pp. 1–77.

Elliott, M., Houde, E.D., Lamberth, S.J., Lonsdale, J.-A., Tweedley, J.R., 2022. Chapter 12. Management of Fishes and Fisheries in Estuaries. In: Whitfield, A.K., Able, K.W., Blaber, S.J.M., Elliott, M. (Ed.), Fish and Fisheries in Estuaries—A Global Perspective. John Wiley & Sons, Oxford, pp. 706–797, ISBN 9781444336672.

Perez-Dominguez, R., Maci, S., Courrat, A., Lepage, M., Borja, A., Uriarte, A., Neto, J.M., Cabral, H., Raykov, V.S., Franco, A., Alvarez, M.C., Elliott, M., 2012. Current developments on fish-based indices to assess ecological-quality status of estuaries and lagoons. Ecol. Indic. 23, 34–45.

Quintino, V., Elliott, M., Rodrigues, A.M., 2006. The derivation, performance and role of univariate and multivariate indicators of benthic change: case studies at differing spatial scales. J. Exp. Mar. Biol. Ecol. 330, 368–382.

www.rspb.org.uk/Images/Indicators_tcm9-132910.pdf.
www.scotland.gov.uk/Resource/Doc/295194/0116507.doc.

Linked

Discharge consents, permits, licenses and authorizations; Integrated management of catchment, transitional and coastal waters

Impacts and assessment: 46. Temporal and spatial physical scales

All coastal industries will influence their local physical environment, whether by occupying space, having marine structures, or having discharges which may influence hydrodynamic structure, processes, and patterns. The duration of most natural processes is constrained by the physical drivers such as tides, the seasonal changes in temperature, or the passage of a weather system, and each of these cycles will have a time and a length scale associated with it. At different time and extent scales, different processes will be important, and to complete a cycle of change there is a progression from small-scale processes that last only a short time, to large-scale processes that take longer. The timescale refers to the time required for the effects of a perturbation of a process to be expressed or a cycle to be completed, and the spatial scale is the distance over which the perturbation is experienced. There may be considerable lags in time between the cause and effect because of inertia. Therefore, largely empirical relationships have been sought between the scales for oceanic and atmospheric processes, planetary phenomena, and biological and sediment distributions and processes.

The figure shows the diagrammatic relationship for physical processes in the sea, and it can be seen that a comparatively narrow band of time and space scales is occupied by each naturally occurring feature. Thus, it is not possible for natural processes that cover large spatial scales to have only a short timescale, or vice versa. Turbulence has timescales of a few seconds to minutes, whereas ocean circulation is bound to have timescales of the order of centuries because of the inertia involved in the movement of large water bodies. These principles extend to the effects of the physics on dependent variables, such as sediment and biology. Tides on the continental shelf, for instance, have a wavelength of several hundred kilometers. The resulting water flows have trajectories of tens of kilometers and produce bedform dunes with a few hundred meters wavelength. Surface wind waves, in contrast, have wavelengths of a few hundred meters or so and generate bed ripples a few meters in wavelength; an expression of this is a relationship established between the near-bed orbital velocity and ripple wavelength (Miller and Komar, 1980).

As a further example, studies of phytoplankton production show patchiness related to duration of light exposure and vertical water movements. Studies in the ocean have shown phytoplankton cycling is about 0.5 h to hundreds of hours for vertical displacements of 10 m. Where stratification exists and mixing is reduced, phytoplankton cycling is longer (Denman and Gargett, 1983).

In the case of an industrial discharge plume interacting with tidal receiving water, the tide will be important in the far field in advecting discharged effluents without mixing them, with any mixing occurring within the water column at smaller scales. In the near field, the mixing, involving short period waves and turbulence, takes place immediately on discharge at smaller time and space scales.

For contaminants, their response will depend upon their decay timescales and whether they are particulate or in solution, or adsorbed onto or absorbed into particles

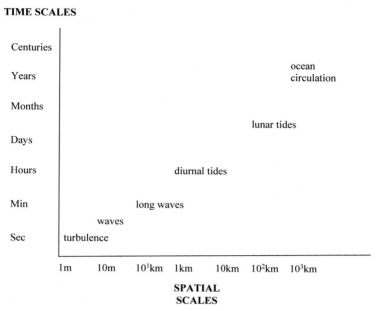

Fig. 1 Temporal and spatial scales for physical processes in the sea.

in suspension. For instance, ammonia may have a toxicity lasting a few hours, and so will affect spatial scales in a tidal flow of up to a few kilometers. Persistent contaminants, such as radionuclides, may become transported through ocean circulation. Short-lived organisms that do not have a pelagic component in their life are likely to have very restricted spatial scales. However, those with a planktonic stage can respond to the tidal timescales and become much more widespread; many benthic organisms may have planktonic stages lasting several weeks. This would result in transport along tidal excursion pathways and being carried in the tidal oscillation (e.g., perhaps moving further in a flood direction than an ebb tidal direction) which can result in transport of 40–80 km during the period to metamorphosis and settlement (Fig. 1).

Since it is the physical processes that largely control the distribution of sediments, contaminants, and biology, much can be concluded from the style of analysis discussed before. Examples are given in Rekacewicz and Bournay (2005) and Steele (1977). An example of application of the relationship for coastal geomorphological features is shown in Dronkers (2005).

References and further reading

Denman, K.L., Gargett, A.E., 1983. Time and space scales of vertical mixing and advection of phytoplankton in the upper ocean. Limnol. Oceanogr. 28 (5) 801–815.
Dronkers, J., 2005. Dynamics of Coastal Systems. World Scientific, New Jersey, p. 519.

Miller, R.L., Komar, P.D., 1980. A field investigation of the relationship between oscillation ripple spacing and the near-bottom water orbital motions. J. Sediment. Petrol. 50, 183–191.
Rekacewicz, P., Bournay, E., 2005. Millenium Ecosystem Assessment. UNEP/GRID-Arendal.
Steele, J.H., 1977. Spatial Patterns in Plankton Communities. NATO Conference Series, Series IV.

Linked

Phenology; Plume characteristics and behavior

Impacts and assessment: 47. Appropriate assessment—Habitats regulations assessment, habitat risk assessment

The regulation, maintenance, protection, and recovery of nature conservation areas will be a major challenge for coastal industries. In particular, such industries will have to consider the effects of their activities on specified habitat and species and, conversely, the effects of those habitats and species on the activities of the coastal industry. Such assessments may be analogous to Environmental Impact Assessments but may be much more focused in that regulatory bodies will identify named components (habitats and species) and then describe these as the conservation objectives for an area. Once these components have been identified, then the industrial activity is regarded as a "plan or project" whose effects on those components require special attention.

Most countries therefore have nature conservation and protection legislation enacted by nature conservation statutory bodies, so-called competent authorities, which requires the industries to assess these aspects. As an example, in Europe, the EU Habitats and Species Directive, 1992, which is transposed into national legal instruments, for example the Habitats and Species Regulations, 2010, for the United Kingdom. These require competent authorities to carry out Appropriate Assessments in certain circumstances where a plan or project affects, or has the potential to affect, the conservation objectives of a designated Natura 2000 (European) site. This is any site designated under the EU Habitats and Species Directive or the EU Wild Birds Directive.

As a further example at the level of a country, in the United Kingdom, under the 2010 Regulations, Habitats Regulations Assessment (HRA) refers to the whole appraisal process, including the Appropriate Assessment step. An Appropriate Assessment is required when a plan or project affecting a Natura site:

- is not connected with management of the site for nature conservation, and
- is likely to have a significant effect on the site and/or its conservation objectives (either alone or in combination with other plans or projects).

This approach applies to any plan or project which has the potential to affect a Natura site, no matter how far away from that site. This requires the relevant spatial and temporal scales to be considered. For example, when the Port of Rotterdam in the Netherlands was considering expanding its size, it had to demonstrate the likelihood of an effect on the nature conservation sites in the Wadden Sea, at the opposite extreme of the country.

It is emphasized that in employing the "precautionary principle," a competent authority must not authorize a plan or project unless, by means of the Appropriate Assessment, they can ascertain that it will not adversely affect the integrity of a Natura[a] (conservation) site. An appropriate assessment should focus exclusively on the qualifying interests of

---

[a] Natura 2000 is a network of sites selected to ensure the long-term survival of Europe's most valuable and threatened species and habitats. Sites are selected according to precise, scientific criteria, but the selection procedure varies depending on which of the two European nature directives—Birds or Habitats—warrants the creation of a particular site.

the Natura site affected and must consider any impacts on the conservation objectives of the site. The only exceptions where authorization might proceed in the context of potential adverse impacts are if there are no alternative solutions and there are imperative reasons of overriding public interest (IROPI) for the plan or project to go ahead. Habitat and species risk assessments, by identifying specific risks to individual designated habitats and species and the pathways of effects, are conducted as part of the appropriate assessment.

Internationally, the global Natural Capital Project is developing Habitat Risk Assessment models that quantify and map the values of ecosystem services. The modeling suite is best suited for analyses of multiple services and multiple objectives. Current models, which require relatively little data input, can identify areas where investment may enhance human well-being and nature.

References and further reading

EC, 1992. Council Directive 92/43/EEC on the Conservation of Natural Habitats and of Wild Flora and Fauna. Official Journal No. L 206.
http://ec.europa.eu/environment/nature/natura2000/management/docs/art6/provision_of_art6_en.pdf.
The Conservation of Habitats and Species Regulations 2010, no 490. http://www.legislation.gov.uk/uksi/2010/490/contents/made.
JNCC Joint Nature Conservation Committee. http://jncc.defra.gov.uk/page-1379.
Natural England, nature conservation objectives. http://www.naturalengland.org.uk/ourwork/conservation/designatedareas/sac/conservationobjectives.aspx.

Linked

Habitats and species directive; Mitigation, amelioration, enhancement, compensation; Nature conservation designations; Precautionary principal and approach

Impacts and assessment: 48. Ecotoxicology assessment

Ecotoxicology is the study of the effects of potentially toxic chemicals on organisms and integrates (non-human) toxicology with ecology. The aim is to assess what changes in the biology or ecology (of the target or receptor organism(s) or the biological community) can be expected with a specified emission of a chemical or mixture of chemicals. In the past, regulation was merely focused on emission levels of potentially toxic compounds, and the concentrations of chemicals found in the discharge water were the basis of a discharge consent. A regulator would require a list of chemicals being discharged and would assess the toxicity of each according to national or international databases. This furthermore focused on single chemicals being discharged in that it was relatively easy for an industry to determine the materials resulting from its processes, but this had the disadvantage of not including chemicals that were created in the wastewater discharge and treatment plants. At present, the focus has shifted to impacts and will therefore include an increased consideration of ecotoxicology.

The key concerns of the regulator will be:

- are the chemicals dispersed to the environment, whether to land, air, or water;
- will the chemicals degrade or disperse or will they be persistent, bioaccumulate, and be toxic to an organism and/or its predator;
- how serious are the effects, and for how long (months, years, or decades) will they persist?
- are the effects reversible, or are we losing certain species or groups of animals in that area?

Hence there is the need to speak not only of ecotoxicology but also of "stressors" or pressures (the mechanism of impact) in the environment and its biology or ecology. Ecotoxicology can be summarized as the assessment and weighting of the effects of anthropogenic activities so the acceptability of that activity may be determined (Van Straalen, 2003). There is the need to consider whether the chemicals are naturally occurring but may be present in unnaturally elevated concentrations, or whether the chemicals are unnatural (synthetic) compounds, for which organisms have not developed a tolerance at even low concentrations. It is also necessary to consider both the *concentration*, the amount of a chemical in a unit volume or weight, and the *content*, the total amount accumulated into a component (environmental or an organism) or discharged into an area, often referred to as the load or loading.

There are some related issues which help the understanding of ecotoxicological data:

- When assessing toxic effects, a clear division can be made between acute and chronic toxicity. Acute toxicity describes the adverse effects within a short period (typically 24–48 h) after exposure. An example is copper which occurs naturally in marine waters at a low level. Being an essential element, all marine aquatic organisms will accumulate as much as possible due to overall shortage of copper in the marine environment but have limited control, so a spill of copper sulfate will result in rapid and severe effects for all marine species within 24 h. The following terms are often used in this context: LC_{50} (median lethal concentration, the amount to kill half the organisms exposed to it); EC_{50} (median effect concentration, the

concentration which will cause some defined response in an organism); LD_{50} (median lethal dose); LT_{50} (median lethal time, the time taken to kill half the test organisms); and ET_{50} (median effects time, the time taken to create an undesirable effect). In these cases, the defined effects may relate, for example, to the behavior, physiology, biochemistry, growth, productivity, genetics, and pathology of the organisms under test.
- Chronic toxicity describes the longer term negative effects after perhaps weeks, months, and sometimes years, generally after repeated exposure to low (not acute) doses of a compound. Chronic toxicity often implies accumulation of the compounds or a buildup of the consequences over time; often this is linked to the mechanism of bioaccumulation where an organism actively absorbs the toxic compound. In contrast to acute toxicity, chronic toxicity frequently results in sub-lethal effects. Chronic toxicity for metal accumulation from pulverized fuel ash, as example, can differ widely among exposed species (Jenner and Bowmer, 1990). Following the example of copper, fish, that is, carp (*Cyprinus carpio*), show a high sensitivity for copper which was reflected in cellular responses in the skin (Iger et al., 1994). In an extended study with sea bass into the bioaccumulation effects of chlorination by-products, it was found that at low-level chlorination (1–2-mg L^{-1} total residual chlorine), there was no significant ecotoxicological stress. In this instance the fish were sensitive to low levels of chlorine and avoided areas with higher concentrations.
- Biomagnification in aquatic organisms refers to the process whereby certain compounds initially at non-toxic concentrations move up in the food chain (e.g., plankton → fish/bivalves → birds) with increasing concentrations and/or content in the higher trophic levels. The substances, as they move up the food chain, can then become concentrated in tissues or internal organs eventually breaching a toxic threshold.

The above describes only the general mechanisms and individual mechanisms may be far more complex. In Europe all discharges of cooling water have to meet IPPC guidelines (integrated pollution prevention and control), based on article 22 paragraph 6 of the EU Water Framework Directive (WFD). The priority (hazardous) substances (PS/PHS) identified under the WFD must be taken into account. Where possible the chemicals must be assessed against the appropriate water quality standards. For this, an integrated approach, as determined in an "emission-evaluation," must be applied. Priority substances are those where discharges must be minimized and priority hazardous substances are those where the aim is to phase out their discharges. Environmental quality standards are available for virtually all priority and priority hazardous substances. An IPPC authorization will be required for all discharges to land, water, and air from complex industrial processes and plant.

Coastal industries may be subject to an Environmental Quality Standards/Environmental Quality Objectives (EQS/EQO) approach in which the amount they can discharge depends on the capacity of the receiving waters to assimilate the materials without adverse effects. Hence, a larger body or water with greater dilution characteristics may be expected to assimilate large quantities of discharge. Conversely, they may be subject to the Uniform Emissions Standards approach (UES) in which the same industry anywhere, irrespective of the receiving body of water, will be allowed to discharge the same concentrations. The EQO/EQS approach allows for the fact that a water body will already have residual chemicals from upstream industries and so those industries further downstream may regard themselves as penalized for their location. In the UES approach, industries are treated the same irrespective of location (McLusky and Elliott, 2004).

Coastal industries may be required to either carry out a toxicity assessment of each material they are discharging or cross-refer to toxicity databases in order to obtain an indication of the likely fate and effects of the materials. However, although such information will give the LC_{50} and LT_{50}, and possibly the EC_{50} and ET_{50}, of the materials, it is likely that the data will be based on organisms not found near the industry receiving area. For example, there is a large amount of toxicity data based on the common goldfish or freshwater microcrustaceans which will be not relevant to coastal and estuarine discharges. Consequently, a recognized series of test organisms has been developed and agreed which then allows direct comparisons between different types of materials; this includes named crustaceans (e.g., the common brown shrimp *Crangon crangon*), mysid shrimps (*Neomysis* sp.), common planktonic crustaceans such as water fleas (*Daphnia* sp., *Acartia* sp.), and fishes such as mummichog/killifish *Fundulus heteroclitus* or rainbow trout *Oncorhynchus mykiss*.

In order to create an even more standardized toxicity testing system, luminescing bacteria are used in the which the light output reduction of a bacterium (*Aliivibrio fischeri*) exposed to toxic materials is determined (the Microtox® system). Once calibrated against more conventional ecotox methods, this chemical spectrophotometric technique can limit behavioral aspects involved in using multicellular organisms.

However, the toxicity of individual components in a wastewater effluent may be a poor indication of the overall toxicity of the whole effluent given the mixture of chemicals together with the nature of the discharged medium, whether it is freshwater or brackish water. Synergistic (additive) and antagonistic (cancelling) interactions between the constituents will affect the overall toxicity of the wastewater. For example, the input of organic materials with halogens in heated effluent will favor the production of organohalogens in the wastewater stream which may be more toxic and/or persistent than either of the constituent parts. Conversely, while discharged acids and alkalis may each be toxic, their effects may be neutralized in a wastewater stream.

Given this potential for synergistic or antagonistic interactions has led to the use of whole effluent testing (also called direct toxicity assessment) for effluents prior to discharge. This requires the industry, or in rare cases their regulators, to carry out toxicological testing on the final effluent and preferably using local organisms. In order to increase the environmental and site-specific relevance of the toxicological testing, local receiving water will be used as the dilution medium, thereby mimicking the dilution process after discharge and providing a more realistic indication of the likely fate and effects of the materials discharged.

While most of the ecotoxicological testing relates to materials in solution, it is well regarded that after discharge many chemicals will attach to particles, flocculate, and eventually be sequestered in bed sediments. The chemical nature of those sediments will then influence the fate and effects of the contaminants discharged, so-called diagenesis. Hence, in some cases there is the requirement to have sediment-based toxicological testing, whether using water or acid extraction techniques to remove the contaminants from the sediment matrix or by using behavioral bioassays. The latter may involve recording the response of organisms, such as lugworms (*Arenicola* sp.) or amphipods (*Corophium* sp.) burrowing into contaminated sediments.

The ecotoxicology information will then be used in permitting the wastewater discharges. As the same the toxicological evidence may eventually be used in any legal proceedings, the testing should have the same rigor and AQC/QA (analytical quality control/quality assurance) of any chemical testing. For example, the EpiSuite discharge elimination process used by the US EPA or the H' process used by the UK environmental protection agencies in their Integrated Pollution Prevention and Control (IPPC) systems will require toxicity information on materials discharged to land, air, or waters. These systems need to have rigorous analytical guidelines and procedures.

References and further reading

IPPC, 2001. BAT-Cooling: European IPPC bureau Sevilla, Document on the application of Best Available Techniques to Industrial Cooling Systems. November 2000 (http://eippcb.jrc.es). BREF (11.00) Cooling Systems.

Iger, Y., Lock, R.A.C., Jenner, H.A., Bonga, W.S.E., 1994. Cellular responses in the skin of carp (*Cyprinus carpio*) exposed to copper. Aquat. Toxicol. 29, 49–64.

Jenner, H.A., Bowmer, T., 1990. The accumulation of metals and their toxicity in the marine intertidal invertebrates *Cerastoderma edule*, *Macoma balthica*, *Arenicola marina* exposed to pulverised fuel ash. Environ. Pollut. 66, 139–156.

Khalanski, M., Jenner, H.A., 2012. Chlorination chemistry and ecotoxicology of the marine cooling water systems. In: Rajagopal, S., Jenner, H.A., Venugopalan, V.P. (Eds.), Operational and Environmental Consequences of Large Industrial Cooling Water Systems. Springer, Heidelberg/New York, pp. 183–226.

McLusky, D.S., Elliott, M., 2004. The Estuarine Ecosystem; Ecology, Threats and Management, third ed. OUP, Oxford, p. 216.

Newman, M.C., 2009. Fundamentals of Ecotoxicology. ISBN-13:9781420067040, CRC Press Inc.

Rand, G.M., 1995. Fundamentals of aquatic toxicology: effects, environmental fate and risk assessment. Taylor & Francis, NY, p. 1148.

Rajagopal, S., Jenner, H.A., Venugopalan, V.P. (Eds.), 2012. Operational and Environmental Consequences of Large Industrial Cooling Water Systems. Springer, Heidelberg/New York.

Van Straalen, N., 2003. Ecotoxicology becomes stress ecology. Environ. Sci. Technol. 37, 324A–329A.

Linked

Biocides; Chemicals which may be prohibited for discharge; Chlorination chemistry; Cooling water and direct cooling: Discharge plumes; Electro-chlorination plants (ECPs)

Impacts and assessment: 49. Underwater sound

The term "underwater sound" incorporates both natural and anthropogenic sounds in water, whether in an ocean, a lake, or a tank. Frequencies studied in underwater acoustics are typically between 10 Hz and 1 MHz, anything lower generally will be penetrating the seabed and anything higher being absorbed very quickly. Natural physical underwater sounds result from, for example, movement of water and waves, beach shingle movement, the release of gases; biological sounds come from many different living organisms, ranging from the snapping claws of the mantis shrimp (*Gonodactylus* sp.), the mating clicks of the haddock (*Melanogrammus aeglefinus*), and the calls and echolocation signals of whales and dolphins.

Anthropogenic noise might derive from shipping, marine construction, or seismic exploration. Pile driving activity generates very high sound pressure levels and relatively broadband signals (20 Hz to >20 kHz) (Fig. 1; Thomsen et al., 2006), while seismic airguns sound at 226 dB re 1 μPa @ 1 m for a single airgun and 248 dB re 1 μPa @ 1 m for an array, with fundamental frequencies between 0 and 120 Hz. Turnpenny and Nedwell (1994) note that airguns are seldom seen to cause direct, physical injuries but can repel fish over distances of several kilometers from source. A recent review of the effects of anthropogenic underwater sound on fish is provided by Popper and Hawkins (2019) and on epibenthos by Roberts and Elliott (2017).

Given the speed of sound in water (~1500 vs 343 ms^{-1} in air at 25°C and 760 mmHg), it is a good medium for acoustic communication, especially over long distances, and many marine animals have evolved acoustic signaling systems to this

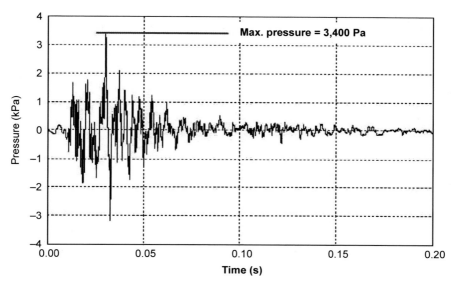

Fig. 1 Waveform from an impact pile (Thomsen et al., 2006).

Table 1 Summary of predicted behavioral avoidance range (based on a 90-dB$_{ht}$ loudness criterion) from measured piling noise during construction of the Burbo Bank Offshore Wind Farm.

Species	90-dB$_{ht}$ range
Bass	500 m
Dab	500 m
Cod	2 km
Herring	2.6 km
Harbor porpoise	5 km
Bottlenose dolphin	4 km
Striped dolphin	4 km
Common seal	3 km

After Parvin, S.J., Nedwell, J.R., 2006. Underwater noise survey during impact piling to construct the Burbo Bank Offshore Wind Farm. Response, pp. 1–5.

end. Sound may also be used where visual signals would be ineffective, such as in water with high sediment suspension, algal growth, or turbidity. Many cetaceans use sound in the form of clicks to echolocate, enabling foraging and navigation around obstacles. They also use whistles, among other sounds, to communicate. Some fishes too have good hearing (see Table 1 and Fig. 2) and research into the importance of sound for fish—and the impacts of anthropogenic noise on fish—is increasing. Scholik and Yan (2001) discovered that long-term impacts (>14 days) on fish hearing

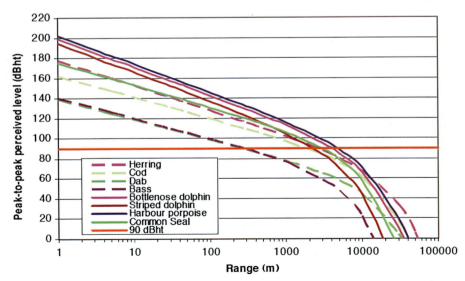

Fig. 2 Peak to peak perceived sound level (dB$_{ht}$) for various marine species during impact piling operations to construct the Burbo Bank Offshore Wind Farm (Parvin and Nedwell, 2006).

can take place when exposed to white noise. Slabbekoorn et al. (2010) give a review of potential impacts of anthropogenic sounds on fish.

As a precise industrial example, during the process of building a power station, various noise and vibration-producing construction activities take place in the marine environment, or on adjacent land or beneath the seabed, creating underwater noise. These may include rock blasting, tunnel boring and piling, associated for example with construction of marine off-loading facilities (MOLFs), cooling water intake/outfall structures and tunnels, and the cooling water pumphouse. Other marine construction projects may create similar noise pollution. Without proper controls, these noise emissions could injure fish, diving birds, or marine mammals; cause them to deviate from established migration routes; or have more subtle long-range effects such as masking communication between animals. Various mitigation measures can be used to limit the risk of environmental harm. They range from selecting methods with the lowest noise emissions, and the use of bubble-curtain sound jackets around construction sites (Reyff, 2009), to timing activities to avoid critical biological seasons.

The sensitivity of fish to sound is used to an advantage for environmental protection at power stations. Acoustic fish deterrents (AFDs: 57), a repellent system developed to exclude fish from certain areas, are now used to reduce the incidence of hearing-sensitive fish entering power station cooling water intakes. Even these have to be designed carefully to prevent unwanted noise "pollution" from affecting a wider area. AFDs and soft-start procedures (piling) are also now commonly used to drive fish and marine mammals from areas of highest noise risk at marine blasting and construction sites (Gordon et al., 2007).

The terms sound and noise tend to be used interchangeably, but a useful distinction is to use "noise" only when the presence of sound energy is (or is at the risk of) causing adverse effects.

References and further reading

Gordon, J., Thompson, D., Gillespie, D., Lonergan, M., Calderan, S., Jaffey, B., Todd, V., 2007. Assessment of the Potential for Acoustic Deterrents to Mitigate the Impact on Marine Mammals of Underwater Noise Arising from the Construction of Offshore Windfarms. (Report No. COWRIE DETER-01-2007). Report by SMRU Consulting.

Hawkins, A.D., Popper, A.N., 2014. Assessing the impacts of underwater sounds on fishes and other forms of marine life. Acoust. Today 10 (2) 30–41.

Parvin, S.J., Nedwell, J.R., 2006. Underwater noise survey during impact piling to construct the Burbo Bank Offshore Wind Farm. Response, pp. 1–5.

Popper, A.N., Hawkins, A.D., 2019. An overview of fish bioacoustics and the impacts of anthropogenic sounds on fishes. J. Fish Biol. https://doi.org/10.1111/jfb.13948. 30864159. (Accessed 10 July 2022).

Reyff, J.A., 2009. Reducing underwater sounds with air bubble curtains [electronic resource] : protecting fish and marine mammals from pile-driving noise. Transportation Research Board, Washington, DC. OCM1bookssj0000939212.

Roberts, L., Elliott, M., 2017. Good or bad vibrations? Impacts of anthropogenic vibration on the marine epibenthos. Sci. Total Environ. 595, 255–268. https://doi.org/10.1016/j.scitotenv.2017.03.117.

Scholik, A.R., Yan, H.Y., 2001. Effects of underwater noise on auditory sensitivity of a cyprinid fish. Hear. Res. 152, 17–24. https://doi.org/10.1016/S0378-5955(00)00213-6.

Slabbekoorn, H., Bouton, N., van Opzeeland, I., Coers, A., ten Cate, C., Popper, A.N., 2010. A noisy spring: the impact of globally rising underwater sound levels on fish. Trends Ecol. Evol. 25 (7) 419–427. Elsevier Ltd https://doi.org/10.1016/j.tree.2010.04.005.

Thomsen, F., Lüdemann, K., Kafemann, R., Piper, W., 2006. Effects of offshore wind farm noise on marine mammals and fish. biola, Hamburg, Germany on behalf of COWRIE Ltd.

Turnpenny, A.W.H., Nedwell, J.R., 1994. The effects on marine fish, diving mammals and birds of underwater sound generated by seismic surveys. Fawley Aquatic Research Laboratories Ltd., Contractors Report prepared for the UK Offshore Operators' Association (UKOOA).

Linked

Acoustic fish deterrent; Abstraction

Section 4

Cooling water

Introduction

Thermal power stations require steam produced by an energy source (nuclear, coal, oil, etc.) to power a turbine; the steam then has to be condensed and recycled as liquid water before being turned back into steam. Direct cooling involves cold water being passed over condenser tubes thereby absorbing the heat from the steam. That external previously cold water then either has to be cooled for recirculating using cooling towers on land or it is part of a once-through (direct) cooling system in which the warmer water is discharged into coastal locations. Owing to limitations implicit in the laws of thermodynamics, 60%–65% of the energy produced in a thermal power station (nuclear or fossil fueled) is rejected as waste heat. With the use of cooling ponds or towers, much of the energy is lost as latent heat of evaporation, but in direct cooled power stations (where water is abstracted, used for cooling and discharged on a once-through loop) all the energy is dissipated in raising the temperature of the cooling water which is then discharged to the receiving water body. Efficiency and safety considerations dictate that direct cooling is used at the majority of European nuclear power stations and is considered "best practice" in the United Kingdom for New Nuclear Build (Environment Agency, 2010, confirmed in a later EA review in 2019) while in the Netherlands direct cooling is favored as it reduces overall chemical use. Cooling water is also used in the chemical and steel industries, and much of the information relating to power stations will also be relevant to these industries.

For coastal industries, and especially coastal power plants using direct cooled systems, the source quantity, quality, and discharge of cooling water are major environmental, production, and operational considerations. Hence, these aspects are given additional emphasis in this volume. It is of note that other industries requiring large volumes of abstracted water, such as desalination plants, will face many of the same problems as coastal, direct cooled power stations.

Cooling water: 50. Cooling water and direct cooling

In a thermal power station, nuclear reactors or fossil fuels are used to raise the steam which then generates electricity using alternators coupled to multi-stage steam turbines. To increase thermal efficiency, the steam is condensed after the final low-pressure turbine stage by passing it through small-bore metal tubes (usually titanium or copper) around which water is pumped. Condensing the low-pressure steam creates a partial vacuum which helps to pull steam through the turbine, increasing efficiency. The cooling water will be sourced locally and may be freshwater, estuarine water, or taken from the sea.

The typical thermal efficiency of a nuclear station is 30%–37% (compared to around 40% for a modern coal-fired station) and hence a 1000-megawatt electric (MWe) nuclear station would typically generate up to 2000 MW of low-grade waste heat, which is removed by the cooling water and discharged into the receiving environment.

Thermal power stations may be directly cooled, using a once-through stream of cooling water to remove waste heat (Fig. 1), or tower cooled (i.e., cooling towers),

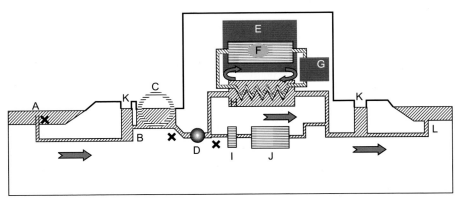

Fig. 1. Basic elements of the cooling water system of a direct-cooled thermal power station, typical of coastal plant. Direct cooled thermal stations reject ∼50 cumecs (cubic meters per second) of cooling water per 1000 MW g (megawatts generation) with a ΔT of +10°C, while indirect-cooled thermal stations (most riverine or inland power stations) reject ∼5 cumecs of purged water from recirculating systems per 1000 MW g, with a ΔT of +10°C or higher. Arrows show direction of cooling water flow in both open (seawater) and closed (boiler water) circuits. A: offshore cooling water intake; B: coarse bar screens; C: 10 mm mesh rotating drum screens (4 of, in parallel); D: main cooling water pump (4 of, in parallel); E: reactor; F: steam generators; G: turbines and electricity generators; H: steam condensers (4 of, in parallel); I: 2 mm mesh pressure strainers (4 of); J: auxiliary and emergency system oil coolers (16 of); K: intake and outfall surge shafts; L: offshore outfall; X: biocidal injection points.

Table 1 Electrical output and cooling water demand for selected current reactor designs.

Reactor	Electrical power output (MW)	CW demand ($m^3\ s^{-1}$ for 12°C rise)
AP 1000	1117	57
EPR	1600	60
HPR 1000	1170	40
UKABWR	1350	53

in which the water is recirculated and heat is mostly lost as latent heat of evaporation. Once-through systems require very large amounts of cooling water (CW) (Table 1), which is why direct cooled stations are frequently sited on the coast, where abundant seawater can be used. The abstracted water is run in a single pass through the condensers, removing the surplus heat, and is then discharged to the original or adjacent water body at a higher temperature (typically +8–12°C). This is the simplest and most thermally efficient method, with the lowest carbon footprint, and is classed under European law as Best Available Technology (BAT) for large nuclear power stations (Environment Agency, 2010).

Some smaller power stations located near to urban or industrial areas are able to use the waste heat in local combined heat and power (CHP) schemes, but the more remote locations selected for larger (particularly nuclear) generation sites preclude this as an option. There have been attempts to use warmed water from nuclear power stations in aquaculture and horticulture but these ventures have rarely succeeded commercially (Turnpenny and Coughlan, 2003). In a combined cycle gas turbine (CCGT) station the fuel is burnt in a gas turbine and the waste heat from this is used to generate steam for a steam turbine. The cooling water requirement applies only to this latter (steam) stage.

The cooling water discharged by a nuclear power station is typically 8–12°C above the inlet temperature and results in a buoyant plume when returned to the source water body (estuary or sea). It may contain residues of biocide chemicals such as chlorine-produced compounds that are added to the water to prevent biological growth within the cooling circuit, which would otherwise further reduce generating efficiency (entry 62).

Many nuclear and conventional stations have used direct cooling to take advantage of the abundant supplies of water available in estuaries and coastal waters. For sites on the upper reaches of estuaries, a combination of direct cooling and tidal storage ponds has been adopted. In those parts of the world where rivers and lakes are larger, freshwater may be used. The use of seawater necessitates the use of higher-grade pipework materials to prevent corrosion but cooling is frequently more efficient. Other "indirect" methods used where water is not plentiful are wet tower cooling, a recirculating system relying on evaporative cooling within cooling towers, and dry tower cooling. Environmental issues associated with direct cooling are reviewed by Rajagopal et al. (2012).

References and further reading

Bamber, R.N., Seaby, R.M.H., 2004. The effects of power station entrainment passage on three species of marine planktonic crustacean, *Acartia tonsa* (Copepoda), *Crangon crangon* (Decapoda) and *Homarus gammarus* (Decapoda). Mar. Environ. Res. 57 (4) 281–294.

Environment Agency, 2010. Cooling water options for the new generation of nuclear power stations in the UK. SC070015/SR3, Bristol, UK, p. 214.

Environment Agency, 2019. Nuclear power station cooling waters: protecting biota. SC 18004/R1 190 pp. Bristol, UK.

Rajagopal, S., Jenner, H.A., Venugopalan, V.P. (Eds.), 2012. Operational and Environmental Consequences of Large Industrial Cooling Water Systems. Springer, Heidelberg/New York.

Taylor, C.J.L., 2006. The effects of biological fouling control at coastal and estuarine power stations. Mar. Pollut. Bull. 53, 30–48.

Turnpenny, A.W.H., Coughlan, J., 2003. Using Water Well? Studies of power stations and the aquatic environment. RWE nPower, Windmill Hill Business Park, Whitehill Way, Swindon SN5 6PB, p. 140.

Turnpenny, A.W.H., O'Keeffe, N., 2005. Screening for intake and outfalls: a best practice guide. Environment Agency Science Report SC030231, Bristol, UK.

Turnpenny, A.W.H., et al., 2010. Cooling Water Options for the New Generation of Nuclear Power Stations in the UK. Environment Agency Project No. SC070015/SR3. ISBN: 978-1-84911-192-8.

Wither, A., Bamber, R., Colclough, S., Dyer, K., Elliott, M., Holmes, P., Jenner, H., Taylor, C., Turnpenny, A., 2012. Setting new Thermal Standards for Transitional and Coastal (TraC) Waters. Mar. Pollut. Bull. 64, 1564–1579.

Linked

Acoustic fish deterrent: Cooling water and direct cooling; Entrainment (biota); Fish recovery and return: Impingement of biota; Source and receiving waters

Cooling water: 51. Water abstraction

Abstraction describes the withdrawal of water from the sea, estuaries, rivers, or lakes. Siting a thermal power station on the coast allows access to large volumes of water. The largest abstractors are thermal power stations that are "direct cooled," which use most of the water for steam turbine condenser cooling. Water may be abstracted for the opposite purpose of giving up heat as in the gasification in liquid natural gas (LNG) plants, resulting in discharges below ambient water body temperatures.

Power plant cooling water intake structures may be situated either offshore or onshore. Offshore structures at modern nuclear facilities may be sited several kilometers offshore to ensure complete immersion at all states of tide and expected wave heights (Environment Agency, 2010). This can make maintenance access difficult in hostile sea conditions. Onshore intake structures, usually built into a wharf bordering reasonably deep water, make maintenance access more straightforward. This can affect the feasibility of operating and maintaining certain environmental protection measures such as acoustic fish deterrents. Long cooling water tunnels from offshore intakes also present greater challenges for biofouling control as the concrete walls and sheltered conditions provide suitable substrata for the settlement of marine fouling organisms.

Water is abstracted and taken into the plant after being filtered by rotating drum- or band-screens which remove some of the larger debris, plants, and animals. Direct cooled or "once-through" nuclear power stations may abstract up to $135 \, m^3 \, s^{-1}$ (cumecs) of cooling water, all but a small percentage of which is returned to the source water body. However, many thermal plants presently operating, with a smaller generating capacity, will typically require only between 10 and $50 \, m^3 \, s^{-1}$.

The key environmental issues associated with water abstraction are the effects of the fish and other biota impingement (the act of being caught on screens) and entrainment (the act of passing through screens and around the plant cooling system) of living animals and plants that are present in the source water and are drawn in incidentally (Turnpenny and Coughlan, 2003). Careful siting of water intakes away from fish and shellfish spawning and nursery areas, fish migration routes, and notable fish and shellfish habitats is therefore required to minimize this risk. The position of the water intakes in the water column will dictate the organisms at risk—for example surface floating seaweed, jellyfish, and ctenophores will be taken by surface intakes; midwater pelagic fish such as herring, sprat, mackerel, and swimming crabs and squid will be taken by midwater intakes, and intakes on, in, or near the bed will take demersal fish species such as cod and benthic species such as crabs, lobsters, and flatfishes. A combination of low-velocity intake designs that allow larger fish to swim away (Fig. 1), acoustic fish deterrents (AFDs) fitted to the water intakes, and fish recovery and return (FRR) systems fitted to the onshore band- or drum-screens will all help to mitigate these effects.

As well as the power station potentially having an effect on the environment, the environment can impede the day-to-day operation of the station. Again, depending

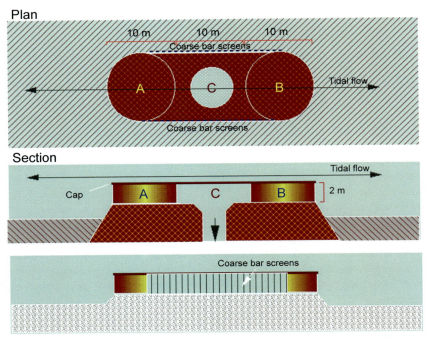

Fig. 1 Low-velocity side-entry (LVSE) intake structure, designed to reduce the effect of tidal flow on intake entrance velocity, improving prospects for fish escape. This type of intake should also be fitted with acoustic fish deterrent (AFD) devices to meet Environment Agency best practice guidance (Turnpenny and O'Keeffe, 2005).

upon the positioning of the intake, stations may have to deal with large agglomerations of organisms (often called "plagues"), such as large shoals of sprat or accumulations of jellyfish which can block the filter screens and starve the plant of cooling water (Environment Agency, 2010). Some sites may experience large amounts of seaweed, particularly after storm events, with the same operational risk.

References and further reading

Bamber, R.N., Seaby, R.M.H., 2004. The effects of power station entrainment passage on three species of marine planktonic crustacean, *Acartia tonsa* (Copepoda), *Crangon crangon* (Decapoda) and *Homarus gammarus* (Decapoda). Mar. Environ. Res. 57 (4) 281–294.

Environment Agency, 2010. Cooling water options for the new generation of nuclear power stations in the UK. SC070015/SR3, Bristol, UK, p. 214.

Environment Agency, 2019. Nuclear power station cooling waters: protecting biota. SC 18004/R1 190 pp., Bristol, UK.

Turnpenny, A.W.H., Coughlan, J., 2003. Using water well? Studies of power stations and the aquatic environment. RWE nPower, Windmill Hill Business Park, Whitehill Way, Swindon SN5 6PB, p. 140.

Turnpenny, A.W.H., O'Keeffe, N., 2005. Screening for intake and outfalls: a best practice guide. Environment Agency Science Report SC030231, Bristol, UK.

Linked

Acoustic fish deterrent: Cooling water and direct cooling; Entrainment (biota); Fish recovery and return: Impingement of biota; Source and receiving waters

Cooling water: 52. Cooling water discharge guidelines

Cooling water discharges must meet a range of Environmental Regulations which will vary from site to site depending on whether the discharge is to a coastal or estuarine water body (termed transitional waters in the European Water Framework Directive), and whether there are sites of conservation interest close by. The core discharge regulations derive from legal instruments such as the Water Framework Directive. This requires no deterioration in ecological status of a set of biological quality elements (BQE), and specifies a range of chemical and physical standards that should be met. The BQE of transitional and coastal (TraC) waters are phytoplankton, macroalgae (seaweed), macrophytes (saltmarsh, seagrasses), and benthic invertebrates together with fish in transitional waters. Where chemical standards are not specifically prescribed, the relevant regulatory agency may, with justification, determine appropriate standards unique to the site.

To date there are no formal temperature standards for TraC waters under European law or UK regulation although there are guidance values for specific cases, but these may be conflicting. In many countries, direct cooling is accepted as an efficient and environmentally sound method of cooling especially in the case of the high-volume discharges (in excess of $100 \, m^3 \, s^{-1}$ (cumecs)) which are planned for the current and next generation of coastal power stations. Typically, in temperate regions the design (and statutory licensing) is for an uplift in temperature over ambient of between 8°C and 12°C at the point of discharge. Most large plants discharging cooling water are situated in estuaries or on coastal sites to use the large volume of water available but this means the most vulnerable areas exposed to warm water are the intertidal and shallow water seabed. Conservation sites are frequently located in the intertidal and in estuaries and near-shore areas such that the discharge would lead to the need for an appropriate assessment or a comprehensive assessment using the outputs of the modeling used for mixing zone predictions.

The permission to discharge (which may be termed a permit, consent, license, or authorization) is usually written as a discharge from an agreed point, with standards that are to be met at the end of the pipe. Most regulatory standards are based on a specific environmental need which means that a key element of getting approval is understanding the extent of a mixing zone with the help of computer models based on measured hydrographic and physicochemical data. Conventionally most industrial discharges are designed to allow rapid dilution with the receiving water in the initial or primary mixing phase, the so-called mixing zone, and regulators may decide that the mixing zone is an area of allowable impact as long as the surrounding area is protected.

In the case of cooling waters, a different approach is required as the most rapid loss of heat load is to design for a buoyant plume which maximizes heat loss to the atmosphere. Such an approach gives slower dilution to any chemical constituents that may be in the discharge and assumes that the temperature is the critical parameter. The dilution at the edge of the mixing zone needs to be calculated to ensure that relevant chemical standards are met. No two sites are the same and an appropriate mixing zone and the related end of pipe standards can only be calculated after adequate local physical and chemical data are collected and analyzed. Cooling water guidelines must take this into account but focus on the critical elements that affect ecosystem structure and functioning and the means of achieving regulatory standards.

References and further reading

EC, 2010. Technical Guidelines for the Identification of Mixing Zones pursuant to Art. 4(4) of the Directive 2008/105/EC C(2010)9369. European Commission, Brussels.
Wither, A., Bamber, R., Colclough, S., Dyer, K., Elliott, M., Holmes, P., Jenner, H., Taylor, C., Turnpenny, A., 2012. Setting new thermal standards for transitional and coastal (TraC) waters. Mar. Pollut. Bull. 64, 1564–1579.

Linked

Plumes; Regulatory mixing zones

Section 5

Impingement and entrainment

Introduction

Many coastal industries require seawater for their industrial processes which, as often emphasized in this volume, lead to problems caused by the industry on the local receiving environment, and problems caused by the local receiving environment on the industry and its plant. Hence the plant is designed with tackling these problems in mind.

Seawater used for condensing the steam circuit cooling in power stations is filtered by coarse bar screens then fine-meshed (2–10 mm) drum screens to remove litter, seaweed, fish, and other marine organisms and materials that might otherwise block the heat exchangers. The impingement on the screens of fish and other organisms is undesirable from both environmental and plant operational points of view. This issue and its mitigation are therefore of high importance to the industry.

The residual animals, plants, and material in the water flow which pass through the cooling water system include adults and larvae of aquatic animals, predominantly members of the plankton. The planktonic organisms, together with microorganisms such as bacteria, diatoms, fungi, have the propensity to settle and grow within the system, with the potential of causing fouling problems. This occurs given that once weathered (i.e., with any toxic materials in the concrete being either dispersed or covered by organisms), the concrete surfaces inside the plant have characteristics no different from rocky shore and hard substrata in the marine environment. It is as a result of this risk to the operation of the system that power plants commonly operate anti-fouling procedures such as biocidal applications.

Equally, the impinged and entrained assemblage of organisms will include species of commercial or conservation significance, such as juveniles and planktonic stages of commercial fish, shellfish, and crustaceans. These organisms are exposed to a combination of potentially deleterious impacts such as pressure stress, physical damage, thermal shock, and biocide toxicity during their passage through the system. Assessments of the potential environmental impact of large cooling water systems have to account for any possible mortality of, or damage to, these organisms as a result of impingement and entrainment.

Finally, the material discharged from the cooling water system includes the remains of the organisms impinged and entrained. This includes the residues of the entrained planktonic organisms such as phytoplankton and gelatinous animals, which can create unsightly foaming at the outfall; the whole or parts of fishes and invertebrates taken into the plant, and hence aggregations of predators or scavengers taking advantage of a readily available food resource.

Impingement and entrainment: 53. Source and receiving waters

A discharge of cooling water from a power station or other industry will be into a receiving water body and will either be immediately mixed or it will create a plume in which the characteristics of the water differ from the surrounding waters. The extent of this water body will depend upon the characteristics of the hydrodynamics of the system, as well as the discharge rate. For a continuous discharge, a plume is likely to be created and initially extend and influence the water body for at least the trajectory of the tidal excursion (the maximum distance the water could travel between low and high water). For high cooling water discharge rates, the effects may not dissipate sufficiently rapidly to avoid accumulation and influencing a wider area. This is particularly true for semi-enclosed estuarine water bodies.

Many large fossil-fueled and most nuclear power stations using direct (once-through) water cooling achieve greater thermal efficiency than power stations with cooling towers but use many times more water. The latest generation of nuclear stations, depending on reactor type, will use around 60–70 $m^3 s^{-1}$ per reactor, so that a site which has two reactors would require a throughput of at least 120 $m^3 s^{-1}$ of cooling water. These large quantities are seldom available from lakes and rivers and the only practicable sources are often estuaries and coastal waters, hence the need to site nuclear power stations in particular in these areas.

The cooling water is abstracted, passed through the plant cooling system, and returned to source at an elevated temperature (typically +8°C to +12°C above ambient). In some cases, residues of anti-fouling biocide (usually halogenated compounds) will also be contained in the cooling water (entry 74). Cooling water supply and receiving waters are usually the same or a contiguous water body; for example water may be abstracted and discharged into the same body of coastal water or abstracted from an estuary and discharged on the adjacent open coast (e.g., the 2000-MW combined cycle gas turbine station at Pembroke in South Wales, United Kingdom which abstracts from the Pembroke River and returns cooling water to the large estuarine water body of Milford Haven).

The choice of supply and receiving waters used for direct cooling needs careful consideration during the planning stages of a power station development. Environmental factors are paramount in the selection process and the subject of detailed assessment studies. For example, abstraction from locations close to fish or shellfish spawning and nursery areas is generally avoided to reduce risks of drawing young life stages of these species into the cooling system where they might suffer impingement or entrainment mortalities. Equally, although warmer waters will be buoyant in comparison to cooler waters in the receiving area, the discharge of heated water has to avoid sensitive benthic habitats and fish migration routes that might be disturbed by elevated temperatures. The position of the intake and discharge pipes in the water column (in the seabed, on the seabed, in midwaters, or at the surface) dictates the effects in terms of material taken in and the influence on the receiving areas. This involves extensive mathematical modeling of how the thermal plume disperses for a representative range of tidal and meteorological conditions.

Many surface water bodies are now subject to international, national, or local conservation designations, which must be accommodated when planning a new power station, for example, special areas of conservation (SACs), special protection areas (SPAs), or sites of special scientific interest (SSSIs).

References and further reading

Environment Agency, 2010. Cooling Water Options for the New Generation of Nuclear Power Stations in the UK. SC070015/SR3, Bristol, UK, p. 214.

Environment Agency, 2019. Nuclear power station cooling waters: protecting biota. SC 18004/R1 19 0pp, Bristol, UK.

Rajagopal, S., Jenner, H.A., Venugopalan, V.P. (Eds.), 2012. Operational and Environmental Consequences of Large Industrial Cooling Water Systems. Springer, Heidelberg/New York.

Linked

Biocides; Cooling water; Direct cooling; Entrainment of biota; Habitats and species legislation; Plume characteristics and behavior

Impingement and entrainment: 54. Impingement of biota

The term "biota" refers to the animals, plants, and other organisms in a given environment or ecosystem and hence the biota in the receiving waters adjacent to a coastal industry is likely to be taken in to any coastal plant requiring cooling water. "Impingement" is the term given to the catching of these organisms on screens designed to filter abstracted water, removing larger debris (termed "trash") before it is used as process or cooling water. Turnpenny (1988) lists some common reasons for fish to be drawn into the intakes:

- high abstraction velocities exceeding the swimming abilities of the fish;
- older designs for offshore intake structures, for example, the "bath plug-hole" effect caused by vertical flows;
- fish not recognizing a danger, thereby not exhibiting avoidance behavior.

Other organisms such as seaweed, jellyfish, and shrimps, with little or no swimming ability, in comparison to the intake suction rates, that happen to come close to the intake are sucked in passively.

Drum- (Fig. 1) and band-screens are commonly used to filter out the trash that enters with seawater via the intake structure, preventing it from entering the fine tubes of the condensers (power plant heat exchangers). Individuals too small to be retained by the screens are instead "entrained," passing through the screens and into the plant before being released back into the original water body. Large quantities of seaweed detached and blown inshore by storms can block screens, restricting cooling water supply, forcing power stations to shut down. Fig. 2 shows a bar screen from the seaward entrance to the cooling system blocked by weed. Summer invasions of jellyfish and winter invasions of sprats have also caused problems of this type at coastal power stations. Such inundations have cost the industry very large amounts of money in lost generation and repairs (Turnpenny and Coughlan, 2003).

Fig. 3 illustrates the quantities of fish that have been recorded at coastal power stations in United Kingdom and France. The catch rate can vary on a seasonal, daily,

Fig. 1 Rotating drum screens are used to filter the cooling water before it enters the plant. After Environment Agency, 2010. Cooling water options for the new generation of nuclear power stations in the UK. SC070015/SR3, Bristol, UK, 214 pp.

Fig. 2 Weed buildup can block screens, reducing cooling water flow and damaging the screens. After Turnpenny, A.W.H., Coughlan, J., 2003. Using Water Well. Studies of Power Stations and the Aquatic Environment. Innogy plc., ISBN 095171726X.

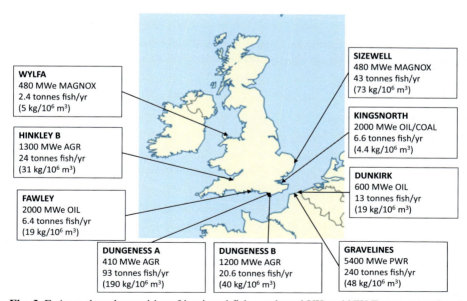

Fig. 3 Estimated total quantities of impinged fish at selected UK and NW France coastal and estuarine power stations.
Adapted from Turnpenny, A.W.H., Coughlan, J., 2003. Using Water Well. Studies of Power Stations and the Aquatic Environment. Innogy plc, ISBN: 095171726X.

or hourly basis, but seasonal migrations are an important factor bringing fish into the vicinity of the intake (Turnpenny and Coughlan, 2003). These fish quantities reflect older power stations which were not designed with the environmental protection features presently available. Nowadays, the cooling water intakes and screening systems of new power stations follow best-practice design criteria, for example as set out in Environment Agency (2010) guidance. This requires fitting of, e.g. acoustic fish deterrents, intake water currents low enough to allow fish to escape, and for screens to be fitted with fish recovery and return systems that allow any trapped fish to be put back to sea. At older power stations, it is/was common practice to either macerate the debris and discharge it as a permitted industrial waste or to send it to landfill.

References and further reading

Environment Agency, 2010. Cooling water options for the new generation of nuclear power stations in the UK. SC070015/SR3, Bristol, UK, p. 214.

Environment Agency, 2019. Nuclear power station cooling waters: protecting biota. SC 18004/R1, Bristol, UK, p. 190.

Turnpenny, A.W.H., 1988. Fish impingement at estuarine power stations and its significance to commercial fishing. J. Fish Biol. 33 (suppl. A), 103–110.

Turnpenny, A.W.H., Coughlan, J., 2003. Using Water Well. Studies of Power Stations and the Aquatic Environment. Innogy plc, ISBN: 095171726X.

Linked

Abstraction; Acoustic fish deterrent; Cooling water and direct cooling; Entrainment (biota); Equivalent adult value; Fish recovery and return (FRR); Source and receiving waters

Impingement and entrainment: 55. Stationary trawlers

Older seawater cooled power stations did not generally incorporate advanced fish protection measures, and their continuous sieving of large volumes of water, and the consequent recovery of large quantities of fish, has been likened to the action of a "stationary trawler." Despite this, power stations have provided a very powerful adjunct to conventional fish survey methods. The early recovery of the tidal Thames, passing through London, from a state of virtual fishlessness caused by organic pollution in the middle of the 20th century, to the position today where more than 124 fish species have now been recorded, has been closely followed by monitoring the incidental catches of fish impinged on power station screens at Thames-side stations such as West Thurrock, Tilbury, and Barking. More recently commissioned stations are often required to monitor impingement and entrainment for a number of years or indefinitely and submit findings to environmental regulators, and therefore they continue to facilitate biological monitoring.

Losses of fish to impingement have not escaped the notice of commercial fishermen, who have often expressed concerns over the "unfair competition" and potential losses of revenue of power stations. The industry has employed fishery scientists to examine this issue and a number of accounts have been published in the scientific literature. A series of studies carried out over a decade at the Sizewell A and B station are summarized in Turnpenny and Taylor (2000). These incidental catches at power stations comprise mainly juvenile fish that fall below the minimum statutory landing size allowable for commercial fishers, and therefore the fishing industry has expressed concerns that removing juveniles may pose a threat to future stocks. However, Turnpenny and Taylor (2000) demonstrated that the catches at Sizewell (United Kingdom) were very small compared with, for example, commercial discards of undersized fish (Table 1). The annual "catch" at Sizewell A power station apparently equaled that of a single, small inefficient trawler.

The development of new, more fish-friendly intake designs, combined with more sensitive intake siting and the deployment of effective fish deterrent and fish return systems have all helped to minimize fish entrapment by power stations.

Table 1 Sizewell a power station, 1981–82 study.

Species	Immediate loss (tonnes year^{-1})	Consequential loss (tonnes year^{-1})	Percent of North Sea stock taken by power station
Plaice	0.03	1.0	0.00072
Sole	0.63	0.9	0.013
Dab	0.41	3.5	0.00034
Cod	1.8	2.8	0.00044
Whiting	1.5	43	0.0087
Herring	0.24	15	0.0017
Total	4.6	66	

Estimated annual loss to the fishery of commercial-sized fish due to CW abstraction.

References and further reading

Turnpenny, A.W.H., 1988. Fish impingement at estuarine power stations and its significance to commercial fishing. J. Fish Biol. 33 (suppl. A), 103–110.

Turnpenny, A.W.H., Coughlan, J., 2003. Using Water Well. Studies of Power Stations and the Aquatic Environment. Innogy plc, ISBN: 095171726X.

Turnpenny, A.W.H., Taylor, C.J.L., 2000. An assessment of the effect of the Sizewell power stations on fish populations. Hydroecologie Appl. 12 (1–2), 87–134.

Wheeler, A.C., 1979. The Tidal Thames: The History of a River and its Fishes. Routledge & Kegan Paul, London.

Whitfield, A.K., Able, K.W., Blaber, S.J.M., Elliott, M., 2022. Fish and Fisheries in Estuaries—A Global Perspective. vols. 1 and 2 John Wiley & Sons, Oxford, ISBN: 9781444336672, p. 1056.

Linked

Abstraction; Acoustic fish deterrent; Cooling water and direct cooling; Entrainment (biota); Fish recovery and return; Source and receiving waters

Impingement and entrainment: 56. Equivalent adult value (EAV)

Given the importance of the impingement of large quantities of fishes in cooling water intakes, it is important to place the resulting data in context of the total available fish populations in either the adjacent area or the movement of populations through coastal areas into estuaries and catchments. Therefore, the equivalent adult value (EAV) technique is a biological accounting procedure used to predict population-level effects of "fishing" impacts where juvenile life stages are affected. This can be whether "fished" by a trawl net (e.g., as juvenile bycatch) or a power station intake and equates to the terms "catchability" and "gear efficiency" of scientific and commercial fishing methods and gear (Franco et al., 2022).

The EAV estimates what the future value to the population of early life stage fishes would have been had they been allowed to grow to maturity. The EAV of an egg, larva, or juvenile fish is defined as the fraction of average lifetime fecundity of a newly mature adult that is required to replace that juvenile, in a balanced population where the birth rate is equal to the death rate (Turnpenny, 1988). This allows comparison of the biological value of fish at different life stages. Mathematically the EAV is calculated using the equation:

$$EAV = 1/(S_t \cdot F_a)$$

where S_t is the probability of survival from birth to any future time and F_a is the average lifetime egg production of an adult. Examples of EAV curves for common commercial fish species are shown in Fig. 1.

The context of EAV is the breeding strategy of fishes. K-selected species, such as mammals or birds, produce few young but offer a high amount of care for each individual, whereas r-selected species, including most fish species, produce large numbers of young to allow for the very high natural mortality that will occur without this parental care. While K-strategists invest in quality, r-strategists invest in quantity.

Estimation of EAV requires the construction of detailed life tables containing life history parameters such as natural mortality, fecundity, sex ratio, and age at sexual maturity for each age class in the population. Such tables usually rely on years of study. The EAV does not consider density-dependent effects (Turnpenny and O'Keeffe, 2005), that is, when the population is larger, there is more competition for resources (such as food and shelter) and a smaller proportion will survive; consequently, it tends to present the "worst-case" estimate of impingement and entrainment impacts.

Turnpenny (1988) examined fish impingement rates at a number of UK estuarine power stations and showed how catches of young fish standardized to adult equivalents using the EAV procedure could be compared with commercial fish landings as a measure of the significance of "competition" from power stations to the fishing industry. Turnpenny and Taylor (2000) describe how the technique was applied to the Sizewell power stations on the east coast of England.

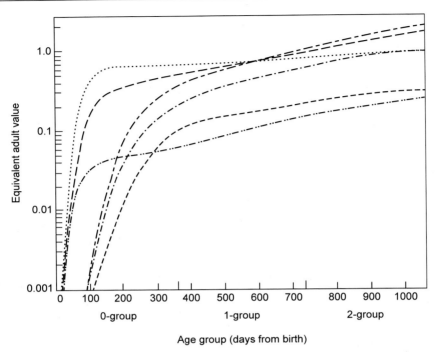

Fig. 1 "Adult equivalent" curves for six fish species, showing values from shortly after spawning to the end of the third year of life. ..., Herring; — — —, whiting; — - —, dab; — · —, sole; - - -, plaice; — ·· —, cod.
After Turnpenny, A.W.H., 1988. Fish impingement at estuarine power stations and its significance to commercial fishing. J. Fish Biol. 33, 103–110. https://doi.org/10.1111/j.1095-8649.1988.tb05564.x.

Whereas life tables can be constructed from available stock management data for the more commercially important fish species, this is less straightforward for some other species which may be of notable ecological importance, for example gobies and other small inshore fish species. Figures for natural mortality rates are particularly hard to estimate without years of study. In order to extend the technique to such species as part of the environmental impact assessment for UK sites for new power stations, CEFAS (BEEMS Technical Report TR383) introduced a revised EAV methodology which employs known relationships between natural mortality rates and average life span of the species in an area (Gislason et al., 2010). This enables a more harmonized approach across commercial and non-commercial fishes (Table 1).

Table 1 Captures of fish by age group at estuarine power stations and equivalent adult values (EAV).

Species		Heysham I No. caught	Heysham I EAV	Hinkley B No. caught	Hinkley B EAV	Fawley No. caught	Fawley EAV	Kingsnorth No. caught	Kingsnorth EAV
Plaice	0-group	3622	172	182	1	21	2	3835	138
	I-group	3336	439	182	30	161	24	6427	853
	II-group	1129	273	0	0	0	0	1822	420
	Total	8087	884	364	31	182	26	12,084	1411
Sole	0-group	1490	121	8531	132	203	9	672	27
	I-group	117	34	1089	825	448	147	11,555	3191
	II-group	0	0	364	381	21	17	2397	1590
	Total	1607	145	9983	1338	672	173	14,624	4808
Dab	0-group	1174	258	9075	1767	0	0	2876	412
	I-group	2013	1099	1815	1202	35	17	5898	4173
	II-group	150	155	0	0	0	0	8823	9199
	Total	3337	1512	10,890	2969	35	17	17,597	13,784
Cod	0-group	445	21	6376	281	0	0	215	10
	I-group	30	2	10,036	760	0	0	0	0
	II-group	0	0	1634	246	0	0	0	0
	Total	475	23	18,046	1287	0	0	215	10
Whiting	0-group	9850	3759	71,655	25,994	42	17	48,905	19,567
	I-group	214	110	65,522	34,360	791	419	10,309	5378
	II-group	0	0	182	231	0	0	0	0
	Total	10,064	3869	137,359	60,585	833	436	59,214	24,945
Herring	0-group	8576	6121	3630	2342	3402	2292	22,536	15,810
	I-group	11,722	8859	364	138	53,291	40,292	20,330	15,325
	II-group	128	178	364	163	224	279	767	690
	Total	20,426	15,158	4358	2643	56,917	42,863	43,633	31,825

After Turnpenny, A.W.H., 1988. Fish impingement at estuarine power stations and its significance to commercial fishing. J. Fish Biol. 33, 103–110. https://doi.org/10.1111/j.1095-8649.1988.tb05564.x.

References and further reading

BEEMS Technical Report TR383, 2016. A new approach for calculating equivalent adult value (EAV) metrics. Cefas, Lowestoft.

Franco, A., Elliott, M., Franzoi, P., Nunn, A., Hänfling, B., Colclough, S., Young, M., 2022. Appendix a study methods: field equipment, sampling and methods. In: Whitfield, A.K., Able, K.W., Blaber, S.J.M., Elliott, M. (Eds.), Fish and Fisheries in Estuaries—A Global Perspective. John Wiley & Sons, Oxford, ISBN: 9781444336672, pp. 874–940.

Gislason, H., Daan, N., Rice, J.C., Pope, J.G., 2010. Size, growth, temperature and the natural mortality of marine fish. Fish Fish. 11, 149–158.

Turnpenny, A.W.H., 1988. Fish impingement at estuarine power stations and its significance to commercial fishing. J. Fish Biol. 33, 103–110. https://doi.org/10.1111/j.1095-8649.1988.tb05564.x.

Turnpenny, A.W.H., Coughlan, J., 2003. Using Water Well. Studies of Power Stations and the Aquatic Environment. Innogy plc, ISBN: 095171726X.

Turnpenny, A.W.H., O'Keeffe, N., 2005. Screening for intake and outfalls : a best practice guide. Environment Agency Science Report SC030231, 154 pp. Bristol, UK.

Turnpenny, A.W.H., Taylor, C.J.L., 2000. An assessment of the effect of the Sizewell power stations on fish populations. Hydroécologie Appl. 12, 87–134.

Linked

Biota impingement; Cooling water and direct cooling; Entrainment (biota); Fish recovery and return

Impingement and entrainment: 57. Acoustic fish deterrent (AFD)

As indicated in this volume, coastal industry managers will be charged with trying to minimize the fish impinged in a coastal water intake, and hence will be concerned to introduce any measures to achieve that reduction of fish, both for environmental and plant logistic reasons. An acoustic fish deterrent (AFD) is a type of behavioral fish barrier that can be installed at a water intake such as the cooling water intake entrance of a power station to deter fish from becoming drawn into the cooling circuit. In some cases the system is designed to simply block fish entry but elsewhere to guide fish around structures, such as offshore water intakes or hydropower turbines.

The fitting of AFDs is specified by the local statutory environmental protection bodies, for example the English Environment Agency, as "best available technology" (BAT) for large coastal or estuarine power stations; this was included in 2010 regulatory advice for nuclear new-build power stations and other large thermal stations (Environment Agency, 2010) and reconfirmed in a further EA review in 2019. AFDs can be used in the form of a sound projector array (SPA) system, effectively an array of underwater loudspeakers, or a Bio-Acoustic Fish Fence (BAFF™) (Turnpenny and O'Keeffe, 2005) (Figs. 1 and 2). The SPA relies upon producing a repellent sound

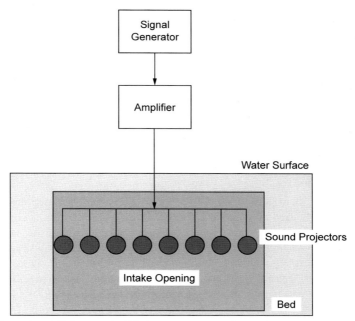

Fig. 1 Schematic of SPA acoustic fish deterrent system (front elevation) showing sound projectors arranged in front of an intake opening (Turnpenny and O'Keeffe, 2005).

Fig. 2 AFD sound projectors shown in raised position (out of water) at Pembroke CCGT power station. In use, they are lowered down the vertical rails that can be seen below the walkway. Courtesy Fish Guidance Systems Ltd.

field to elicit avoidance behavior from hearing-sensitive fish species, whereas the BAFF™ uses an air bubble curtain to confine the sound signal, creating a guiding "wall of sound" that can be used like a net or screen to divert fish past an intake or into a purpose-designed bypass channel. SPAs are therefore purely acoustic, while BAFFs™ combine acoustic and visual cues.

Power stations and other industries positioned on coasts or estuaries most commonly use the SPA-type system, as the main aim is to repel fish from the intake entrance, pushing them out into a tidal flow that will carry them to safety. BAFFs™ are more frequently used in rivers, to divert fish around obstructions, such as turbines, or to force them to pass the dam or weir through a narrow bypass, for example.

The SPA creates its repellent acoustic field in front of a water intake using underwater sound projectors powered by audio amplifiers and electronic signal generators. The arrangement of sound projectors must be customized to each site to allow for such physical differences as bathymetry, water current, and tidal range, but also the hearing thresholds of the species to be targeted. Once in situ, the AFD can be fine-tuned in order to increase the efficiency of the signal and may be retuned at different times of the year to optimize efficiency for the different species of fish that occur seasonally. The more sophisticated systems incorporate diagnostic telemetry systems that allow web-based monitoring of system integrity and performance. Continuous access to operating parameters enables shore-based monitoring of system condition and remote control of AFD signals and settings.

Installation and servicing of AFD equipment is undoubtedly more challenging and has potential staff safety implications at distant offshore intake locations. This is exacerbated by limited access opportunities on continuously running plant where maintenance can only be undertaken during planned shutdowns. These issues may be overcome by new developments that allow service intervals to be extended to 24 months and by the latest generation of remotely operating diving vehicles which could remove the need for human divers mechanically cleaning the plant.

AFDs find application in a variety of settings. In addition to fish exclusion from water intakes, they have found use in clearing fish from areas where construction blasting is about to take place and as barriers to the spread of invasive fish species.

The English Environment Agency (2010) BAT guidance for large direct cooled power stations recommends the use of AFDs in conjunction with a fish recovery and return (FRR) system. The two systems are complementary, as the AFD is most effective against hearing-sensitive pelagic species, while the FRR system will safely return to the sea the more robust bottom-dwelling species that can tolerate handing.

References and further reading

Environment Agency, 2010. Cooling water options for the new generation of nuclear power stations in the UK. SC070015/SR3, Bristol, UK, p. 214.

Environment Agency, 2019. Nuclear power station cooling waters: protecting biota. SC 18004/R1 190 pp., Bristol, UK.

Turnpenny, A.W.H., O'Keeffe, N., 2005. Screening for intake and outfalls: a best practice guide. Environment Agency Science Report SC030231, 154 pp. Bristol. UK.

Linked

Biota impingement; Cooling water and direct cooling; Entrainment of biota; Fish recovery and return; Underwater sound

Impingement and entrainment: 58. Fish recovery and return (FRR)

A fish recovery and return (FRR) system at a coastal power station makes it possible to return fish and other biota such as shrimps and crabs to their natural environment, rather than collecting them in "trash baskets" for disposal at landfill as has long occurred at older stations. The English Environment Agency (2010) best practice guidance recommends that direct cooled coastal and estuarine power stations will employ a suite of techniques to mitigate against fish losses, starting with ensuring that the intake is located in a fish-poor area of the seabed and away from key fish migration routes. The plants should ensure that intake water velocities are kept low ($<0.3\,\mathrm{ms}^{-1}$) so that fish are not drawn in involuntarily, hence by fitting acoustic fish deterrents (AFDs) or other fish warning devices and finally by deploying FRR techniques. FRR is a mitigation measure aimed at fish that have not been deterred at the intake entrance and these typically comprise fish of poor hearing ability that do not respond to the AFD system, notably flatfish and other bottom-dwelling fish. A well-designed FRR therefore has the potential to minimize losses of fish, as bottom-dwelling species have been shown to have a high rate of survival after becoming impinged on a band- or drum-screen and returned to sea (see Table 1, after Turnpenny and O'Keeffe, 2005).

Organisms handled by the FRR system enter the plant with the cooling or process water. Larger organisms and other debris are filtered out of the cooling water by drum- or band-screens to prevent condenser or process blockages within the plant circuit. The event of being trapped on filter screens is termed "impingement." The FRR process differs from standard debris screening procedures by ensuring sympathetic handling of delicate fish, lifting them in water-filled "fish buckets," washing them off the screens with a low-pressure spray, and providing a return route to the sea via open launders (troughs) or pipes which carry a continuous flow of water and are covered to exclude predators. The discharge point is carefully chosen to reduce the risk of fish being recycled back to the water intake, usually by particle track modeling methods.

The nature of the capturing gear and the type of fish caught dictate the amount of damage to the fishes. Latest best practice guidance from the English Environment

Table 1 Typical fish survival reported from studies of drum- or band-screens with simple modifications for fish return (e.g., with fish buckets, low-pressure sprays, and continuous screen rotation) (Turnpenny and O'Keeffe, 2005).

Fish group	Survival rate >48 h after impingement
PELAGIC E.g., herring, sprat, smelt	<10%
DEMERSAL E.g., cod, whiting, gurnards	50%–80%
EPIBENTHIC E.g., flatfish, gobies, rocklings, dragonets, and crustacea	>80%

Fig. 1 Fish return system at barking power station, Thames tideway. Left-hand insert shows screen panel with fish buckets inverted. Right-hand insert shows modern fish bucket profile (Turnpenny and O'Keeffe, 2005).

Agency (UK) (2010) provides detailed design requirements, specifying, for example, screen rotation rate, fish bucket design, backwash pressure, pipeline and launder size and slope, bend radius, and other details. System design requires careful hydraulic analysis to meet these requirements, including analyses of pressure and velocity stresses which may impair fish survivability.

An unknown (and largely unmeasurable) factor is the potential losses to bird, fish, and marine mammal predators that rapidly identify such discharge points as a source of easy prey. FRR designs therefore need to minimize any stress or disorientation of the fish in order to maximize survival rates. However, even returning non-viable organisms contributes to the local marine food web as detritus, unlike the process of disposing the waste material in former landfill waste practices.

These studies were carried out at power stations using the older trash elevator, which often failed to retain fish, rather than the latest designs of "fish bucket" (Fig. 1).

References and further reading

Environment Agency, 2010. Cooling water options for the new generation of nuclear power stations in the UK. SC070015/SR3, Bristol, UK, p. 214.
Turnpenny, A.W.H., O'Keeffe, N., 2005. Screening for Intake and Outfalls : a best practice guide. Environment Agency Science Report SC030231, 154 pp. Bristol, UK.

Linked

Cooling water and direct cooling; Entrainment (biota); Impingement of biota

Impingement and entrainment: 59. Disposal of impinged material

Given the potentially large amounts of material such as fish, large invertebrates, seaweed, and other debris and litter being taken in to coastal cooling water intakes, it is then both an environmental and an operational production problem to dispose of the material as sustainably as possible. Depending upon the mesh size of the band- or drum-filter screens (usually between 2 and 10 mm^2 openings), larger organisms and debris that cannot pass through will become impinged on the screens. Smaller elements will pass through the screens and be entrained in the cooling circuit and returned to the water body in the discharge stream. In many older UK power stations, all of the trash (the debris and biota) impinged is washed off and routed to trash skips. Some of the earliest and most valuable fish monitoring data sets have arisen from filtering out and recording the fish and other biota from the trash stream. In some instances, the trash material has been sent direct to landfill whereas in others it has been macerated and returned to the water body via the discharge. In some areas, waste to landfill attracts a Landfill Tax at the standard (higher) rate given the high organic content whereas its discharge back into the adjacent water body is also treated as a highly organic discharge despite the material previously being removed from that water body; hence discharging the material into the water body will require an effluent discharge license.

The English Environment Agency has defined best practice screening arrangements for intakes in two Science Reports (2005 and 2010). Where screens are fitted with fish recovery and return technology (FFR) systems in pursuance of Best Practice requirements, organisms will be returned to the wild in varying condition. While flatfish and crustaceans may survive, those roundfish with swim bladders are less able to withstand the pressure changes during transit through the cooling intakes and so are easily damaged or killed; loss of scales and hemorrhaging of eye are also common problems leading to a lower survivorship of the fish being ejected from the cooling water. Past practice on the treatment of trash including biota as a potential trade waste has varied. The English Environment Agency has no formal policy on this issue, but the 2010 Science Report contains the following advice:

- If the waste is returned to the water in a continuous stream, together with entrained material, there should be no consenting issues. This reduces landfill waste.
- Maceration of screen arisings may be acceptable if discharged continuously. Maceration occurs incidentally on some current, primitive fish return systems.
- Concentrated dumping of accumulated waste could have significant waste and/or water quality issues. This would constitute a land-based discharge and subject to regulation.

References and further reading

Turnpenny, A.W.H., O'Keefe, N., 2005. Best practice guide for intake and outfall fish screening. The Environment Agency, R&D Contract No. W6-103, Final Report, The Environment Agency, Bristol.

Turnpenny, A.W.H., Coughlan, J., Ng, B., Crews, P., Rowles, P., 2010. Cooling water options for the new generation of nuclear power stations in the UK. Science Report SC070015/SR, Environment Agency, Bristol.

Linked

Abstraction; Discharge consent, permit, licence, authorisations; Entrainment of biota; Fish recovery and return; Impingement of biota

Impingement and entrainment: 60. Entrainment (biota)

Entrained organisms (biota) are those organisms entering with the cooling or industrial water supply that are too small to be impinged on the fine band- or drum-screens used to filter the cooling or process water and remove larger debris. Instead, they pass through the screens and continue along the internal cooling water route before being returned in the plant effluent to the original water source. On older installations, the screen meshes are typically of 8–10 mm^2 aperture but newer systems may have smaller meshes of 2–6 mm, allowing more impinged organisms to be recovered and put back to sea by the fish recovery and return facilities.

The entrained biota typically includes phytoplankton and zooplankton, as well as the eggs, larvae, and juveniles of fish and crustaceans (Turnpenny and Coughlan, 2003). The maximum size of organisms entrained depends on the size of the screen meshes used (Fig. 1). Mortalities of organisms at power stations caused by entrainment have to be assessed in combination with impingement mortalities when considering overall impacts of entrainment. For fish this is facilitated using, for example, the equivalent adult value concept, a biological accounting procedure that projects the future value that an early life stage would have to the mature adult stock after allowing for the naturally high mortality rates in larval and juvenile stages. Although this concept is still valid the approach now varies.

Entrained organisms are subject to a variety of stresses while passing through the plant, including abrasion in pipes and tunnels, contact with pump blades, hydrostatic

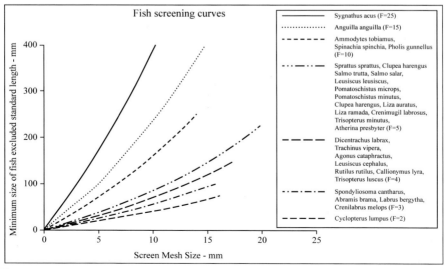

Fig. 1 Mesh size curves for screening fish of different body shape (Turnpenny, 1981). F is the fineness ratio, = fish length divided by maximum body depth.

and hydrodynamic pressure changes (positive and negative), sharp changes of temperature (up to +8–12°C above ambient) as water enters and leaves power plant cooling condensers, and biocide (usually halogenated compounds) toxicity. Plant designers and operators have little control over these effects.

Some organisms are killed by these stresses and some survive and are put back to sea with the cooling water. Without adequate biocide treatment, sessile, filter-feeding species such as mussels (mainly the blue mussel *Mytilus edulis* in Northern temperate areas) and barnacles can settle and thrive within the cooling circuit causing problems with cooling efficiency. An entrainment simulator, one type used in the United Kingdom known as the entrainment mimic unit (EMU), can be programmed to simulate the stress conditions experienced by entrained organism, allowing some indication of potential entrainment mortality. Available results (Table 1) show that survival can be very different according to species and life stage but that significant proportions can pass through the cooling system and returned back to sea alive.

Survival of phytoplankton can be measured in terms of primary productivity or the rate of carbon fixation (Davis and Coughlan, 1983). Seawater samples from the plant intake and outfall are incubated with a radioactive isotope (C^{14}), the uptake of which is then measured. Comparing carbon fixation rates of the samples allows estimation of the rate of survival through the cooling system (Fig. 2A). The chief measure for reducing fish entrainment is to ensure that the cooling water intake is sited away from fish spawning and nursery areas (Environment Agency, 2010). The same would apply avoiding important shellfish grounds.

Table 1 Survival rates of entrained fish and crustacean from EMU CW passage simulation experiments (Bamber and Turnpenny, 2011).

Species	Life stage	Entrainment survival rate at 0.2 ppm TRO and ~10°C ΔT	Prime causes of mortality
Bass (*Dicentrarchus labrax*)	Eggs	54%	Thermal stress
	Larvae	56%	Thermal stress and chlorine toxicity
Lobster (*Homarus gammarus*)	Larvae	92%	Mechanical stress
Shrimp (*Crangon crangon*)	Larvae	75%	Thermal stress and chlorine toxicity
Silver eel (*Anguilla anguilla*)	Larvae[a]	52%	TRO
Sole (*Solea solea*)	Eggs	93%	Pressure, thermal stress
	Post-larvae	8%	Thermal stress and chlorine toxicity
Turbot (*Psetta maxima*)	Eggs	93%	Pressure, thermal stress
	Post-larvae	30%	Thermal, mechanical, and pressure stress

[a] Eel tested at 2 ppm TRO.

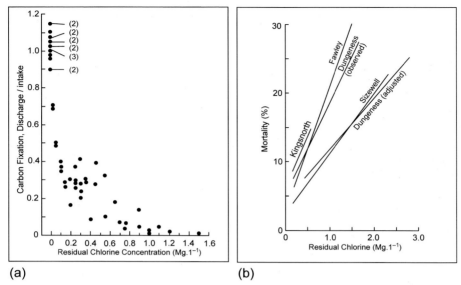

Fig. 2 (A) Effect of chlorine concentration on carbon fixation by phytoplankton at Fawley Power Station, Hampshire (Davis, 1983). (B) Percentage mortality of adult calanoid copepods within one hour of entrainment under various chlorination regimes at four different UK power stations (Davis and Coughlan, 1983).

References and further reading

Bamber, R.N., Turnpenny, A.W.H., 2011. Entrainment of organisms through power station cooling water systems. In: Rajagopal, S., Jenner, H.A., Venugopalan, V.P. (Eds.), Operational and Environmental Consequences of Large Industrial Cooling Water Systems. Springer Science & Business Media, Norwell, MA (Chapter 15).

Davis, M.H., Coughlan, J., 1983. A model for predicting chlorine concentration within marine cooling water circuits and its dissipation at outfalls. In: Jolley, R.L., Brungs, W.A., Cotruvo, J.A., Cumming, R.B., Mattice, J.S., Jacobs, V.A. (Eds.), Fourth Conference on Water Chlorination: Environmental Impact and Health Effects, vol. 4. Ann Arbor Science, Pacific Grove, CA, pp. 347–357.

Environment Agency, 2010. Cooling Water Options for the New Generation of Nuclear Power Stations in the UK. SC070015/SR3, 214 pp, Bristol, UK.

Turnpenny, A.W.H., 1981. An analysis of mesh sizes required for screening fishes at water intakes. Estuaries 4 (4), 363–368.

Turnpenny, A.W.H., Coughlan, J., 2003. Using Water Well. Studies of Power Stations and the Aquatic Environment. Innogy plc, ISBN: 095171726X.

Linked

Abstraction; Cooling water and direct cooling; Equivalent adult value; Impingement of biota; Mechanical, thermal and chemical stressors; Source and receiving waters

Section 6

Biofouling

Introduction

As indicated elsewhere, most marine organisms have planktonic dispersal stages, irrespective of whether the adult is planktonic or benthic. The lifestyle of many sedentary and sessile benthic marine organisms requires searching for suitable settlement surfaces and settling once they reach metamorphosis, the stage at which they change from a larva to a settling adult. Bio(logical) fouling refers to the unwanted settlement and growth of organisms in industrial installations, resulting in performance and operational efficiency reductions of the plant and equipment. Biological fouling is a phenomenon which can be found at all phase boundaries, for example, air to solids (lichens on stones), air to water (a microlayer with its own ecology), and water to substratum (micro- and macrofouling). Most notably, the inside surfaces of an industrial plant often provide suitable protected conditions for settlement as well as an abundant supply of food and oxygen.

The colonization process in sea or estuarine water is reviewed together with the problems caused by biofouling on station and cooling water performance. The sequence of fouling starts with surfaces becoming "weathered," the deposition of organic and inorganic molecules rapidly followed (within a few hours) by the attachment of bacteria, yeasts, and microalgae. At a later stage, and depending on the time of year, these stages are followed by the settling stages of larger animals (oysters, mussels, barnacles, hydroids, bryozoans). In turn, these organisms may then create crevices and sediment accumulation which will support small crabs and mobile crustaceans such as amphipods and isopods.

The final species composition and community within fouling layers will vary with geographical location season. As a general rule, reproduction occurs during spring and summer months leading to settlement of larvae in summer and autumn. The control methods are discussed with the focus on sodium hypochlorite and heat treatment (Thermoshock). A relatively new method "BioBullets" is also discussed. The section ends with a short explanation of "Microbiological influenced corrosion" (MIC). The entries emphasize that while biofouling and its control is non-discernible as an environmental problem, in that the loss of planktonic and settlement stages would not be detected among the billions of stages in the plankton, it is an operational and production problem for the industrial plant. Therefore, control of biofouling is necessary and proven antifouling measures are discussed here with the focus on once-through cooling water systems.

Biofouling: 61. Microbial and macrobial fouling

Fouling is a generic term which can be divided into three different forms: trash, scaling, and biological fouling. Trash is the unwanted debris arriving on the trash racks in front of the intake bay, consisting of driftwood, algae, seaweeds, plastic bags, etc. On rare occasions sea gooseberries (ctenophores), fish (sprat, pipefish), and jellyfish can enter the cooling system in quantities sufficient to force the shutdown of a power station. Scaling in the general form of calcium carbonate deposition is a chemical process and mostly restricted to heat exchangers in seawater cooled systems.

Sea or estuarine water will cause deposition of (in)organic material in all pipework of a cooling water system (CWS) changing the surface and enabling the initiation of colonization by biota (living organisms). This colonization proceeds in a predictable pattern once any chemicals exuding from the surface have been lost (the "weathering" process). At first, organic and inorganic molecules are deposited, rapidly followed (within a few hours) by the attachment of bacteria and the establishment of a chemical layer forming a film of polypeptides, polysaccharides, and lipids. Planktonic bacteria will settle and change their structure and physiology by forming fibrils, leading to the biofilm. They produce slime or extracellular polymeric substances (xPS), mostly polysaccharides, and this biofilm forms the housing for bacteria and creating their optimal environment (Flemming, 2009).

This process of forming a biofilm takes less than 12h and, together with settled fungi (yeasts) and microalgae, is the basis for macrofouling settlement by bivalves (mussels and oysters), barnacles, tube worms, bryozoans (sea mat), hydroids (white weed), etc. by making the settlement surfaces attractive and resembling natural hard substrata. All of these fouling species are common to natural marine hard structures, such as rocky shores and outcrops, and they usually have a specialized fixation system to withstand the sheer forces of the cooling water flow although they may have different flow tolerances. For example, barnacles, hydroids, and bryozoans require fast flowing conditions and so may settle on clean, unrestricted surfaces, whereas mussels will settle in the crevices between barnacles as they prefer more turbulent conditions. Once these early colonizers have settled then tube worms, and crustaceans such as small crabs, amphipods, and isopods may occupy the increasingly complex community and building a fouling ecology (Jenner et al., 1998).

Many of the initial settling organisms such as mussels, barnacles, hydroids, bryozoans, and tube worms are suspension feeders which obtain their food by filtering particles and plankton from the water column. It is widely accepted that whereas other types of organisms, such as carnivores and detrital feeders, have their populations regulated by the amount of food, in the case of suspension feeders it is space which regulates the population size. Hence, the availability of large areas for colonization inside coastal plants gives an attractive environment for these sessile forms.

The nature of the surface provided by the plant dictates its suitability for settlement. Within the condensers, the speed of colonization depends to a large extent on the metals used in construction: inert titanium condenser tubes are rapidly colonized, whereas copper alloys, cupronickel and brass, which have an inherent acute toxicity to bacteria, are colonized at much slower rates. Although the fouling community constitutes both the microbial bacterial slimes and the larger animals or macro-invertebrates (macrofouling), the latter are often most prevalent in the incoming stages of the cooling water system up to the main condensers where the temperatures will be similar to ambient conditions. In general and in contrast, the increased temperature (ΔT) over the condensers discourages the settlement of fouling organisms in the outgoing parts of the CWS.

It is the colonizing power of mussels and barnacles (1 million m^{-2} is possible) in the incoming CWS of direct cooled power stations which constitutes the main threat to the safety and operational efficiency of the plant. As indicated before, in a CWS, the level of settlement is regulated by the availability of suitable substratum, for example, the surface areas of the inner walls of pipework, areas in which particulate food and oxygen will be continuously delivered; this may be similar to the natural environment where nutrients and available space for colonization are available but where, in contrast to inside the plant, predation regulates the biomass of the sessile forms.

Inside a CWS, nutrients are not limited, surface space is available, and predators are lacking. Consequently, the growth of mussels can be rapid, up to 3–4 cm in the first year. Mussel fouling commonly takes the form of mattresses of mussels on an underlying substratum of barnacles (Anil et al., 2012). The barnacles would have settled first and change the boundary turbulent conditions allowing mussel spat (the juvenile stages) to settle in between barnacles or in dead barnacle exoskeletons. After the barnacles have died, large rafts of mussels can break off from the barnacle substratum with subsequent clogging of condenser tubes or plate coolers. Fouling problems with oysters can also be severe as they cement themselves to the substratum with one (ventral) valve still fixed at the wall after death of the oyster. The remaining valve will act as a suitable substratum for other bivalves and barnacles with settlement aided by the increased turbulence at the remaining valve.

The sequence of settlement depends on the spawning period and time of larvae in the plankton until metamorphosis leading to settlement; for NW Europe, Rasmussen (1973) gives a large amount of information for many of the inshore and estuarine species regarding their maturation rate and season, time of spawning, planktonic phase, and settlement period. As such, successful settlement depends on encountering a suitable substratum at the appropriate time (i.e., following metamorphosis from a larva to an adult of young phase). Consequently, an adequate control of all fouling is a necessary design and operational requirement for all coastal and estuarine power stations.

Mussels produce byssus threads to fix themselves to the substratum or to their neighbors. Barnacles are cemented to their substratum and, owing to their internal fertilization, they must be settled as close as possible to each other (Khandeparker and Anil, 2007; Anil et al., 2012). At dense populations, barnacles grow vertically owing to a lack of space. The growth season of barnacles starts in early spring before the settlement of mussels, and they form a good substratum for mussels to settle. Barnacles are able to settle at water velocities of ≤ 3 m s^{-1} and they roughen the walls and change the turbulent boundary conditions thereby enabling mussels to settle (Fig. 1).

Fig. 1 Mussel settlement at a flow $>2.5\,\mathrm{m\,s^{-1}}$, note the mold seam at which the smaller mussel spat has settled.

Without fouling control, the mussels will grow on the barnacle layer as thick mattresses which detach after the barnacle layer dies off. A power station is then at risk from the sudden clogging of the condenser tubes. Inside a cooling water system, the preferred sites for barnacle and bivalve settlement are the inside bends of the conduits and near inspection hatches. Here the turbulence is higher with pockets of low flow conditions which enable the organisms to settle. The molded seams of concrete conduits and the crevices on inspection hatches form other spots for settlement owing to the local turbulence (see Fig. 2).

In the case of oysters, the attached shell can remain fixed for many years and can only be removed by mechanical means which in turn can damage the finish of a concrete conduit. Tube worms (such as *Spirobranchus* spp.) are less troublesome, but encrustations of both old oyster shells and tube worms form an excellent substratum for mussel growth owing to the increased turbulence along the surfaces of the conduits. Hydroids (white weed) are colonial "micro" anemones (polyps) forming tufts of branched threads at lower water velocities. If not removed, they can grow rapidly and clogging problems will arise; removal of the fouling organisms in the condenser tubes can be achieved using rubber "Taprogge®" balls, although these balls can get stuck on remains of barnacles thereby increasing the clogging of tubes (see Fig. 4).

There are four main factors or driving forces for settlement: (i) relatively high turbulence along the surfaces, juveniles need to touch the wall or substratum to secure themselves, already settled organisms or a molded seam in the concrete wall will increase the turbulence and encourage settlement; (ii) owing to the relatively high velocity and turbulence in the intake bay, oxygen levels are always optimal; (iii) the intake system and conduits, including water velocity, prevent predators (e.g.,

Fig. 2 Barnacle fouling and at the right the vertical "tulip growth" forms when space is limited and growth high.

water fowl, fish, and crabs) from preying on the settled biofouling; owing to the lack of sunlight, algae are not able to overgrow the biofouling; (iv) nutrients are constantly available during operation of the station.

In the natural environment there are usually periods of low (tidal) flow or exposure whereas the environment inside the CWS may be more uniform. Consequently, in a CWS growth of mussels can be high, with large specimens reaching 4 cm at the end of September in their first year. Often the shells are thinner and more yellow as the result of a combination of rapid growth and lack of sunlight. Too high water velocities will preclude settlement and at water velocities of 1.5–2.5 m/s, substratum is the limiting factor. In general, the abundance of mussel spat ready to settle is generally so high that after the settlement period a density of $10^6 \, m^{-2}$ of mussel spat is possible. The mortality rate inside a CWS is lower compared to that of naturally settled mussels so without biofouling control problems quickly arise.

Biofouling also occurs on external structures in the sea, such as cooling water intake structures, fuel-supply or waste-disposal jetties, and seawalls. Fouling on static structures tends only to become a problem if the weight of the fouling threatens the physical integrity of the structure itself. While this can be a serious issue with offshore oil and gas platforms, it is not documented as a significant problem at power stations and other industrial users (Figs. 3–5).

Fig. 3 Stolen cars recovered from Amsterdam-Rhine canal at Utrecht (the Netherlands) serving as substratum for mussel growth.

Fig. 4 Inlet water box with severe barnacle/mussel fouling and trapped blue "Taprogge" balls.

Fig. 5 Debris fouling at a trash rack and mussel fouling at a traveling screen (note the sacrificial anodes).

References and further reading

Anil, A.C., Desai, D., Khandeparker, L., Gaonkar, C., 2012. Barnacles and their significance in biofouling. In: Rajagopal, S., Jenner, H.A., Venugopalan, V.P. (Eds.), Operational and Environmental Consequences of Large Industrial Cooling Water Systems. Springer, New York, p. 522.

Bruijs, M.C.M., Jenner, H.A., 2012. Cooling water system design in relation to fouling pressure. In: Rajagopal, S., Jenner, H.A., Venugopalan, V.P. (Eds.), Operational and Environmental Consequences of Large Industrial Cooling Water Systems. Springer Science & Business Media, Inc., Norwell, MA, pp. 45–693.

Flemming, H.C., 2009. Why micro-organisms live in biofilms and the problem of biofouling. In: Flemming, H.C., Murthy, P.S., Venkatesan, R., Cooksey, K. (Eds.), Marine and Industrial Biofouling. Springer, Berlin.

Jenner, H.A., Whitehouse, J.W., Taylor, C.J.L., Khalanski, M., 1998. Cooling water management in European power stations: biology and control of fouling. Hydroécologie Appl. 10. Electricité de France (EdF) Chatou Paris: vols. 1–2, 225 pp.

Khandeparker, L., Anil, A.C., 2007. Underwater adhesive: the barnacle way. Int. J. Adhes. 27, 165–172.

Rao, T.S., 2012. Microbial fouling and corrosion: fundamentals and mechanisms. In: Rajagopal, S., Jenner, H.A., Venugopalan, V.P. (Eds.), Operational and Environmental Consequences of Large Industrial Cooling Water Systems. Springer Science & Business Media, Inc., Norwell, MA, pp. 95–126.

Rasmussen, E., 1973. Systematics and ecology of the Isefjord marine fauna (Denmark). Ophelia 11, 1–507.

Turnpenny, A.W.H., Coughlan, J., 1992. Power generation on the British coast: thirty years of marine biological research. Hydroecol. Appl. 1, 1–11.

Whitehouse, J.W., Khalanski, M., Saroglia, M.G., Jenner, H.A., 1985. The control of biofouling in marine and estuarine power stations. Joint report of CEGB, EDF, ENEL and KEMA, 48 pp; EDF Energy, Barnwood, Gloucs., UK.

Linked

Antifouling measures; Biofouling control by chlorine; Cooling water and direct cooling

Biofouling: 62. Settlement of planktonic organisms

It is emphasized throughout this volume that the coastal manager requires a good knowledge of the local ecological and physicochemical conditions in order to determine the effect of the industry on the natural system and the effect of the natural system on the industry. This is particularly the case for the organisms in the vicinity of the plant, in the water column, and on the bed of the estuary or coast. The majority of invertebrate species in the meroplankton are larval stages of taxa which, as adults, live as benthic forms, commonly sessile (e.g., barnacles, mussels, bryozoans) or species with limited benthic mobility (e.g., some polychaetes). After a period of planktonic dispersion, which may be as brief as a few days but usually up to several weeks, these larvae must metamorphose and settle to develop further (Rasmussen, 1973; Levinton, 2021).

In many cases, larval stages adapted to a planktonic life metamorphose into a secondary larval stage whose role is settlement. Thus decapod crustacean zoea larvae metamorphose into megalopae prior to settlement, barnacle nauplius larvae metamorphose into cypris larvae prior to settlement, bivalve veliger larvae metamorphose into a pediveliger (by developing a foot) prior to settlement. These secondary larval stages commonly have a much more restricted potential life span than do their obligately planktonic primary stages and must settle within that limited time or die.

Settlement of most species is selective. The larvae seek appropriate substrata, using chemosensory cues, and appropriate depth levels (e.g., the littoral or the sublittoral) using baroperception, light cues, and behaviorally by remaining inshore. Particularly favorable chemical cues include the presence of adults of the same species, as is found with barnacle cypris larvae which preferentially settle on or adjacent to other barnacles or on the calcareous bases left by dead barnacles. Secondary cues for settlement, particularly of species which live (as adults) in soft sediments, are sediment organic content and granulometry.

Some species show more complex patterns of settlement. Mussel pediveliger larvae settle initially into soft sediments or on filamentous red algae, where they feed and grow briefly. They subsequently resuspend (by byssal drifting—the byssus being the "hair" by which the mussel is fixed to substrata or other mussels) and secondarily seek out hard substrata on which to settle. Byssal drifting may be very widespread in bivalve mollusks, including species in which the adult has no byssus.

Settlement of planktonic organisms is thus substratum selective and is highly seasonally restricted in temperate and polar waters, but all year round in subtropical and tropical waters (although the abundance of settling larvae may vary seasonally, e.g., in response to monsoon effects).

References and further reading

Levinton, J.S., 2021. Marine Biology: Function, Biodiversity, Ecology, sixth ed. OUP, Oxford.
Rasmussen, E., 1973. Systematics and ecology of the Isefjord marine fauna (Denmark). Ophelia 11, 507.

Raymont, J.E.G., 1983. Plankton & Productivity in the Oceans: Volume 2 Zooplankton. Pergamon, Oxford.

Sigurdsson, J.B., Titman, C.W., Davies, P.A., 1976. The dispersal of young post-larval bivalve molluscs by byssus threads. Nature 262, 386–387.

Linked

Phenology; Plankton

Biofouling: 63. Peak times of settlement by fouling organisms

The principal metazoan (multi-cellular) species involved in fouling of cooling water intake systems colonize via dispersive larvae, mainly planktonic. In temperate waters, reproduction, and thus the production of such larvae, is commonly seasonal. As a general rule, reproduction occurs during spring and summer months leading to settlement of larvae in summer-autumn although many species have temperature thresholds for spawning (see Rasmussen, 1973). Evolutionary selection is for the settlement time most appropriate to the biology of the species involved. Inshore barnacles commonly settle in the late winter/early spring, for example, *Semibalanus balanoides* settling between January and May, but often in a very restricted period in May in NW Europe. The benthic individuals are then able to feed on phytoplankton as it becomes available in the water column in spring. Similarly, species of the honeycomb worm *Sabellaria* settle early in the season, *Sabellaria spinulosa*, for example, settling in southwestern England in March and April.

Climatically, latitude has a restricting effect on settlement seasonality. Blue mussels (*Mytilus edulis*) settle between May and October in the colder waters of the Wadden Sea, while further south, in UK waters *M. edulis* settles all year round. Even then, there can be peak times, some sites showing peak recruitment in summer and winter, respectively, while in the Turkish Mediterranean, *Mytilus galloprovincialis* pediveligers settle predominantly from March to December. Serpulid polychaetes, such as *Spirobranchus* (formerly *Pomatoceros*) spp., a common inhabitant of concrete surfaces inside coastal industry cooling systems, tend to show a single restricted peak of settlement, typically during summer and autumn months in British waters.

Bacteria, which are responsible for creating the surface slimes within coastal power plant condenser tubes (i.e., the production of biofilms), also show seasonal patterns in density and diversity, but studies give conflicting results. Densities have been found to be higher in warmer months, diversity highest in winter or stable over time, and populations limited by phosphorus availability during spring and summer, but by carbon availability in autumn and winter, mitigating against seasonal patterns. It therefore appears that there is little seasonality in bacterial settlement, even though reproduction would be expected to be greatest in warmer months.

References and further reading

Anil, A.C., Desai, D.V., Khandeparker, L., Gaonkar, C.A., 2012. Barnacles and their significance in biofouling. In: Rajagopal, S., Jenner, H.A., Venugopalan, V.P. (Eds.), Operational and Environmental Consequences of Large Industrial Cooling Water Systems. Springer Science & Business Media, Inc., Norwell, MA, pp. 65–93.

Bernbom, N., Ng, Y.Y., Kjelleberg, S., Harder, T., Gram, L., 2011. Marine bacteria from Danish coastal waters show antifouling activity against the marine fouling bacterium *Pseudoalteromonas* S91 and zoospores of the green alga *Ulva australis* independent of bacteriocidal activity. Appl. Environ. Microbiol. https://doi.org/10.1128/AEM.06038-11.

Cotter, E., O'Riordan, R.M., Myers, A.A., 2003. Recruitment patterns of serpulids (Annelida: Polychaeta) in Bantry Bay, Ireland. J. Mar. Biol. Assoc. U.K. 83, 41–48.

Gilbert, J.A., Field, D., Swift, P., Thomas, S., Cummings, D., Temperton, B., Weynberg, K., Huse, S., Hughes, M., Joint, I., Somerfield, P.J., Mühling, M., 2010. The taxonomic and functional diversity of microbes at a temperate coastal site: a 'multi-omic' study of seasonal and diel temporal variation. PLoS One 5 (11), e15545. https://doi.org/10.1371/journal.pone.0015545.

Pinhassi, J., Gómez-Consarnau, L., Alonso-Sáez, L., Sala, M.M., Vidal, M., Pedrós-Alió, C., Gasol, J.M., 2006. Seasonal changes in bacterioplankton nutrient limitation and their effects on bacterial community composition in the NW Mediterranean Sea. Aquat. Microb. Ecol. 44, 241–252.

Pulfrich, A., 1996. Attachment and settlement of post-larval mussels (Mytilus edulis L.) in the Schleswig-Holstein Wadden sea. J. Sea Res. 36, 239–250.

Rasmussen, E., 1973. Systematics and ecology of the Isefjord marine fauna (Denmark). Ophelia 11, 507.

Schauer, M., Balagué, V., Pedrós-Alió, C., Massana, R., 2003. Seasonal changes in the taxonomic composition of bacterioplankton in a coastal oligotrophic system. Aquat. Microb. Ecol. 31 (2), 163–174.

Snodden, L.M., Roberts, D., 1997. Reproductive patterns and tidal effects on spat settlement of *Mytilus edulis* Populations in Dundrum Bay, Northern Ireland. J. Mar. Biol. Assoc. U. K. 77, 229–243.

Wilson, D.P., 1970. The larvae of *Sabellaria spinulosa* and their settlement behaviour. J. Mar. Biol. Assoc. U. K. 50, 33–52.

Yildiz, H., Berber, S., 2010. Depth and seasonal effects on the settlement density of *Mytilus galloprovincialis* L. 1819 in the Dardanelles. J. Anim. Vet. Adv. 9, 756–759.

Linked

Biofouling; Settlement of planktonic organisms

Biofouling: 64. Biology of fouling organisms

Fouling communities or settlements are to be found on hard substrata, such as vessels, oil platforms, piers, jetties, within cooling water systems. Fouling in relation to power station cooling water intakes relates predominantly to that last category, although there are also risks of fouling on external structures in the sea, such as cooling water intake structures, fuel-supply or waste-disposal jetties, and seawalls. The plant and animal species involved in fouling settlements or communities in this context are sessile. Being incapable of locomotion in their settled (fouling) stage of life, they will have a dispersive phase, normally a larva or spore (although in the case of hydrozoans and scyphozoans the dispersive phase is the adult) (Levinton, 2021).

The animal species involved are almost invariably filter feeders (suspension feeders), and it is of note that whereas deposit and detritus feeders are limited by food supply, suspension feeders are limited by space; hence they settle in large densities until restricted by space. Overall, the biology of these species while in cooling water systems is similar to that experienced under natural conditions, where these species are living on rocky shores or similar hard substrata; in the case of mussels, the bed may be composed of mussels attaching by byssus threads to stones or other mussels.

In possible contrast to natural environments, which are often tidal involving exposure at regular intervals, within cooling water systems, during operation, the artificial habitat is more stable and normally well oxygenated and provides a constant and adequate supply of organic particulate and dissolved material appropriate as food; however, unlike natural habitats, there is virtually no light present, and thus no plants or microalgae survive. This removes niches which may be afforded by plant and algal shelter in a natural habitat, but conversely removes competition for space by plants and macroalgae (seaweed). Furthermore, most direct predators of such species, for example, asteroids (starfish) or dog whelks (respectively, notable predators of mussels and barnacles), do not gain access to this habitat, either greatly reducing or eliminating any predation pressure.

A further factor within cooling water systems may be the water velocity. Barnacles and zebra mussels (*Dreissena polymorpha*) do not settle or feed at water velocities significantly exceeding $2\,\mathrm{m\,s^{-1}}$; in extremis, most sessile animals are detached by velocities substantially in excess of $5\,\mathrm{m\,s^{-1}}$. Growth of bryozoan colonies is inhibited above $0.6\,\mathrm{m\,s^{-1}}$, and they are detached at velocities $>0.9\,\mathrm{m\,s^{-1}}$. In practice, where organisms more tolerant of higher velocities can settle in the boundary layer of a cooling water culvert, their growth imparts a roughness reducing the boundary layer velocity, which can then allow subsequent settlement of other species. For example, marine mussels prefer slower flowing and more turbulent conditions.

After settlement, suspension feeding bivalves such as mussels capture and excrete suspended sediment particles (as feces and pseudofeces) in niches which then allows other detritus-feeding, scavenging, and even carnivorous invertebrates to colonize the area.

References and further reading

Anil, A.C., Desai, D.V., Khandeparker, L., Gaonkar, C.A., 2012. Barnacles and their significance in biofouling. In: Rajagopal, S., Jenner, H.A., Venugopalan, V.P. (Eds.), Operational and Environmental Consequences of Large Industrial Cooling Water Systems. Springer Science & Business Media, Inc., Norwell, MA, pp. 65–93.
Collins, T.M., 1964. A method for designing seawater culverts using fluid shear for the prevention of marine fouling. CERL Report No. RD/1/N93/64, Central Electricity Research Laboratories, Leatherhead, p. 5.
Hayward, P.J., Ryland, J.S. (Eds.), 2017. Handbook of the Marine Fauna of North-West Europe, second ed. OUP, Oxford, p. 785.
Jenner, H.A., Whitehouse, J.W., Taylor, C.J.L., Khalanski, M., Mattice, J., l'Abeé-Lund, J.H., Ambrogi, R., Bamber, R.N., Coughlan, J., Duvivier, L., Humphris, T.H., Moreteau, J.-C., Bachmann, V., Rajagopal, S., Turnpenny, A.W.H., van der Velde, G., 1998. Cooling water management in European power stations. Biology and control of fouling. Hydroécologie Appl. 10 (1-2). i–v + 1–225.
Levinton, J.S., 2021. Marine Biology: Function, Biodiversity, Ecology, sixth ed. OUP, Oxford.
OECD, 1963. Catalogue of Main Marine Fouling Organisms, Volumes 1-9. Organisation for Economic Co-operation and Development. OECD, Paris.
Synopses of the British Fauna (New Series), nos 10—1998, 14—1999, 33—1985, 34—1985, 57—2008. Published for the Linnean Society of London by the Field Studies Council.

Linked

Microbial and macrobial fouling; Peak times of settlement by fouling organisms; Settlement of planktonic organisms

Biofouling: 65. Antifouling measures

Fouling of cooling water systems (CWS) can be divided into four different processes: (i) biological fouling inside the CWS by settlement of sessile micro- and macroorganisms; (ii) clogging problems by ingress of fish, jellyfish, or weeds; (iii) chemical fouling by scaling and sedimentation inside the system; (iv) fouling by the generic term of trash (timber, plastic etc.) which plays a role in the blockage of the intake trash racks and band screens. This section reviews the first of these: control and mitigation of settled micro- and macrofouling inside the cooling system. Uncontrolled macrofouling will cause severe operational problems, leading to high maintenance costs as well as costly outages and loss of production. Typical CWS conditions provide an optimal environment for many fouling species to settle and develop into large populations. The most important fouling species are bacterial slime, barnacles, chalk worms, bryozoans (sea mats), hydroids (white weed), mussels, and oysters. This section focuses on proven (non-experimental) antifouling measures for once-through CWS. In general, antifouling measures can be divided between physical and chemical measures, with the latter further divided into oxidizing and non-oxidizing compounds.

Methods of control

Physical

Filtration and water pre-treatment techniques: these techniques may be subdivided into macrofiltration and microfiltration. Macrofilters keep debris and adult organisms (macrofouling organisms, fish, and jellyfish) out of the CWS. Commonly employed systems are drum screens and band sieves and automatically cleaning trash racks may be employed. At power stations subject to extensive macrofouling, automatically cleaning mussel sieves are often installed inside the CWS immediately before the main condensers. Microfiltration is a means to reduce the organic and inorganic load of the cooling water, thus reducing the biocide demand and carbon source for biological growth. Existing methods are rotating drum filters and sand filters. Because of the large water volumes involved and the speed with which filters become clogged, microfiltration is currently not a feasible option for large once-through CWS.

Heat treatment or "thermoshock"

This is an established antifouling method that replaces the use of a biocide for macrofouling control. The treatment consists of heating the cooling water to a temperature of 38–42°C by means of partial recirculation and maintaining this temperature for ~30 min to 2 h. This ensures elimination of existing mussels, oysters, and barnacles. In seawater CWS, this should be done 3–4 times a year. The crucial factor here is that mussels and oysters detaching from the walls must be sufficiently small to pass through heat exchanger tubes; otherwise, large detached mussels can cause greater blockages down-flow in passing through the CWS. In plants abstracting from freshwaters, CWS treatments can be limited to once or twice a year. In order to control microfouling with heat treatment, much higher temperatures (60–80°C) are required. These temperatures can only be reached when the condensers are dewatered sequentially. The heat loosens and dries out the microfouling which detaches as thin leaves from the inner side of the tubes.

Sponge rubber balls

The use of recirculating sponge rubber ball systems, such as the "Taprogge®" system, has found worldwide acceptance for the maintenance of heat exchanger efficiency and protection of heat exchanger tubes. The technique is capable of preventing the accumulation of particulate matter, biofilm formation, but is not effective against macrofouling. The slightly oversized sponge rubber balls continuously pass, at random through the tubes of a heat exchanger, propelled by the water flow. The balls are extracted from the main flow by a suitable strainer, drawn off and returned to the inlet section of the heat exchanger to be recycled through the equipment. However, impediments inside the tubes, such as old barnacle exoskeletons, can cause the balls to get stuck thereby exacerbating the blockage problems.

Manual cleaning for macrofouling

An alternative adopted at some nuclear power stations is manual cleaning of the CWS each year during shutdown for fuel exchange. This approach is applicable to any industry where there are planned shutdowns at regular intervals. The whole operation is logistically optimized and the advantage is that no biocides are used. The mussel and other biofouling growth after one year are usually not sufficiently severe to cause operational problems.

Oxidizing and non-oxidizing biocides

Chemicals used in antifouling applications are biocides which can be subdivided between oxidizing and non-oxidizing biocides. The tables below present an inventory of the well-known oxidizing and non-oxidizing biocides that are commercially available and used in once-through CW systems.

Oxidizing biocides (Table 1) do not only react with organisms but also with dissolved and suspended matter, and both living and dead organic material present in the cooling water will be oxidized; this involves usually halogenated biocides and as a result, a complex mixture of halogenated by-products is formed. Some of the constituents of this mixture have been identified (e.g., haloamines, haloforms, haloacetonitriles, halogenated amides, halophenols, haloacetic acids, bromate, and chlorate), but many others remain unidentified. In terms of concentrations, bromoform is the most important halogenated by-product found in seawater with concentration levels between 1 and $25\,\mu g\,L^{-1}$ depending on location and season (White, 2010; Khalanski and Jenner, 2012). Sodium hypochlorite (NaOCl) is the most

Table 1 Common oxidizing biocides for once-through CW systems.

Group	Biocide	Chemical formula	Reaction
Chlorine based	Sodium hypochlorite	NaOCl	Fast
	Chlorine dioxide	ClO$_2$	Very fast
Bromine based	Sodium hypochlorite + NaBr	NaOCl + NaBr	Fast
	1-Bromo-3-chloro-5,5,-dimethylhydanthoide (BCDMH)	C$_5$N$_2$O$_2$H$_6$ClBr	Fast
Other	Ozone	O$_3$	Very fast
	Peracetic acid	C$_2$H$_4$O$_3$	Fast

common oxidizing biocide and it may be supplied in bulk by tanker or generated as Cl$_2$ gas on-site in an electro-chlorination plant (ECP) by electrolysis of brine or seawater. Commercially available sodium hypochlorite has an active chlorine concentration of 10%–15%. The commercial hypochlorite solution may be used as such, but more often it is diluted prior to dosing to about 500–2000 mg L^{-1} Cl$_2$ of free oxidants (FO) to improve mixing with the CW. Hypochlorite (analogous to the oxidizing chemicals in household bleach) remains the most commonly used cooling water biocide because of its effectiveness and relatively low cost. This has resulted in an extensive literature and operational experience. In Europe, hypochlorite is by far the most commonly applied biocide used in once-through coastal CWS.

Non-oxidizing biocides. The most common non-oxidizing biocides are presented in Table 2. These biocides are not suitable for CWS with flows of >25 m^3 s^{-1} but some have found application in cooling tower systems. The best known are the quaternary ammonium compounds, usually referred to by practitioners as QACs or Quats. While oxidizing biocides exert a non-specific biocidal action on the target organism, non-oxidizing biocides have more specific modes of action on the cell surface

Table 2 Non-oxidizing biocides, chemical formula, reaction time, and half-life.

Group	Name of biocide	Chemical formula	Reaction	Half-life
Isothiazolones	2-Methyl-4-isothiazolin-3-one	C$_4$H$_4$NO$_5$	Slow	Long
	5-Chloro-2-methyl-4-isothiazolin-3-one	C$_4$H$_4$ClCNO$_5$	Slow	Long
QACs	Alkyl-dimethylethyl-benzylammonium-chloride	R(CH$_3$)$_2$(C$_8$H$_9$)NCl	Average	Average
	Didecyl-dimethyl-ammoniumchloride	C$_{22}$NH$_{48}$.Cl	Average	Average
	Alkyl-dimethyl-benzyl-ammoniumchloride	R(CH$_3$)$_2$(C$_7$H$_7$)NCl	Average	Average
	Poly[oxyethylene (dimethyliminio)ethylene-	(C$_{10}$H$_{24}$N$_2$O.Cl$_2$)$_n$	Average	Average

Table 2 Continued

Group	Name of biocide	Chemical formula	Reaction	Half-life
	(dimethyl-iminio)-ethylenedichloride]			
	2,2-Dithiobisbenzamide	$C_{14}S_2NH_{10}$	–	–
	2-Bromo-2-nitropropane-1,3-diol (BNPD)	$C_3H_6NO_4Br$	Average	Long
	2,2-Dibromo-3-nitrilo-propionamide (DBNPA)	$C_4N_2H_2OBr_2$	Fast	Short

(e.g., QACs), or inside the cell, affecting metabolic processes. The application of non-oxidizing biocides is only recommended in cases where oxidizing biocides are not able to give sufficient protection, such as in systems with high organic loads, or in recirculating CWS systems where daily control is not practical.

References and further reading

Jenner, H.A., Whitehouse, J.W., Taylor, C.J.L., Khalanski, M., 1998. Cooling water management in European power stations: biology and control of fouling. Hydroécologie Appl. 10, 225. Electricité de France (EdF) Chatou Paris: vol. 1–2.

Jolley, R.L., Carpenter, J.H., 1983. A review of the chemistry and environmental fate of reactive oxidant species in chlorinated waters. In: Jolley, R.L., Bull, R.J., Davis, W.P., Katz, S., Roberts Jr., M.H., Jacobs, V.A. (Eds.), Water Chlorination, Environmental Impact and Health Effects. vol 4. Ann Arbor Science Publishers, Ann Arbor, MI.

Khalanski, M., Jenner, H.A., 2012. Chlorination chemistry and ecotoxicology of the marine cooling water systems. In: Rajagopal, S., Jenner, H.A., Venugopalan, V.P. (Eds.), Operational and Environmental Consequences of Large Industrial Cooling Water Systems. Springer Science & Business Media, Inc., Norwell, MA, pp. 183–227.

White, 2010. White's Handbook of Chlorination and Alternative Disinfectants, fifth ed. Black and Veatch. Wiley & Sons, Chichester, ISBN: 978-0-470-56133-1.

Linked

Biocides; Continuous and pulse dosing; Cooling water and direct cooling; Chlorination chemistry; Electro-chlorination plants; Heat treatment; Pulse or continuous dosing

Biofouling: 66. Biofouling control by chlorination

Chlorine was first used in Europe early in the 20th century for drinking water disinfection and it is still used as such as well as being widely employed in the food and drinks industry. Chlorine is the most extensively studied disinfectant for both its chemistry and (eco)toxicology. However, at cellular level the mode of action has never been fully resolved and is still best regarded as "a black box." The claim that undissociated (non-charged) forms of chlorine and bromine pass across cellular membranes more easily is a hypothesis which has never been scientifically demonstrated. There are indications chlorine mainly destroys the extracellular enzymes outside the cells, while another halogen, bromine, passes the cell walls destroying cell organelles.

Chlorine is not the ideal biocide for cooling water systems (CWS) application because it is non-specific to target organisms. Reaction with many non-target organic and inorganic compounds lowers the concentration available for biocidal action, and these unwanted secondary reactions may produce persistent and sometimes carcinogenic residuals. However, most of the chemistry is well understood, the significant by-products are all identified, and they show low acute and chronic toxicity; chlorine remains a relatively cheap and effective alternative compared to other oxidative biocides. Worldwide, chlorine is the most commonly used antifouling biocide in CWS.

Chlorine (as hypochlorite) for CWS is produced either on site in Electrochlorination Plants (ECP) or delivered in bulk and, with both options, storage tanks are necessary. They require to be insulated as the decay at higher ambient temperatures can be significant (30% at 30°C. in 2 weeks). ECP-produced chlorine will have concentrations of $2-5\,g\,L^{-1}$ and industrial grade bulk supplies around $150-160\,g\,L^{-1}$. In general, the hypochlorite is diluted by cooling water to a concentration of $0.5-2\,g\,L^{-1}$ prior before dosing in the CWS enabling a better mixing. In seawater, with bromide concentrations of $\sim 65\,mg\,L^{-1}$, the dosed hypochlorite will oxidize the bromide to bromine with much of the subsequent chemistry bromine based.

Chlorine is described in the older literature in terms such as free, active, available, combined, residual, or by combination of these adjectives. Currently, the terminology for seawater is simplified to free oxidants (FO) and total residual oxidants (TRO). In Fig. 1 a summary is given of the chlorine/bromine chemistry.

To combat fouling in a CWS, for example, by mussels, oysters, barnacles, and tube worms, a chemical treatment strategy is necessary to ensure adequate control. Some of these treatments are found to work well at one location but not at all at another owing to species differences, location, climate, and water quality. Treatment methods are:

- end-of-season treatment during which all fouling is killed in a relatively short period at the end of the settlement period
- periodic treatment when adult and juvenile fouling biomass has reached a critical value before the end-of-season treatment
- shock dosing. This remains a commonly employed method although its impact is debatable. One commonly employed method of shock dosing entails one or two extra-high-dosing events on top of continuous dosing, although there is poor evidence that this technique is successful despite the assumption that short periods of higher concentrations will be more

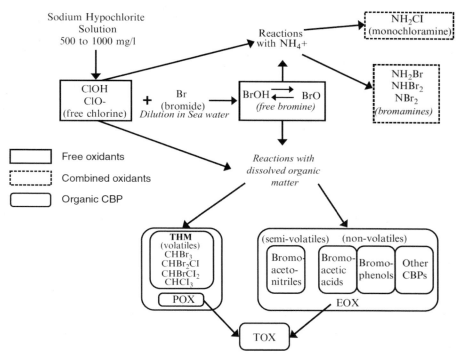

Fig. 1 Chlorine and bromine basal chemistry during chlorination of seawater. *From Taylor, C.J.L., 2006. The effects of biological fouling control at coastal and estuarine power stations. Mar. Poll. Bull. 53, 30–48.*

effective. The mode of action of chlorine against bivalves is on the metabolic level and not by oxidation of internal tissue, and the application of a high concentration of biocide causes bivalves to close their shells very tightly. Mussels easily survive a full week chlorination at concentrations of $\sim 10\,\mathrm{mg\,L^{-1}}$. Their condition may even be improved as the biocide has killed all bacteria and parasites on the outer surfaces of the mussels
- Continuous dosing is a simple dosing strategy with low concentrations applied during the whole year or at least several months consecutively. Although effective and simple from an operator standpoint, the use of chlorine will be much higher, more chlorination by-products will be formed, and costs will be higher.
- Intermittent treatment is a regime with frequent dosing changing from minutes to hours with and without chlorine. A specific mode of dosing is Pulse-Chlorination® at which the opening and closure behavior of bivalves are monitored during dosing and not dosing periods. The principle here is that the continuous changes in metabolic activity (aerobic-anaerobic) stress the mussels and oysters and mortality is quicker compared to continuous dosing alone.

Mussels and oysters appear to be the most resistant fouling species to chlorination so ensuring complete mortality of those bivalves means that all other species like tube worms and barnacles will also be eradicated.

Hypochlorite dosing is more complicated than dosing with chlorine given that the mixing of hypochlorite is slow and it takes time before adequate chlorine concentrations are found along the wall boundary layer where mussel spat are found. Spat enter the system with their foot extruded about three times the length of the spat as it tries to touch the wall where, after some time, the first byssus is produced. Patchiness in chlorine concentration can be found in the CWS even after 500 m beyond the dosing point. Therefore, an effective dosing system by a rack or ring is necessary to ensure uniform dilution and mixing.

Measuring the active products, the free oxidant (FO) and total residual oxidant (TRO) levels can be done by several techniques including amperometric, potentiometric titration, and colorimetric/spectrophotometric. The latter is the well-known technique using DPD (*N,N*-diethyl-*P*-phenylene-diamine) and portable or fixed spectrophotometric equipment which can measure a lowest concentration of $\sim 0.05\,\text{mg L}^{-1}$ FO or TRO with a resolution of $0.01\,\text{mg L}^{-1}$. In general, (semi-)continuous measurement equipment is installed close to the inlet water boxes of the main condensers. In seawater, bromine and eventually monobromamine are formed due to oxidation of bromide. The acute toxicity of bromine and bromamines is higher than chlorine and so TRO has an efficacy (toxic action) nearly equal to FO.

References and further reading

Khalanski, M., Jenner, H.A., 2012. Chlorination chemistry and ecotoxicology of the marine cooling water systems. In: Rajagopal, S., Jenner, H.A., Venugopalan, V.P. (Eds.), Operational and Environmental Consequences of Large Industrial Cooling Water Systems. Springer, New York, pp. 183–227.

Rajagopal, S., Venugopalan, V.P., Van der Velde, G., Jenner, H.A., 2005. Dose response of mussels to chlorine. In: Lehr, J.H., Keeley, J. (Eds.), Water Encyclopedia. Vol. V Water Quality and Resource Development. John Wiley & Sons, Inc., New York, pp. 401–406.

Taylor, C.J.L., 2006. The effects of biological fouling control at coastal and estuarine power stations. Mar. Poll. Bull. 53, 30–48.

White, 2010. White's Handbook of Chlorination and Alternative Disinfectants, fifth ed. Black & Veatch. Wiley & Sons, Chichester, ISBN: 978-0-470-56133-1.

Linked

Antifouling measures biocides; Biofouling control by chlorination; Chlorination chemistry; Cooling water and direct cooling

Biofouling: 67. BioBullets

An innovative method for the control of mussel growth in industrial cooling water systems is marketed under the trade name of BioBullets® (BioBullets Ltd., London, United Kingdom). The method consists of encapsulated crystals of an active biocidal compound with a nutritional coating attractive to target biota. The active compounds used include quaternary ammonium compounds (QACs) and potassium chloride (KCl). Dosing QAC loaded BioBullets in a cooling system might ultimately discharge concentrations of QACs which were unacceptable for the regulator. A solution is parallel dosing of fine mineral clays such as bentonite to neutralize the QACs, through adsorption and sedimentation, while retaining an adequate toxic impact to control benthic fauna (Aldridge et al., 2006; Costa et al., 2008, 2011).

Bivalves (mussels, oysters, and clams) will ingest the dosed BioBullets, in the same way as their natural food particles, and the BioBullets will also be transported via the gills through the food groove to the mouth. In the intestine the coating is removed and the biocidal action is released. A potential shortcoming is that the bivalve needs to be of adequate size to ingest particles of ~40–80 µm. An advantage over chlorination is that the bivalves do not close their shells as they do with hypochlorite in detecting potentially harmful materials. Hence, with BioBullets the toxic effect will be more direct and benefits will be seen in days instead of weeks.

BioBullets may be an attractive alternative for mussel fouling mitigation in terms of effectiveness and dosing time. However, large-scale application of QAC loaded BioBullets for cooling water systems of $20–50 m^3 s^{-1}$ needs further experimental assessment, especially to determine the need for detoxification by bentonite-like materials which in turn may have effects on the benthic communities (Fig. 1).

Summary

Advantages

- targeted dosing for bivalves;
- no excessive dosing/discharge of unused chemicals;
- will not affect other non-target organisms in the receiving water bodies.

Disadvantages

- only larger specimens take up sufficient amounts of BioBullets;
- the quantity of the preparations needed to treat the high volumes in once-through system;
- relatively costly compared to chlorine.

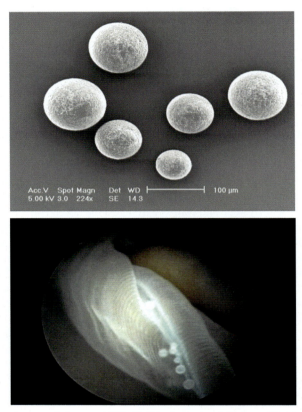

Fig. 1 Electron microscopic images of BioBullets (upper) and their position in the food groove by endoscopy (lower).

References and further reading

Aldridge, D.C., Elliot, P., Moggridge, G.D., 2006. Microencapsulated BioBullets for the control of biofouling zebra mussels. Environ. Sci. Technol. 40, 975–979.

Costa, R., Elliott, P., Saraiva, P.M., Aldridge, D., Moggridge, G.D., 2008. Development of sustainable solutions for zebra mussel control through chemical product engineering. Chin. J. Chem. Eng. 16 (3), 435–440.

Costa, R., Moggridge, G.D., Aldridge, D.C., 2011. Improved mussel control through microencapsulated BioBullets. In: Rajagopal, S., Jenner, H.A., Venugopalan, V.P. (Eds.), Operational and Environmental Consequences of Large Industrial Cooling Water Systems. Springer, New York, pp. 273–287.

Linked

Anti-fouling measures; Biocides, biofouling control by chlorination; Cooling water and direct cooling

Biofouling: 68. Heat treatment

Thermal treatment or "Thermoshock" is an established and widely accepted antifouling method that can completely replace the use of biocides for the control of macrofouling, although its practical application is currently limited to a limited number of locations in Europe. It is also used at freshwater cooled locations against Zebra and Quagga (non-indigenous) mussels. The application of thermal treatment requires a special design of the cooling water system (CWS), so the decision to use heat treatment has to be implemented in an early stage of plant design. Adaptations afterward are often technically difficult and expensive.

"Thermoshock" or heat treatment consists of heating the cooling water to a temperature of 38–45°C by means of partial recirculation or steam injection and maintaining this for half an hour (in case of mussel fouling). This guarantees elimination of existing growth of blue mussels, oysters, and barnacles. In seawater cooling water systems (CWS) treatment should be repeated 3–4 times a year. It is especially important to ensure that the mussels, or other bivalves, detaching from the walls are still sufficiently small to pass through the condenser tubes and avoid blockages. Therefore monitoring (e.g., by a bypass monitor over the condensers or with settlement plates or tubes at the inlet) of growth rate of settled mussels and oysters in the CWS is essential. A "rule of thumb" is that when the largest animals from a sample of ~200 individuals are around 10 mm the treatment becomes necessary.

In order to control microfouling by heat treatment, much higher temperatures (70–80°C) are required. A disadvantage is that regrowth starts immediately after ending the treatment and "Thermoshock" is not an option against microfouling in general. However, as an example, the Borssele nuclear station in the Netherlands installed an innovative variation on heat treatment to clean the condenser tubes of biofilm. The condensers were dewatered during operation one by one during which the heat loosens and dries out the microfouling which detaches as thin leaves from the inner side of the tubes.

Particular attention is needed for the control of the invasive (in NW Europe) Japanese oyster *Crassostrea gigas*, a species naturally more adapted to higher temperatures. Owing to their growth in the whole CWS, temperatures of at least 45°C are necessary for a period of ≥ 2 h in order to ensure 100% mortality. This elevated temperature must be present in the whole system including the outlet conduits. At temperatures higher than 45°C, the treatment time is reduced. In Fig. 1, the representative temperature, tolerance, and mortality graphs for seawater and brackish water species of bivalves are shown.

Summary

Heat treatment or "Thermoshock" is an elegant method for control of mussel fouling and has already been applied in the Netherlands at two coastal and two freshwater cooled stations. A specific problem is the rapid colonization in certain locations by the Japanese oyster which requires a temperature up to 45°C for control. The latter reduces the potential of "Thermoshock" as a relatively environmentally friendly method. With a prospective station life of 40–50 years for the new generation of nuclear

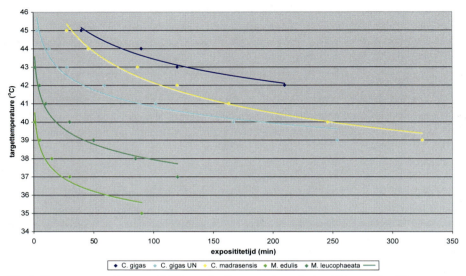

Fig. 1 Temperature tolerance of different fouling organisms during "Thermoshock" treatment. The Pacific oyster *Crassostrea gigas*; *C. gigas* UN (lab. experiments done at the university); *Crassostrea madrasensis* (tropical species); *Mytilus edulis* (common blue mussel); *Mytilopsis leucophaeata* (brackish water invasive species) (vertical axis: Target temperature [°C]; horizontal axis: Exposure time [min]).
From Rajagopal, S., van der Velde, G., van der Gaag, M., Jenner, H.A., 2005. *Factors influencing the upper temperature tolerances of three mussel species in a brackish canal: size, season and laboratory protocols.* Biofouling 21, 87–97.

stations and with the forecasted change in fouling species composition at the station sites in both the United Kingdom and elsewhere in Europe (reflecting climate change), the complete replacement of chlorine by heat treatment is considered unlikely.

References and further reading

Jenner, H.A., Whitehouse, J.W., Taylor, C.J.L., Khalanski, M., 1998. Cooling water management in European power stations. Biology and control of fouling. Hydroécologie Appl. 10 (1-2), 225.

Rajagopal, S., van der Velde, G., Jansen, J., van der Gaag, M., Atsma, G., Janssen-Mommen, J.P.M., Polman, H., Jenner, H.A., 2005a. Thermal tolerance of the invasive oyster *Crassostrea gigas*: feasibility of heat treatment as an antifouling option. Water Res. 39, 4335–4342.

Rajagopal, S., van der Velde, G., van der Gaag, M., Jenner, H.A., 2005b. Factors influencing the upper temperature tolerances of three mussel species in a brackish canal: size, season and laboratory protocols. Biofouling 21, 87–97.

Linked

Control of biofouling control by chlorination; Non-indigenous, alien, invasive, and other non-native species (NIS, AIS)

Biofouling: 69. Microbially influenced corrosion (MIC)

Microbial fouling in heat exchangers and cooling towers can lead to reduced heat transfer and damage to plant components (e.g., valves) by microbial influenced corrosion (MIC), as well as an increased risk of infection from human pathogens. Human pathogens commonly associated with cooling towers are *Naegleria* spp., *Acanthamoeba* spp., and *Legionella pneumophila*. Numerous outbreaks of Legionnaire's disease have been associated with aerosols from cooling towers and air conditioning units.

All of the three items mentioned before result in increased operating costs. For example, Siemens Kraftwerk Union reports a 26-μm-thick biofilm results in a loss of 2.0 MWe (megawatt electric) for a 740-MWe unit and 5.1 MWe for a 1300-MWe unit. Mitigation of microbial fouling (biofilm) is therefore especially necessary although adequate monitoring of the biofilm is rarely practiced.

MIC starts with biofilm formation, and formation of a slime layer or biofilm starts immediately when the cooling system is operational. The first stage is the formation of a conditioning film (the surface is regarded as being "weathered"); after initial attachment, bacteria adhere irreversibly to the surface and proliferate to transform into bacterial microcolonies; other microbes may also develop such as yeasts. Subsequently, bacteria generate a coating of xPS (extracellular substances mainly consisting of polysaccharides), which is essential for the development of the architecture of any biofilm matrix; it provides a framework into which microbial cells are inserted. Confocal laser scanning microscopy indicates that microcolonies within a biofilm are three-dimensional structures of mushroom-like bacterial growth with water channels running between them providing a constant supply of nutrients. A biofilm consists mostly of water, with the microbial cells forming only a thin layer (Fig. 1).

The deposition of microorganisms on metallic surfaces promotes microbial induced corrosion (MIC) with an increase in the overall rate of corrosion. Application of noble materials (such as monel) and biotoxic materials (such as copper) does not in practice give any guarantee of MIC resistance, with the exception of titanium. Approximately 20% of all corrosion damage in heat exchangers is caused or influenced by microorganisms. MIC involves the interaction between microbial activity and corrosion processes and has been observed in virtually all surface waters. Microbial activity can influence the initiation and propagation of corrosion and induce corrosion processes where they would otherwise not occur (Figs. 2 and 3).

Accelerated corrosion may occur as the result of corrosive metabolic products such as sulfides, ammonia, organic acids, or mineral acids. Microorganisms may actively participate in the reactions that produce corrosion, for example, by catalyzing the reduction of oxygen on surfaces or by utilizing chemical species such as nascent hydrogen that otherwise would polarize the corrosion cell and cause the overall corrosion process to stop or slow down significantly. The xPS that holds the biofilm together can also participate directly in corrosion process by complexing metal ions

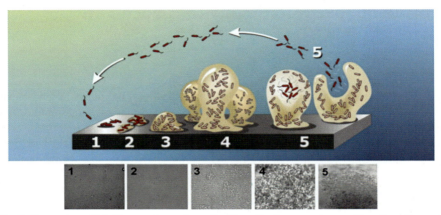

Fig. 1 Formation of biofilm on the surface. Sessile cells form mushroom like slimy "houses" with small water channels which act as a protection shield against biocides; nutrients are gathered by the sticky, slime layer and inside the protection shield bacteria can grow and multiply. A biofilm is an ecosystem on itself.

that would otherwise inhibit further corrosion. Mitigation and control can be achieved by dosing with biocides such as hypochlorite for at least 2 h a day. For heat exchangers and condenser tubes, the "Taprogge®" rubber ball cleaning is a widely adopted option. These balls are slightly larger in diameter than the tubes in the condenser and they clean all tubes at random. Both simple and more sophisticated measurement techniques for bacterial activity are commercially available.

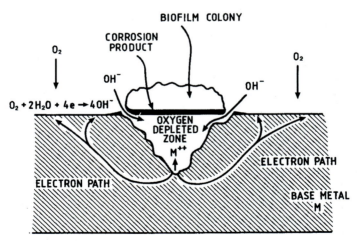

Fig. 2 MIC induced "pit" corrosion. Corrosion spot = anodic and metal surface = cathodic. Corrosion occurs as $I_{anodic} > I_{cathodic}$.
After Videla, H.A., 2001. Microbially induced corrosion: an updated overview. Int. Biodeterior. Biodegrad. 48, 176–201.

Fig. 3 Pit corrosion in the *light blue* material with metal deposits and debris from the biofilm along the metal surface. Electron microscope recording by NV KEMA (the Netherlands).

References and further reading

Bruijs, M.C.M., Venhuis, L.P., Jenner, H.A., Licina, G.J., Daniels, D., 2001. Biocide optimization using an on-line biofilm monitor. Power Plant Chem. 3 (7), 400–405.
Flemming, H.C., 1995. Biofouling und Biokorrosion—die folgen unerwunschter Biofilme. Chem. Ing. Tech. 67.
Licina, G.C., Nekoksa, G., 1993. An electrochemical method for on-line monitoring of biofilm activity in cooling water using the BIoGEORGE™ probe. In: Kearns, J.R., Little, B.J. (Eds.), Microbiologically Influenced Corrosion Testing, pp. 118–127. ASTM STP-1232, American Society for Testing and Materials, 1993.
Videla, H.A., 2001. Microbially induced corrosion: an updated overview. Int. Biodeterior. Biodegrad. 48, 176–201.

Linked

Microbial and microbial fouling

Section 7

Chemicals

Introduction

Process chemicals are used primarily to limit biological fouling in cooling water systems and to inhibit corrosion in high-pressure steam circuits. Corrosion chemicals, in particular, tend to be discharged at non-uniform rates with an increased potential during or after a shutdown. This section describes the main chemical encountered, their behavior in the receiving water, and potential environmental consequences.

There is the potential for many other chemicals to enter the effluent in low concentrations. These may originate from trace impurities in process chemicals, the products of corrosion, or in treated sewage. Notably, given the many chemicals that can emanate from complex industrial plants and their interactions in the wastewater discharge pipes and treatment plants, there is the potential for chemicals to be formed unknown to the operators of the industrial plant.

Many of the chemicals used in the coastal industries and discharged in their wastewater or process water are commonly used and their chemistry and fate and effects are well known. These chemicals will have long been subject to discharge permits and the use of toxicological assessments to determine their ability to degrade, disperse, and be assimilated or to accumulate in biota. Discharge permits will therefore give both the concentrations and contents (loadings) of such materials discharged to the adjacent waters. However, brief details are also given here of those chemicals for which discharge permits are difficult or impossible to obtain, albeit few of these are likely to be encountered by a power station operator.

Chemicals: 70. Biocides

The large volumes of water passing through the cooling circuit of direct cooled power stations provide an ideal environment for the settlement and growth of a wide range of biota, a process known as biofouling. The use of biocides (acutely toxic and mostly oxidizing compounds) is therefore frequently necessary to control settlement and growth of organisms so preventing blockage of the condensers and associated plant parts. The role of a biocide is to control settlement and growth. Without this control, large numbers of hard mussel shells, for example, will soon block condenser tubes. The goal of the plant operator is to obtain the effective mitigation and control of biofouling but to avoid excessive dose rates which may leave unwanted residuals in the discharge which will threaten the wider marine ecosystem.

The most common biocide is "chlorine." This may be generated in situ by electrolysis of seawater or added as a solution of sodium hypochlorite. In general, an electrochlorination plant (ECP) produces between 2 and $5\,g\,L^{-1}$ chlorine, and bulk sodium hypochlorite contains between 140 and $160\,g\,L^{-1}$ as Cl_2. Safety considerations preclude the use of gaseous chlorine for large industrial cooling water applications. A solution of sodium hypochlorite is an equilibrium mixture of the hypochlorite ion and hypochlorous acid:

$$HOCl \leftrightarrow H^+ + OCl^- \tag{1}$$

The position of the equilibrium is determined by the temperature and pH of the solution. In seawater, hypochlorite reacts almost instantaneously with the bromide naturally present ($\sim 65\,mg\,L^{-1}$) to produce a hypobromite/hypobromous acid mixture

$$OCl^- + Br^- \leftrightarrow OBr^- + Cl^- \tag{2}$$

$$OBr^- + H_2O \leftrightarrow HOBr + OH^- \tag{3}$$

The result is an analogous mixture of hypobromite ion and hypobromous acid:

$$HOBr \leftrightarrow H^+ + OBr^- \tag{4}$$

The free acid is reported to be more acutely toxic than the hypobromite ion (see Fig. 1). Although hypobromous acid is the principal biocide in "chlorinated" seawater, many other secondary reactions may occur (e.g., with ammonia to produce chloramines and bromamines), which themselves are oxidizing agents with a biocidal action.

An effective biocide will be highly reactive, and so this mixture of oxidizing agents is by its nature unstable. The determination of chemical speciation is difficult and the complex mixture of oxidizing agents is usually measured as free oxidants (FO) for reactions (1) and (4), or as total residual oxidant (TRO) by which all oxidizing agents including amines and organic residuals are expressed as a chlorine equivalent.

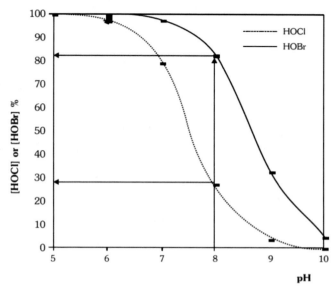

Fig. 1 pH and T dependence of chlorine and bromine.

Although chlorine dominates the market for biocides, a perceived environmental risk from the discharge of chlorination residuals has led to the development of alternatives, although their use remains limited particularly on direct cooled applications.

The main group of alternatives is the quaternary ammonium compounds (referred to variously as "QACs," "Quats," or "polyquats"). They have strong bactericidal properties and some have been shown to be effective against mollusks. Unlike chlorine, they are not oxidizing agents nor do they decay through reaction with other compounds. Regulators therefore may require treatment inside the plant before discharge, for example, by an injection of bentonite, which adsorbs QACs and promotes anaerobic digestion, which is the usual technique. They are marketed as a variety of propriety mixtures, and preparations used for cooling water treatment have included the following active ingredients:

- alkyl dimethyl benzyl ammonium chloride/dodecyl guanidine mixture
- didecyl dimethyl ammonium chloride
- poly [oxyethylene (dimethyliminio) ethylene (dimethyliminio) ethylene dichloride]

QACs are surfactants and their toxic action is by destruction of the cellular membranes and specifically the branchiae (gill filaments) of mollusks. Their effectiveness is species dependent and highly sensitive to temperature.

Many other methods of biocidal control have been proposed, including potassium permanganate, low frequency agitation, thermal shock, controlled hypoxia, and CO_2 injection. However, none of these have proved suitable for large-scale direct cooled applications.

The non-oxidizing biocide (chloro)methylisothiazolinone/methylisothiazolinone (CMIT/MIT) has proved effective in cooling towers particularly if used in conjunction with a dispersant, but it has not so far been used in direct cooled systems.

References and further reading

Jenner, H.A., 2021. A novel combination of CMIT/MIT with a new non-biocide dispersant in cooling tower biofilm control power plant. Chemistry 23 (4), 74–86.
Jenner, H.A., Whitehouse, J.W., Taylor, C.J.T., Khalanski, M. (Eds.), 1998. Cooling water management in European power stations: biology and control of fouling. In: Hydroécologie Appliquée. Electricité de France (EdF), Chatou, Paris, p. 225.
Khalanski, M., Jenner, H.A., 2012. Chlorination chemistry and ecotoxicology of the marine cooling water systems. In: Rajagopal, S., Jenner, H.A., Venugopalan, V.P. (Eds.), Operational and Environmental Consequences of Large Industrial Cooling Water Systems. Springer, New York.
Rajagopal, S., Jenner, H.A., Venugopalan, V.P. (Eds.), 2012. Operational and environmental consequences of large industrial cooling water systems. Springer, Heidelberg/New York.
Whitehouse, J.W., Khalanski, M., Saroglia, M.G., Jenner, H.A., 1985. The control of biofouling in marine and estuarine power stations. A collaborative research working group report. GECB, EDF, ENEL & KEMA: 48 pp., Barnwood, UK.
Anon., 2010. White's Handbook of Chlorination and Alternative Disinfectants, fifth ed. Wiley and Sons, Chichester, ISBN: 978-0-470-18098-3. Black & Veatch Corporation.

Linked

Antifouling measures; Biofouling control by chlorination; Chlorination chemistry; Cooling water and direct cooling; Electro-chlorination plants

Chemicals: 71. Chlorination chemistry

Historically, the application of chlorine has been the universal method to combat planktonic (floating) and settled fouling organisms in cooling water systems (CWS). Dosing with sodium hypochlorite for the control of fouling species in larger cooling water systems ($>5 \text{ m}^3 \text{ s}^{-1}$) is well established and chlorination remains the best known and most widely applied anti-fouling method (White, 1972). Currently its benefits are seen to outweigh its disadvantages.

Terminology

Dissolved chlorine gas, or a solution of sodium hypochlorite, produces many different oxidizing compounds. In seawater free chlorine/hypochlorite reacts rapidly with the bromide naturally present in saltwater to produce a hypobromous acid/hypobromite mixture. Reaction with the ammonia also present in natural waters produces a range of chloramines or bromamines. The oxidants also react with organic matter to produce chlorinated or brominated organics. As a result, the chemistry of chlorinated water involves many molecular and ionic species or groups, both oxidants and non-oxidants, whose terminology must be precisely defined. Chlorine is described in the literature as "free," "active," "available," "combined," or "residual" or by a combination of these terms (Jenner et al., 1998):

1. Free chlorine/free available chlorine (FAC) describes an equilibrium mixture of hypochlorous acid (HOCl) and hypochlorite ions (OCl$^-$). Both are oxidants, but OCl$^-$ is a less effective disinfectant than HOCl. Fugitive elemental chlorine can be ignored.
2. Combined chlorine/combined available chlorine represent chloramines or other compounds having an N-C link that are also oxidants.
3. Total available chlorine (TAC) is the sum of (1) and (2).
4. It is becoming more common and is, in many instances (e.g., seawater applications) more correct, to refer to "oxidants" rather than to chlorine.
5. Residual is analogous to available, but serves to emphasize the concept of a pool of oxy-disinfectant capacity that persists after the initial demand (discussed later) has been met. Hence, using terms such as free residual chlorine, combined residual chlorine, and total residual chlorine (TRC).
6. Chlorine demand is defined as the difference between the amount of chlorine added (dosage) and the useful (residual) chlorine remaining at the end of a specified contact period. Demand is best considered as a one-off reaction that "mops up" a finite quantity of oxidizable substrate.
7. Chlorine decay is a continuing series of reactions that will in time lead to the complete disappearance of all measurable chlorine. Decay, unlike demand, is not substrate limited. The toxicity of chlorinated water, and particularly its acute toxicity, is correlated with its oxidizing potential. It is generally found that a chlorine species, if it is to have any significant efficacy as a biocide, must have an oxidation-reduction or redox potential (ORP) sufficient to oxidize iodide to iodine at pH 7.

8. Total residual oxidant (TRO). Most methods for determining "chlorine" actually measure oxidant capacity via the stoichiometric iodide/iodine route. For this reason, in seawater where brominated oxidants are the dominant species, it is preferable to refer to total free and combined residual oxidant rather than to total, free, and combined residual chlorine. TRO is numerically and operationally equivalent to TRC and TAC as defined before.
9. Chlorine-produced oxidants (CPO) is a self-explanatory term, indicating that oxidants occur naturally in many waters and that some will be measured as though they were CPOs. These will appear as "background" in blank determinations but for most practical purposes their concentrations are so low that they can be ignored. When using oxidant terminology, it is important to remember that the instrument or method used for measurement was almost certainly developed and calibrated for chlorine and that (strictly) the result should be expressed as "mg L^{-1} TRO as Cl$_2$." Generally, the calibration curve is drawn for free chlorine in HOCl-dominant solutions, which means that chloramines, bromamines, and brominated oxidants are measured as free chlorine equivalents.
10. The concept of available chlorine (as distinct from "free available" and "total available") is confusing and should be avoided wherever possible. It originated as a means of comparing bleaching agents, much as water hardness may be measured "as CaCO$_3$." The concept can be applied equally well to oxidants containing no chlorine at all (e.g., ozone) and to those whose chlorine does not hydrolyze to produce hypochlorous acid (e.g., chlorine dioxide), but the calculations are heavily dependent upon knowledge of valence state during reaction conditions.
11. Neither ozone nor chlorine dioxide are "available chlorine compounds," specifically chlorine compounds that hydrolyze to hypochlorous acid. This is therefore another term to be avoided.
12. A common source of error is the expression of biocide concentrations as, for example, mg L^{-1} bromine. Such values must be factored by the atomic weight ratio 33.5:80 to be directly comparable with chlorine oxidants.

The recommended term suitable for use in most situations is "residual oxidant" prefaced by "free," "combined," or "total" as appropriate, with "as Cl$_2$" implicitly understood.

Free oxidant chemistry

The addition of chlorine (or sodium hypochlorite) to water can be viewed as an instantaneous reaction resulting in an equilibrium mixture of hypochlorous acid (HOCl) and hypochlorite ions (OCl$^-$):

Cl$_2$ + H$_2$O → HCl + HOCl

HOCl → H$^+$ + OCl$^-$

Both are oxidants but the negatively charged OCl$^-$ is far less effective as a biocide than HOCl. Decreasing the concentration of H$^+$ ions (i.e., increasing the pH) shifts the equilibrium toward OCl$^-$, which is substantially less acutely toxic compared to HOCl. The ionization constant also varies with temperature. Comparison of the percentage of un-dissociated HOCl as a function of pH is shown in Fig. 1.

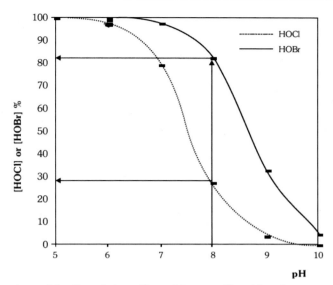

Fig. 1 Comparison of the dissociation of hypochlorous acid and hypobromous acid with changing pH at 20 degrees.

Nitrogenous compounds

Chlorination would follow the simple and predictable chemistry outlined before, were it not for the presence of nitrogenous compounds in the water. Chlorine reacting with any amine will form a molecule broadly classified as a chloramine or *N*-chloro-compound. The term "combined (residual) chlorine" or "combined (residual) oxidant" generally refers to such halogenated nitrogen compounds of which the most significant are the inorganic chloramines, formed by reaction with free ammonia. The reaction with ammonia proceeds by sequential substitution of each of its hydrogen atoms. The reactions are competitive and, for practical purposes, may be considered as irreversible:

$$NH_3 + HOCl \rightarrow NH_2Cl + H_2O \quad \text{(monochloramine)} \quad (1)$$

$$NH_2Cl + HOCl \rightarrow NHCl_2 + H_2O \quad \text{(dichloramine)} \quad (2)$$

$$NHCl_2 + HOCl \rightarrow NCl_3 + H_2O \quad \text{(trichloramine/nitrogen trichloride)} \quad (3)$$

The reactions are dominated by five factors: pH, temperature, contact time, the initial ratio of chlorine to ammonia and, most importantly, by the initial chlorine and ammonia concentrations. The first reaction (1) is almost instantaneous at pH 8.3 and will convert all the free chlorine to monochloramine when the initial ratio (of chlorine to NH_3-N) is equimolar (5.1:1 w/w). The other reactions (2) and (3) are slower (several minutes), consistent with the decreased basicity of the chloramines. At higher chlorine:ammonia ratios, chloramines disappear and the TRC decreases to a minimum so-called breakpoint value, theoretically located at the molar ratio of 1.5:1 (7.6:1 w/w). Under normal conditions encountered in cooling water systems, the breakpoint value is not reached and the formation of trichloramine does not occur. The combined forms are less toxic (and less effective) but are more persistent than free chlorine (Dotson and Helz, 1984; Khalanski and Jenner, 2012).

When chlorine is added to seawater (full seawater contains ~65 mg L^{-1} bromide) bromine is displaced, yielding hypobromous acid (HOBr). This reaction is rapid, with 99% conversion within 10 s at full seawater salinity and within 15 s at half seawater salinity. There is a competing, similar rapid reaction between HOCl and ammonia to produce monochloramine (NH$_2$Cl). Where ammonia levels are high with respect to chlorine dose, NH$_2$Cl is a major product, resulting in a chlorine-dominated chemistry, since NH$_2$Cl, although an oxidant, cannot oxidize bromide. Conversely, where the ammonia level is low in relation to chlorine dose and bromide content, HOBr formation will predominate. Moreover, HOBr reacts rapidly with any NH$_2$Cl that was formed, yielding dibromamine and a bromine-based chemistry. More often than not, a mixed chlorine/bromine system prevails, particularly at estuarine sites with varying salinity and elevated ammonia levels.

Hypohalous acids are more powerful oxidizing agents than hypohalite anions, so bromine should be more effective than chlorine at seawater pH and, incidentally, in most freshwater cooling tower circuits (see Fig. 1). A further feature of bromine

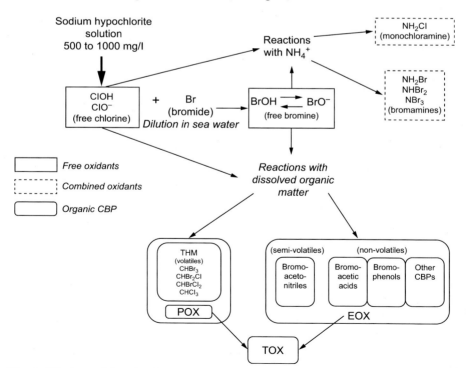

Fig. 2 Chlorine and bromine basal chemistry during chlorination of seawater.
From Taylor, C.J.L., 2006. The effects of biological fouling control at coastal and estuarine power stations. Mar. Pollut. Bull. 53, 30–48.

chemistry is that bromamines, unlike chloramines, inter-convert rapidly and reversibly. $NHBr_2$ decays rapidly, as does $NHCl_2$, but unlike $NHCl_2$ it does not form a persistent monohalamine at seawater pH. This high oxidative reactivity leads to claims that bromo-compounds are more potent biocides than their chlorine analogues.

Even if there is clear evidence that the residuals are wholly bromine-based derivatives, there is little merit in expressing them "as bromine" rather than "as chlorine" because the former would be numerically 2.25 times higher (80/35.5) and there is scope for misinterpretation and confusion. In summary, chlorination of seawater results in a bromine rather than a chlorine dominated chemistry.

Fig. 2 relates chlorine and bromine chemistry for seawater.

References and further reading

Dotson, D., Helz, G.R., 1984. Chlorine decay chemistry in natural waters. In: Jolley, R.L., Bull, R.J., Davis, W.P., Katz, S., Roberts, M.H., Jacobs, V.A. (Eds.), Water Chlorination: Environmental Impact and Health Effects. vol. 5. Lewis Publishers Inc., Boca Raton, FL, pp. 713–722.

Jenner, H.A., Whitehouse, J.W., Taylor, C.J.L., Khalanski, M., 1998. Cooling water management in European power stations. Biology and control of fouling. Hydroécologie Appl. 10 (1-2), 225.

Khalanski, M., Jenner, H.A., 2012. Chlorination chemistry and ecotoxicology of the marine cooling water systems. In: Rajagopal, S., Jenner, H.A., Venugopalan, V.P. (Eds.), Operational and Environmental Consequences of Large Industrial Cooling Water Systems. Springer, New York, pp. 183–227.

Anon., 1972. White's Handbook of Chlorination for Potable Water, Wastewater, Cooling Water, Industrial Processes, and Swimming Pools, fifth ed. John Wiley & Sons, Chichester, ISBN: 978-0-470-56133-1, p. 447. (2010) Black & Veatch.

Linked

Antifouling measures; Biocides; Biology of fouling organisms; Chlorination chemistry; Cooling water and direct cooling

Chemicals: 72. Electro-chlorination plants (ECPs)

When antifouling control is required as at a direct cooled power station, chlorine is generally the antifouling agent of choice. Use of gaseous chlorine is impractical and considered unsuitable by coastal industries because of the hazards involved, and hence the most common alternative is to add chlorine in the form of sodium hypochlorite (HOCl). The use of bulk sodium hypochlorite is not without its own issues and involves the provision of large storage tanks with associated delivery and dosing equipment. There will be some natural decay of sodium hypochlorite during storage reducing its oxidizing potential, which in summer can reach a 50% reduction in 4–6 weeks. At coastal sites an alternative is to produce sodium hypochlorite in situ by the electrolysis of seawater.

Standard seawater contains c.19,000 mg L^{-1} chloride, and so passing an electric current through seawater electrolyzes the chloride-producing reactive chlorine.

The reactions are complex but can be represented by:

$2Cl^- \rightarrow Cl_2 + 2e^-$ anode reaction

$2Na^+ + 2H_2O + 2e^- \rightarrow 2NaOH + H_2$ cathode reaction

$2NaOH + Cl_2 \rightarrow NaCl + NaOCl + H_2O$ reaction within cell

The previous three reactions may be represented by:

$NaCl + H_2O + energy \rightarrow NaOCl + H_2 \uparrow$

Seawater ECPs are used for a wide variety of applications other than cooling water (including wastewater treatment, ballast water treatment, and treatment of produced water from oil and gas wells) and hence the technology is well developed.

Traces of other oxidizing compounds will be produced both by electrolysis and also reaction of hypochlorite with, *inter alia*, the bromide in seawater. By regulating the voltage to 3.9 V, the reduction of chloride to chlorine is favored over competing reactions. The output from the ECP is predominantly an equilibrium mixture of sodium hypochlorite and hypochlorous acid with the relative proportions being governed largely by the pH and temperature of the solution. Adverse side reactions include the production of Mg^{++} and Ca^{++} compounds which precipitate and require periodic acid cleaning to prevent blockages within the ECP.

Numerous proprietary ECPs units are available with a variety of designs but are all based on the same principles. Commercial units include parallel plate and concentric tube configurations, while one common design uses multiple cells connected hydraulically in series where the concentration of hypochlorite increases as the flow passes through the cells. A recent development is the reverse polarity electrolyzer which is claimed to significantly reduce maintenance requirements. Electro-chlorination plants for seawater produce typically 2–4 g L^{-1} "free chlorine," although concentrations up to 5 g L^{-1} are theoretically possible. The power requirement depends on the temperature and salinity of the incoming seawater as well as the strength of the hypochlorite solution produced.

There is some variation in the power consumption quoted by the various manufacturers but 4–5 kWh kg^{-1} chlorine equivalent is typical. Gaseous hydrogen is a necessary by-product of the electrolytic reaction. If the hypochlorite solution is used immediately on production, natural dilution with air may provide adequate protection, but some degassing will be required if ECP-produced hypochlorite is to be held in buffer or storage tanks. The aim is to reduce the hydrogen concentration to <1% in the surrounding atmosphere. Most proprietary designs incorporate gas release tanks, although hydrocyclones are also used (Figs. 1–3).

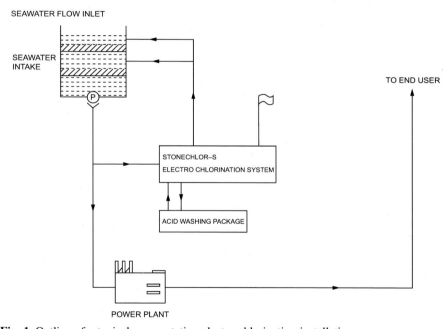

Fig. 1 Outline of a typical power station electro-chlorination installation.

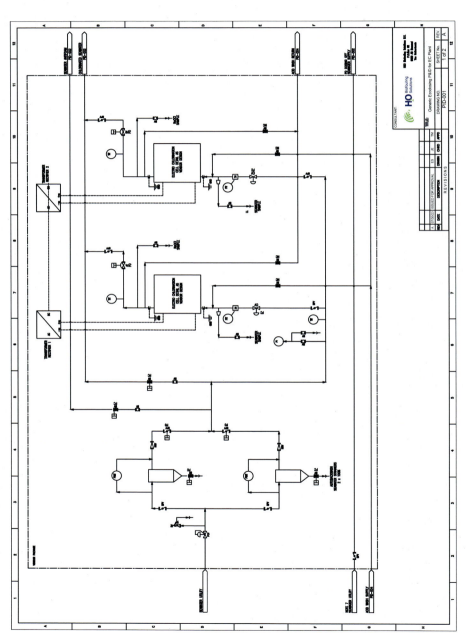

Fig. 2 Design for a modern electro-chlorination plant. Reproduced with permission from H₂O Biofouling Solutions B.V.

Fig. 2, Continued

 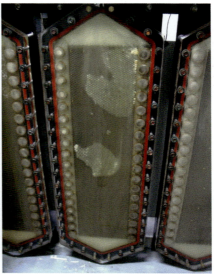

Fig. 3 (A) (Upper) a commercial electro-chlorination plant and (B) An electro-chlorination plant showing Ca^{++} salt deposition.
From H.J.G. Polman.

References and further reading

Särkkä, H., Bhatnagar, A., Sillanpää, M., 2015. Recent developments of electro-oxidation in water treatment. J. Electroanal. Chem. 754, 46–56.

Scialdone, O., Proietto, F., Galia, A., 2021. Electrochemical production and use of chlorinated oxidants for the treatment of wastewater contaminated by organic pollutants and disinfection. Curr. Opin. Electrochem. 27, 10068.

Anon., 2010. White's Handbook of Chlorination and Alternative Disinfectants, fifth ed. John Wiley & Sons, Chichester, ISBN: 978-0-470-56133-1. Black & Veatch Corporation.

Linked

Anti-fouling measures; Chlorination chemistry

Chemicals: 73. Continuous and pulse dosing

Sodium hypochlorite (chlorine) is commonly used as the disinfectant employed to control biofouling in power station cooling water systems (CWS) and there are several alternative techniques and dosing regimes available. The required level of dosing of sodium hypochlorite (or any other oxidizing antifouling compound) is dependent on the reproduction and settlement period of bivalve spat and their growth rate. Spat is the collective noun for ready-to-settle larvae with a length of $\sim$200–300 µm. These biological processes are influenced *inter alia* by water temperature; for example, growth of natural mussel populations requires a minimum temperature of about 12°C and so there is little winter growth. Normally, a specific seawater temperature is used as the trigger for starting the biocide dosing regime.

Treatment regimes may be continuous (year around or seasonal), discontinuous or intermittent (based on hours), and short period or pulsed dosing (based on minutes):

- Continuous: Year around continuous dosing is a simple procedure that guarantees success provided concentrations are sufficiently high and the equipment is reliable. The main disadvantage is the large quantity of hypochlorite used, which results in a high operating cost and the potential for increased environmental impact around the station. To reduce "chlorine" consumption, the concentration level is set at minimum effective levels, so the process is described as low-level continuous chlorination. Physiologically, mussels that are forced to close their shells then change their metabolism from aerobic to anaerobic which enables them to overcome prolonged unfavorable periods. A disadvantage is that they produce organic by-products which are toxic in higher concentrations (succinate, propionate, and butyrate); as a result, they must release those by-products which requires them to open their shells for a short period. At very low chlorine levels the mussels can survive >10 weeks owing to their anaerobic metabolism and the low toxicity of the chlorine during the necessary short opening periods.
- Discontinuous: An improved method with lower chlorine usage is "discontinuous dosing" which alternates regular periods of dosing and no dosing. These specific regimes will be site specific and are determined on the basis of water temperature, settlement of spat, and history. As an example, a station on the North Sea coast experiencing heavy mussel growth was controlled by a regime of 4 h on and 4 h off, although mechanical removal of many tonnes of mussels was still required during planned station outages. That station changed over to pulsed dosing and the CWS has remained clean for over 10 years.
- Pulsed dosing: A refinement to intermittent dosing is pulsed (chlorine) dosing (PD) which is based on the observation that bivalves (e.g., mussels, oysters, and clams) show a recovery period after exposure to chlorination. Only after this period do they open their valves fully before restarting filtering water for oxygen and nutrients. PD exploits a cyclic mode of hypochlorite dosing (on/off dosing regime on minute timescales), based on the behavioral responses of the specific bivalve to chlorine, thereby taking advantage of this recovery period to delay the restart of PD.
- After dosing, the bivalves close and further dosing is unnecessary during the subsequent recovery period. As soon as the bivalves have opened fully and start filtration again, dosing is restarted resulting in the immediate closing of the valves. PD is thus based on the repetitive dosing at intervals too short for the bivalves to recover completely; PD forces the bivalves to switch continuously between aerobic and anaerobic metabolism, leading to physiological exhaustion. In this situation, bivalves deplete their energy reserves (fat, muscles, and glycogen) and so their body condition decreases (Gosling, 2015). They may survive for several

weeks but ultimately the population will collapse. This dosing procedure does not use the "chlorine" as a toxicant to control bivalves, but rather to force bivalves to switch their metabolic modes and to prevent them from restarting filtration. The result is a high mortality rate of the bivalves, compared to the conventional low-level continuous chlorination method but with reduced chlorine usage (Figs. 1 and 2). Pulse dosing is more complex technique but, if correctly employed, can lead to significant cost reductions (>50% is possible), lower environmental impact, and reduced maintenance. In Europe, pulse dosing is accepted by regulators as best available technology (BAT).

Fig. 1 In halant (lower side) and exhalant siphons of two mussels.

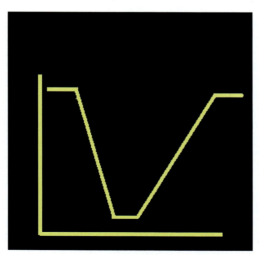

Fig. 2 x-axis time; y-axis opening valves; upper left: start chlorination and change to anaerobic metabolism; down left: stop chlorination; right upper: end of the recuperation period, the shells are open and filtration (aerobic metabolism) starts again.

References and further reading

Anon., November 2000. BAT-cooling: European IPPC bureau Sevilla, Document on the application of Best Available Techniques to Industrial Cooling Systems. http://eippcb.jrc.es. BREF (11.00) Cooling systems.

Bayne, B.L. (Ed.), 1976. Marine Mussels: Their Ecology and Physiology. Cambridge University Press.

Gosling, E., 2015. Marine Bivalve Molluscs, second ed. Wiley Blackwell, Oxford, p. 524.

Macdonald, I.A., Polman, H.J.G., Jenner, H.A., Quyam, S.Q.B.M., 2012. Pulse-Chlorination®: anti-fouling optimization in seawater cooling systems. In: Rajagopal, S., Jenner, H.A., Venugopalan, V.P. (Eds.), Operational and Environmental Consequences of Large Industrial Cooling Water Systems. Springer, New York, p. 522.

Polman, H.J.G., Jenner, H.A., 2002. Pulse-Chlorination®, the best available technique in macrofouling mitigation using chlorine. Power Plant Chem. 4 (2), 93–97.

Polman, H.J.G., Bruijs, M.C.M., Jenner, H.A., 2008. Optimisation of the cooling water chlorination strategy at Verve energy's power plants Cockburn and Kwinana. Presented at the API.

POWERCHEM, 2008. Power Station Chemistry Solutions for the 21st Century. Sunday 25 May to Friday 30 May 2008.

Linked

Antifouling measures; Biocides; Chlorination chemistry

Chemicals: 74. Non-oxidizing residuals

Chlorination of cooling water is still considered to be one of the best available technologies for biofouling control in seawater cooling systems. After dosing, chlorine produces a mixture of hypochlorous acid and the hypochlorite ion. In seawater (which contains $\sim 65\,\text{mg}\,\text{L}^{-1}$ bromide) this rapidly reacts with the bromide ion forming a mixture of hypobromous acid and hypobromite. With the exception of some low salinity estuarine and brackish waters, most of the subsequent chemistry is bromine based. However, if ammonia levels are high there can be a competing reaction to form a variety of chloramines and bromamines. The disinfecting action in "chlorinated" seawater will be expressed through a complex mixture of halogenated compounds comprising principally hypobromous acid and monobromamine.

The acute oxidants formed by chlorination are (by definition) short-lived toxic compounds and are not persistent in natural waters. The major environmental concern of "chlorination" is the production of numerous more persistent compounds, formed by complex reactions between chlorine and bromine and mineral or organic constituents of natural waters collectively described as chlorine by-products (CBPs). In seawater, CBPs will be mainly brominated compounds. A summary of the main chemical pathways is given in Fig. 1. Many CBPs are persistent and have been proven or suspected to be toxic, mutagenic, or carcinogenic for animals and humans when subject to long-term exposure. Fortunately, in seawater at initial dosing concentrations of about $2\,\text{mg}\,\text{L}^{-1}$ as chlorine, CBPs, are not formed in high concentrations.

Among the great number (>200 different compounds) of brominated and chlorinated chemical species that could be formed in chlorinated seawater, the four major groups in terms of detection frequency and quantity are as follows:

- trihalomethanes (THMs)
- haloacetic acids
- haloacetonitriles
- halophenols

Trihalomethanes: chlorine and bromine react with many organic substrates (phenolic compounds, aromatic acids, ketones) to produce chlorinated and brominated methane: chloroform, $CHCl_3$; bromodichloromethane (BDCM), $CHBrCl_2$; chlorodibromomethane (CDBM), $CHBr_2Cl$; and bromoform, $CHBr_3$. A significant consumer of total residual oxidant (TRO) which goes to produce THMs will be the resorcinol structures that make up a large part of humic materials. All THMs are more or less volatile.

Haloacetic acids are, with the THMs, among the most frequently found CBPs in chlorinated waters. They are formed not only by reaction of TRO with organic compounds but also by hydrolysis of haloacetonitriles. In chlorinated seawater, two brominated acetic acids have been frequently identified: monobromoacetic acid (MBAA), $BrCH_2COOH$ and dibromoacetic acid (DBAA), $Br_2CHCOOH$. They are non-volatile.

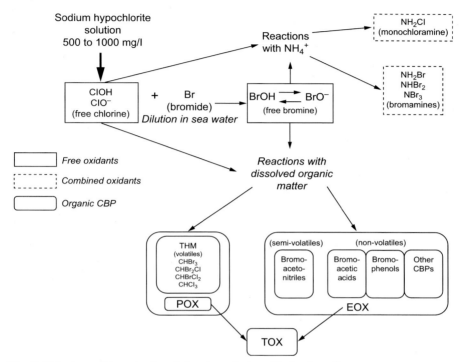

Fig. 1 Chlorine and bromine basal chemistry during chlorination of seawater. From Taylor, C.J.L., 2006. The effects of biological fouling control at coastal and estuarine power stations. Mar. Pollut. Bull. 53, 30–48.

Haloacetonitriles (HANs) are formed by oxidation of amino acids and the weak nitrogen bonds in protein chains. In chlorinated seawater, dibromoacetonitrile (DBAN), Br_2CHCN, is most commonly found. They are semi-volatile.

Halophenols: 2,4,6-tribromophenol (TBP) is the most commonly encountered, though dibromophenol (DBP) is occasionally reported. Halophenols are potentially unstable in cooling water plumes. Humic materials are broken down by chlorine to dihydroxyphenols which are then halogenated. In the presence of TRO, halogenated phenols are further oxidized to quinones which will then be cleaved to CO_2 and haloforms. They are non-volatile.

Bromate (BrO_3^-) may also be present, but it has often been ignored, probably because of the poor sensitivity of the classical analytical methods in the presence of high chloride matrix (detection limit $= 5\,mg\,L^{-1}$ as BrO_3^-). Recently IC-ICP-MS methods have been developed with detection limits of $2-3\,\mu g\,L^{-1}$, allowing detection of traces of bromate in seawater. It is thought that bromate is formed mainly in the hypochlorite storage tanks.

Biofouling is a natural phenomenon (e.g., in coral reefs and coastal zones) and many aquatic organisms use halogenated compounds (chloro- and bromoform, methylchloride, etc.) to combat the attack of substratum-seeking fouling organisms and grazers/predators. Bacteria, algae, kelps, sponges, mollusks, and worms are all able to produce halogenated compounds for anti-fouling purpose. In a long-term study by EDF on three coastal nuclear power stations, chloroform was intermittently detected. There was no correlation with either chlorine dosage or bromoform concentration, and it was surmised that the origin of the chloroform is natural production, rather than formation in the cooling water.

The presence of CBPs in the receiving waters could form an ecotoxicological hazard following the use of sodium hypochlorite for fouling control. A division should be made between acute and chronic toxicity. Acute toxicity is not considered a problem: in the case of bromoform the acute toxicity threshold lies at $\sim$3 orders of magnitude higher than levels commonly encountered.

A case study of a possible chronic toxicity response at a large-scale commercial sea bass breeding site in chlorinated water showed no indication of negative effects due to (bio)accumulation. However, the potential for chronic impacts must be considered.

References and further reading

Allonier, A.-S., Khalanski, M., Camel, V., Bermond, A., 1999. Characterization of chlorination by-products in cooling effluents of coastal nuclear power stations. Mar. Pollut. Bull. 38 (12), 1232–1241.

Grote, M., et al., 2022. Inputs of disinfection by-products to the marine environment from various industrial activities: comparison to natural production. Water Res. 217, 1–9.

Jenner, H.A., Taylor, C.J.L., van Donk, M., Khalanski, M., 1997. Chlorination by-products in chlorinated cooling water of some European coastal power stations. Mar. Environ. Res. 43 (4), 279–293.

Khalanski, M., Jenner, H.A., 2012. Chlorination chemistry and ecotoxicology of the marine cooling water systems. In: Rajagopal, S., Jenner, H.A., Venugopalan, V.P. (Eds.), Operational and Environmental Consequences of Large Industrial Cooling Water Systems. Springer, New York, p. 522.

Nightingale, P., Malin, G., Liss, P., 1995. Production of chloroform and other low-molecular-weight halocarbons by some species of macroalgae. Limnol. Oceanogr. 40 (4), 680–689.

Prasse, C., von Gunten, U., Sedlak, D., 2020. Chlorination of phenols revisited: unexpected formation of α,β-unsaturated C4-dicarbonyl ring cleavage products. Environ. Sci. Technol. 54 (2), 826–834.

Taylor, C.J.L., 2006. The effects of biological fouling control at coastal and estuarine power stations. Mar. Pollut. Bull. 53, 30–48.

Linked

Antifouling measures; Chlorination chemistry; Electro-chlorination plants

Chemicals: 75. Chemicals which may be prohibited for discharge

Most discharges from power stations and other coastal industries in Europe may be expected to discharge into waters controlled by the Water Framework Directive (WFD) or Marine Strategy Framework Directive (MSFD). Most European countries not in the EU (e.g., United Kingdom) have adopted the same standards and targets in national legislation. The WFD covers estuaries (termed transitional waters) and applies up to 1 nautical mile (nm) from the coast for biological quality and 3 nm for chemical quality (3 nm in Scotland for both ecological and chemical quality), and any pipe or structure outside this 1-nm line will be controlled by regulations made under the MSFD. While specific limits for discharges controlled by the MSFD are within the discretion of a Member State, in order to achieve good environmental status, standards set by other directives will need to be met to meet the ecological targets specified but are expected to be similar to those applicable to WFD.

As part of the Water Framework Directive, a European "priority list" of substances posing a threat to or via the aquatic environment was established, with the aim of reducing (or eliminating) pollution of surface water (rivers, lakes, estuaries, and coastal waters) by the pollutants on the list. This list was last updated in 2013 as a daughter directive to the Water Framework: Directive 2013/39/EU of the European Parliament and of the Council. This directive provides environmental quality standards which will need to be met in any discharge.

While the aim is to reduce as far as practical the discharge of all priority substances there is a subset, the Priority Hazardous Substances, where the aim is to eliminate discharges completely.

Priority hazardous substances
    Anthracene
    Brominated diphenylethers
    Cadmium and its compounds
    C_{10-13}-chloroalkanes
    Di(2-ethylhexyl)phthalate (DEHP)
    Dicofol
    Dioxins
    Endosulfan
    Heptachlor and heptachlor epoxide
    Hexabromocyclododecanes
    Hexachlorobenzene
    Hexachlorobutadiene
    Hexachlorocyclohexane
    Mercury and its compounds
    Nonylphenols
    Pentachlorobenzene
    Perfluorooctane sulfonic acid
    Polyaromatic hydrocarbons (PAH)

Quinoxyfen
Tributyltin compound
Trifluralin

Other priority substances
Aclonifen
Alachlor
Anthracene
Atrazine
Benzene
Bifenox
Chlorfenvinphos
Chlorpyrifos
Cybutryne
Cypermethrin
1,2-dichloroethane
Dichloromethane
Dichlorvos
Diuron
Fluoranthene
Isoproturon
Lead and its compounds
Naphthalene
Nickel and its compounds
Octylphenols
Pentachlorophenol
Simazine
Terbutryn
Trichlorobenzenes
Trichloromethane
Trifluralin

The presence of priority hazardous substances in an industrial cooling water discharge is usually relatively low and is not normally considered to be a risk. However, some priority substances (particularly the metals) may be found as corrosion products or impurities in process chemicals.

These lists and their associated quality standards are subject to periodic review and the latest legislation should always be consulted at the planning stage. Although strictly applicable only to Europe, many industrialized countries have similar list of chemicals presenting a risk or hazardous to the environment. It is emphasized that Regional Seas Conventions will also have their priority lists for controlled discharges either from land-based sources (pipelines) or vessel disposal. In Europe, these Regional Seas Conventions are OSPAR for the NE Atlantic, HELCOM for the Baltic, MedPol for the Mediterranean, and the Black Sea Convention.

References and further reading

De Pooter, D., 2019. Flanders Marine Institute: WFD list of priority substances. Available from: http://www.coastalwiki.org/wiki/WFD_list_of_priority_substances.

McLusky, D.S., Elliott, M., 2004. The Estuarine Ecosystem; Ecology, Threats and Management, third ed. OUP, Oxford, p. 216.

Official Journal of the European Union, 2013. Directive 2013/39/EU of the European Parliament and of the Council, Brussels.

Linked

Compliance monitoring; Discharge consents, permits, licenses, authorizations

Chemicals: 76. Microbial pathogens—Chemical interactions

It may seem counterintuitive that a cooling water discharge which has almost certainly been treated with a biocide should pose an increased risk from microbial pathogens. However many, perhaps most, microbial agents (bacteria, viruses, and fungi) respond to temperature and water warmed above ambient. This may favor growth of those pathogenic to humans (frequently most active around body temperature). As discussed in other sections, an effective biocide is a highly reactive compound and will rapidly decay on discharge. Therefore, the footprint of elevated temperature around a discharge may be expected to be significantly greater than that of the residual biocide.

Locations where microbial issues are most liable to become significant are those where the discharge is in the vicinity of either bathing or shellfish waters. In most developed countries, areas are designated for recreational purposes and appropriate microbial standards for the protection of bathers applied. In Europe, the Bathing Waters Directive of 2006 is the appropriate legislation and there are similar acts in North America and elsewhere (McLusky and Elliott, 2004).

Recently the World Health Organisation has published revised recommendations and these are likely to become the basis for all future national legislation. Any future cooling water discharges should therefore be planned with these standards in mind. All administrations will aim to ensure recreational waters achieve at least category "B."

Microbial standards are applied to shellfish intended for human consumption. The European Shellfish Waters Directive of 2006 first set standards but this was repealed in 2013 and all responsibility for legislative protection of shellfish waters was subsumed into the Water Framework Directive. To comply with the European standard, the level of *Escherichia coli* (*E. coli*) must be ≤300 colony-forming units (cfu)/100 mL in the shellfish flesh and intervalvular liquid expressed as a 75th percentile. An interesting divergence between European and North America practice is that European standards are expressed as percentile compliance, whereas in North America a geometric mean is more commonly used. Both approaches have their adherents but in practice the standards tend to align.

In recent years the risk from fungi in beach sand has received increasing attention and again elevated water temperature may favor growth. There is limited information on what constitutes a safe level but the 2021 WHO guidance suggests a preliminary value of $60\,cfu\,gm^{-1}$, with an alternative of $90\,cfu\,gm^{-1}$ wet wt. (cfu = colony-forming units). A disease such as *Legionella* that may be transmitted through aerosols is unlikely to be an issue in direct cooled applications but in tower-assisted seawater cooling systems it will remain a consideration (Table 1).

Table 1 WHO proposed guideline values for microbial quality of coastal and freshwater recreational waters (2021).

Intestinal *Enterococci* (95th percentile value per 100 mL)	Basis of derivation	Estimated risk per exposure
≤40 A	This range is below the NOAEL in most epidemiological studies Low risk or low probability of adverse effects	• <1% GI illness risk • <0.3% AFRI risk • the upper 95th percentile value relates to an average probability of <1 case of gastroenteritis in every 100 exposures. The AFRI burden would be negligible
41–200 B	The 200/100 mL value is above the threshold of illness transmission in most epidemiological studies that have attempted to define a NAOEL or LOAEL for GI illness and AFRI	• 1%–5% GI illness risk • 0.3%–1.9% AFRI risk • The upper 95th percentile relates to an average probability of 1 case of gastroenteritis in 20 exposures. The AFRI illness rates at this upper value would be <19 per 1000 exposures
201–500 C	This range represents a substantial elevation in the probability of all adverse health outcomes for which dose-response data are available	• 5%–10% GI illness risk • 1.9%–3.9% AFRI risk • This range of 95th percentiles represents a probability of 1 in 10 to 1 in 20 of gastroenteritis for a single exposure. Exposure in this category also suggests a risk of AFRI of 19–39 per 1000 exposures
>500 D	Above this level, there may be significant risk of high levels of minor illness transmission	• >10% GI risk • >3.9% AFRI risk • There is a >10% chance of gastroenteritis per single exposure. The AFRI illness rate at the 95th percentile value of >500/100 mL would be >39 per 1000 exposures

AFRI, acute febrile respiratory illness; *GI*, gastrointestinal; *LOAEL*, lowest observed adverse effect level; *NOAEL*, no observed adverse effect level.

References and further reading

Anon., 2016. The Shellfish Water Protected Areas (England and Wales) Directions. https://www.legislation.gov.uk/uksi/2016/138/pdfs/uksiod_20160138_en.pdf.

McLusky, D.S., Elliott, M., 2004. The Estuarine Ecosystem; Ecology, Threats and Management, third ed. OUP, Oxford, 216 pp.

Recreational Water Quality Criteria, 2012. Office of Water document: 820-F-12-058. Office of Science and Technology, United States Environmental Protection Agency. Available at: https://www.epa.gov/sites/default/files/2015-10/documents/rwqc2012.pdf.

WHO, 2021. Guidelines on recreational water quality, Volume 1: coastal and fresh waters. World Health Organization, Geneva, ISBN: 978-92-4-003131-9.

Linked

Compliance monitoring; Microbial and macrobial fouling

Chemicals: 77. Corrosion control: Oxygen scavengers

Corrosion is the reversion of a metal to its ore form; iron, for example, reverts to iron oxide as the result of corrosion. The process of corrosion, however, is a complex electrochemical reaction and it takes many forms (see Fig. 1). Corrosion is a problem for both boilers and elements of the cooling water system (CWS) of coastal industry plants, including condensers, screens, and valves. Two forms of corrosion can be distinguished:

- a general attack over a large metal surface,
- a local pinpoint penetration of metal.

Fig. 1 Electrochemical corrosion causing "pitting."
Photo DNV-KEMA, the Netherlands.

Within electricity generating stations, corrosion is due to the action of dissolved oxygen and chloride in boiler makeup water and chloride from the seawater in CWS. Corrosion may also occur in the feed water system as a consequence of low pH and the presence of dissolved oxygen and carbon dioxide.

Corrosion in a boiler generally occurs when the boiler water alkalinity is low, or when the metal is exposed to oxygen-bearing water either during operation or idle periods. High temperatures (or large temperature differences) and stresses in the boiler metal tend to accelerate the corrosive mechanisms. In both the steam and condensate systems, corrosion is generally the result of the presence of both carbon dioxide and oxygen.

The following general types of corrosion exist:

- uniform corrosion over larger metal surfaces. This type of corrosion occurs when there is an overall breakdown of the passive film formed on the (stainless) steel surface. The entire surface of the metal shows a uniform fine sponge-like appearance. The rate of attack is determined by the oxygen concentration, temperature, and flow velocity. A general rule is that the corrosion speed doubles with each 10°C temperature increase.
- galvanic corrosion by galvanic reaction between different metals in a conductive (seawater) environment (see Fig. 1) with an electric current flowing between the metals. When the current flows, material will be removed from one of the metals or alloys (anode) and dissolved into the electrolyte. The other metal (cathode) will be protected.
- pitting and crevice corrosion by local electrochemical corrosion (see Fig. 2). This is an accelerated form of chemical attack. It occurs when the corrosive environment penetrates the passivated film in only a few areas. Chloride and other halogens will penetrate passivated (coated) stainless steel. Pitting corrosion is therefore simple galvanic corrosion, occurring as a small active area is being attacked by the large passivated area. This difference in relative areas promotes corrosion, causing the pits to penetrate deeper. The (sea) water fills the pits and prevents oxygen from passivating the active metal so the problem becomes worse eventually creating local deep pits and crevices.
- stress corrosion cracking (SCC): an unexpected failure (pipe blow), mainly in boilers, caused by a combination of tensile stress and a corrosive environment (oxygen and/or chloride). Pipe material with severe SCC can appear bright and shiny, while being weakened with microscopic cracks. Conditions for SCC are metals under tensile stress in the presence of dissolved oxygen and chloride ions (see Fig. 3) (Huijbregts, 1986).
- erosion corrosion: this type of corrosion is physically induced corrosion owing to incorrect water flow, that is, turbulence or even cavitation (pump propellers). See Fig. 4.
- microbial influenced corrosion (MIC).

Fig. 2 Pitting corrosion.
Photos by DNV-KEMA, the Netherlands.

Fig. 3 Intergranular SCC of an Inconel heat exchanger tube with the crack following the grain boundaries (500×).
Photo Metallurgical Technologies.

Fig. 4 Erosion corrosion by a clogged mussel in a condenser tube.
Photo Dow Benelux, the Netherlands.

- other more exotic corrosion types exist such as intergranular corrosion caused during chromium carbide welding.

Control of corrosion is primarily through the introduction of oxygen scavengers. Common oxygen scavengers include sodium sulfite, hydrazine, hydroquinone-based derivatives, hydroxylamine derivatives, and ascorbic acid derivatives; some require the presence of a catalyst. Oxygen scavengers reduce oxides and dissolved oxygen and most also passivate metal surfaces.

Hydrazine and sodium sulfite are the most widely used chemical oxygen scavengers for boiler feed water:

- sodium sulfite reacts rapidly with dissolved oxygen even at low temperatures, but it increases the dissolved solids in boiler water necessitating an increased frequency of blowdown.

Decomposition of sodium sulfite also becomes significant at pressures above 65 bar which makes it unsuitable in high pressure boilers.
- hydrazine does not contribute to the dissolved salt content of the boiler water. Hydrazine is slightly volatile and starts to decompose (to ammonia) at high temperatures reducing its efficiency as an oxygen scavenger. In comparison with sodium sulfite, hydrazine reacts with oxygen slowly but reaction rates increase as temperature increases. Hydrazine may be lost from the boiler water by a variety of reactions as follows:

(a) In the presence of oxygen:

$$N_2H_4 + O_2 \rightarrow 2 H_2O + N_2 \tag{1}$$

(b) At high temperature or in the presence of catalysts:

$$3N_2H4 \rightarrow N_2 + 4NH_3 \tag{2}$$

or

$$2N_2H4 \rightarrow N_2 + 2NH_3 + H_2 \tag{3}$$

(c) In the presence of metal oxides:

$$N_2H_4 + 6Fe_2O_3 \rightarrow 4Fe_3O_4 + 2H_2O + N_2 \tag{4}$$

$$2H_2 + 4CuO \rightarrow 2Cu_2O + 2H_2O + N_2 \tag{5}$$

As an alternative to both sodium sulfite and hydrazine, various volatile amines (e.g., diethyl hydroxylamine, cyclohexylamine, morpholine) and some filming (surface active) amines (e.g., octadecylamine) have been successfully used for high pressure boiler water treatments.

Especially during outages (operational shutdowns of coastal industry plant) and, in nuclear stations, fuel changes, there is the possibility of some discharge of hydrazine with the cooling water flow. The fate of hydrazine in the aquatic environment is dependent on dilution/dispersion, chemical and biological degradation, and processes such as volatilization and sedimentation with hydrazine ultimately degrading to nitrogen. Studies have shown that there is a potential toxicity from hydrazine to a range of different marine species. Owing to the low frequency of events during which a discharge of hydrazine may be expected coupled with the high dilution from the cooling water flow, severe environmental problems are not routinely encountered. Nevertheless, operating procedures should be optimized to limit hydrazine discharges and some regulatory regimes require hydrazine to be eliminated prior to discharge.

References and further reading

Anon., 1991. BetzDearborn Handbook of Industrial Water Conditioning., ISBN: 0 913641006.
Cloete, E., Flemming, H.-C., 2012. Environmental impact of cooling water treatment for biofouling and biocorrosion control. In: Rajagopal, S., Jenner, H.A., Venugopalan, V.P.

(Eds.), Operational and Environmental Consequences of Large Industrial Cooling Water Systems. Springer, New York, pp. 303–314.

Huijbregts, W.M.M., 1986. Oxygen and corrosion potential effects on chloride stress corrosion cracking. Corrosion 42 (8), 456–462.

Rao, K.R., 2011. Energy and Power Generation Handbook, Established and Emerging Technologies. ASME, New York, NY, ISBN: 978-0-7918-5955-1, p. 44.

The Nalco Water Handbook, 1988. McGraw-Hill Book Company, ISBN: 0-07-045872-3.

Linked

Chemicals which may be prohibited for discharge; Cooling water and direct cooling; Mechanical, thermal, and chemical stressors

Section 8

Discharge plumes

Introduction

Many coastal industries discharge wastewater or process waters, examples are the cooling waters from coastal power plants, wastewater from petrochemical and chemical plants, and the production and injection water from oil and gas extraction fields. Coastal industries will also discharge effluent from cleaning, laundry, and washrooms as well as runoff from roofs, car parking, and other hard-standing areas. Spillages from loading areas and oil residues from vehicles usually get added to the wastewater runoff from such industries.

Consequently, the wastewater discharges may contain mostly heated cooling waters from power plants, together with the residues from antifouling compounds, and cold water from desalination plants. There is likely to be metals from galvanized roofs and from pipework, organic biodegradable materials in sewage, together with nitrates and phosphates, and particulates from roads and the hard-standing areas. Waste chemicals from the production process and even lost production materials such as nurdles created in plastics manufacture are also discharged. Production and injection waters will contain "fossilized" metals and small amounts of radioactivity, emanating from the oil and gas formation and previously trapped with the fossil fuels. On-site laboratories will release their materials, some emanating from testing and experimentation. In addition, there will be biological materials such as microbes released to or from treatment works.

In addition to the materials known by the industry to be present in the wastewater discharge are also any chemicals or flocculates created by the presence of so many types of materials being present in the wastewater pipes and treatment works. Given that a discharge license can only include the listing of materials known to the industry operator and the regulator than any chemical created post-process inside the pipework, effluent streams or treatment plant may be unknown to both the industry and the regulator. Hence there is the value of whole effluent testing and the need to mimic the behavior of the discharge in the receiving waters.

Given that most wastewater discharges will also include rainwater and runoff from hard surfaces, it has to be remembered that freshwater discharged to a marine coastal area is also a toxicant to those organisms unable to tolerate low salinities. Hence, even in the absence of any other contaminants, a freshwater discharge from a coastal industry requires regulating and licensing.

The characteristics mentioned before indicate that a coastal industry will be required to have a knowledge of the effluent discharge characteristics, especially the physical and chemical nature of the plume formed when the effluent reaches and eventually mixes with the receiving waters. For example, in the particular case of discharges of cooling water by a power station, a plume is formed which undergoes complex movement on mixing with the surrounding water mass creating a zone where the temperature will be elevated above ambient with heat loss principally to the atmosphere. Any other constituents or pollutants in the discharge will be dispersed in a similar way with loss in some instances to bed sediments or atmosphere. Consequently, the discharge will create a zone adjacent to the outfall within which the ecology may be affected. Defining the extent of that zone is a crucial task for the developers and the regulators in especially determining the fate and effects of materials discharged from coastal industries.

Discharge plumes: 78. Plume characteristics and behavior

In this section, cooling water discharges from coastal power plants will be used to illustrate the main considerations for many coastal plumes. In essence, these considerations can be regarded as the knowledge of the behavior of the effluent and its constituents in relation to the behavior of the receiving areas. Of importance here is the density of the discharge in relation to the density of the receiving waters which will control that behavior. Principally, the density is controlled by the temperature and salinity of the effluent given that low salinity and fresh waters are more buoyant than seawater and warmer water is more buoyant than cold water.

The mixing of the effluent with the receiving waters then depends on the force of both water sources, the turbulent and mixing characteristics, the presence of or the ability to form fronts or stratified waters, the nature of the discharge technology (i.e., through nozzles or an open pipe), and even the presence of obstructions and the prevailing wind characteristics. As with an ocean or estuarine front, the plume will be observed as a change in surface water characteristics. These include, among others, the color, smoothness, presence of water vapor rising from the surface, sheen (perhaps with an oily film), and odor. There may even be an ability to hold particles, debris, and even litter and passively floating organisms such as jellyfish along its leading edges.

Apart from a few minor discharges into site surface water, the majority of chemical residuals from power station operation will be discharged with the return cooling water. These residuals may be formed in situ—for example, through chlorination of cooling water to minimize fouling or be independent sources blended into the cooling water stream, for example, water treatment plant effluent. A warm cooling water discharge forms a buoyant plume and the fate and effect of potential contaminants is controlled by the behavior of this plume as it mixes with the ambient water. Unstable materials (e.g., residual oxidants) in the discharge will degrade principally by reaction with dissolved and particulate carbon in the receiving water, but for the more stable contaminants, which are of greater concern, their dispersion is linked directly to plume behavior.

A generalized description of discharge plumes created in shallow tidal water is difficult because of the complications presented by the presence of two boundaries—the seabed and the sea surface—which cause turbulence and frictional shear by the rapidly changing tidal currents and by variable stratification. Nevertheless, observations have confirmed the presence of three zones (Fig. 1):

Cooling water discharged into a receiving water body will be warmer and less dense than its surroundings. The initial characteristics of the plume in terms of density, dimensions, and flow rate will depend crucially on the design of the discharge outlet. If it is discharged from near the seabed, it will rise to the surface and spread out as a thin floating plume. Within the plume there are effectively three zones: a near field, a mid-field, and a far field (see figure).

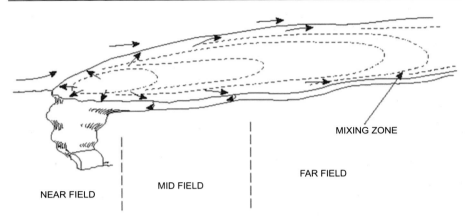

Fig. 1 Behavioral zones within a thermal plume. *Dashed lines* show isotherms, *arrows* show currents. The mixing zone is an isotherm as defined by regulation (see Mixing zone). From Dyer et al. (2017)

1. In the near field, the motion is dominated by the upward momentum of the discharge and buoyancy induced by its lower density. The highly turbulent discharge interacts with the surrounding less turbulent fluid, and the shear produces wave-like motions on the boundary of the rising plume, the breaking of which entrains ambient fluid into the plume. This mixing process known as (physical) entrainment thereby increases its volume and reduces its temperature, so increasing its density and diluting any contained pollutants (Fischer et al., 1979; Lewis, 1997). Once the sea surface is reached, the vertical momentum is translated into horizontal momentum, via an elevation of the surface, and the plume begins to spread out.
2. In the mid-field, radial pressure gradients are caused by the change in the direction of action of the momentum and the horizontal density gradients. The plume spreads outwards and thins as it does so, the spreading even penetrating a short distance into the prevailing current. The spreading involves an outward movement of the near surface water, a down-welling at the edge of the plume, where a front occurs, and a slower return flow above the interface. On the front, there is often a zone of short surface waves and a noticeable change in the water color. The direction of the tidal shear will influence the subsequent movement of the plume. Loss of heat occurs through mixing of the plume with the ambient water and through exchange at the sea surface.
3. In the far field, the velocity shear at the base of the plume will have reduced to the extent that vertical turbulent mixing is minimal, and horizontal spreading and advection dominate. Thus the plume rapidly thins, the rate of spreading diminishes with distance, and the boundaries of the plume become essentially parallel. The interface at the base of the plume is relatively stable, and mixing depends on the degree of turbulence in the water below. The heat loss is predominantly into the atmosphere and depends on such factors as the temperature and humidity of the air and wind speed.

If the discharge is onto the surface of the receiving water, then the near field is constrained by the extent of the exposed surface, the local depression of the ambient water surface, and the lateral boundaries. Mixing will be caused by intense (physical) entrainment, with heat losses mainly direct into the atmosphere. The mid- and far fields will act similarly to that for a buoyant discharge.

In a tidal flow the plume will continually be transported over a wide trajectory, and continuous discharge into the tidal flow leads to complex movement of the plume because of the changing rate and direction of movement of the ambient fluid. A typical sequence described by Dyer et al. (2017) would be:

- Slack water after ebb: there would be an accumulation of discharged water above the outfall which would increase in thickness and diameter with time. Near-, mid-, and far-field zones would form as described before, but with limited dimensions and heat dissipation. As water depth over the outfall would be at a minimum, a boil may occur at the surface for large discharges. This would produce an elevation of the sea surface in the middle of the patch and an outward pressure driving a radial surface flow.
- Flood tide: as soon as the tide begins to flood, the rising plume would be deflected by the current, and the initial dissipation of heat would increase because of the increasing plume length as the patch becomes elongated. In the near field, the patch would spread out, but maintaining some of its integrity as a nose of higher temperature, with high temperature gradients at its leading edge, the following water being relatively cooled. In the mid- and far fields, the discharge becomes influenced by the regional tidal pressure gradient and the plume spreading rate would be related to the tidal velocities. As the ambient fluid is likely to be more turbulent because of shear at the seabed, the velocity differences between the plume and the ambient fluid having decreased, turbulent mixing dominates over entrainment. Heat dissipation is dominated by exchange with the atmosphere. Continued spreading of the plume involves a gradual decrease in thickness and increasing importance of mixing induced by surface waves.
- Slack water after flood: with gradually decreasing tidal velocity, the width and thickness of the plume will increase in the near and mid-fields, and heat dissipation will decrease. In the far field the plume will be extended, wide and thin, with the likelihood of breaking up into ribbons and fingers because of variations in water depth and velocity resulting from bathymetric and seabed frictional differences.
- Ebb tide: during the ebb tide the remnants of the plume will be advected back across both the cooling water intake and the outfall, to interact with the continuing plume discharge, in a similar manner to that occurring on the flood tide.

Under certain conditions the plume may bifurcate during its rise to the surface, in which case two separate lobes of the plume will appear in the far field, with ambient water appearing between them. The mechanism requires that the spreading motion develops into a pair of counter-rotating helical vortices. These create a divergence in the flow at the surface in the middle of the plume, thereby bringing up fluid from below, in addition to creating convergence and downward motion at the outer edges of the plume, where "fronts" form (e.g., Fischer et al., 1979). These vortices are similar to those formed by Langmuir circulation that creates "windrows" on the sea surface. However, the circumstances under which bifurcation occurs are not well understood, but appear to involve the water depth, the discharge rate, the shape and size of the outfall, and the velocity of the ambient fluid (Fig. 2).

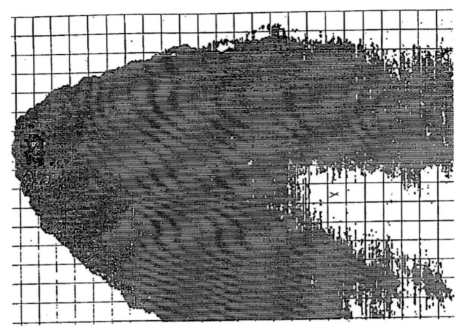

Fig. 2 Infrared image of a bifurcating plume. Discharge at midpoint of left side y-axis. Grid spacing 30.5 m. Flow 21.5 m^3 s^{-1}. ΔT at discharge = 11°C.
Data from MacQueen, J.F., 1978. Vortex pairs and Power Station Cooling Water. Adv. Water Resour. 1 (3), 169–174.

References and further reading

Andreopoulis, J., Praturi, A., Rodi, W., 1986. Experiments on vertical plane buoyant jets in shallow water. J. Fluid Mech. 168, 305–336.
BEEMS Science Advisory Report 007, 2011. Methodology for the measurement of plumes. CEFAS Lowestoft UK.
Dyer, K., Holmes, P., Roast, S., Taylor, C.J.L., Wither, A., 2017. Challenges in the management and regulation of large cooling water discharges. Estuar. Coast. Shelf Sci. 190, 23–30.
Fischer, H.B., List, E.J., Koh, R.C.Y., Imberger, J., Brooks, N.H., 1979. Mixing in Inland and Coastal Waters. Academic Press, New York, p. 483.
Jirka, G., Doneker, R., 1991. Hydrodynamic classification of submerged single-port discharges. J. Hydraul. Eng. 117, 1095–1112.
Lewis, R., 1997. Dispersion in Estuaries and Coastal Waters. John Wiley & Sons, Chichester, p. 312.
MacQueen, J.F., 1978. Vortex pairs and Power Station Cooling Water. Adv. Water Resour. 1 (3), 169–174.
Smith, S.H., Mungal, M.G., 1998. Mixing, structure and scaling of the jet in a crossflow. J. Fluid Mech. 357, 83–122.

Linked

Source and receiving waters

Discharge plumes: 79. Scouring of the seabed

The flow of any fluid either over another water body or over the bed of the water body will create friction depending on the nature of the two media and the force of the water movement. Under normal conditions the sediment on the seabed will have a threshold of erosion in excess of the shear stress the water applies to its upper surface, and it will be stable. It is only during extreme events that the shear stress will exceed the threshold of erosion, and the surface layers will erode until the shear stress diminishes or a previously covered level is exposed that has a higher erosion threshold. Erosion then ceases, but movement of the eroded sediment may still continue. The erosion characteristics of bed sediments are well known, and potential bed level variations can be calculated for changing shear stresses.

Scour is the term used for local erosion caused by the presence of a bed-penetrating or near bed structure, such as an outfall, or a feature on the bed surface, such as a boulder, a pipeline, or a wreck. The structure will cause a local acceleration of the water flow as it diverts around the obstacle and an increase in the flow velocities close to the structure. Regular vortices are shed from around the structure, depending on its shape and roughness. These cause an enhancement of the shear stress at the bed, and potential for scour, provided the increase is sufficient to exceed the threshold of erosion of the bed. The scour creates a local increase in flow depth which decreases the acceleration of the flow, as well as exposing bed material with higher erosion properties, and a new stable situation can result.

Scour effects are well known for unidirectional flow, and the principles can be illustrated by reference to the Fig. 1 below. A regular horseshoe vortex is generated at the upstream side of the structure and spreads around each side, extending a distance downstream. This creates a scour hole on the upstream of the outfall, which also continues around the sides. Separation, or wake vortices, acting mainly in the horizontal plane also arise on either side in the lee of the structure. These produce deeper scour holes, which can extend downstream by several diameters of the obstruction, sometimes becoming crescent shaped.

Normally an effluent discharge outfall would be streamlined, according to the maximum flow direction, to reduce the size and depth of the scour hole. This will depend upon the size and shape of the obstruction, the velocity and depth of the water flow, and the erosion properties of the seabed. Sharp corners or abrupt curves particularly need to be avoided.

For tidal flow, the scour will be more complicated because scour will potentially occur on both phases of the tide, and the scour holes will interact in their effects on the flow. Erosion generally occurs within a few diameters of the obstruction, with equivalent depths of erosion. The side on which deeper scour is produced should indicate the direction of the maximum current. However, for some features, such as wrecks and beach groynes, there may be a buildup of sediment on one side, indicating the direction of sediment transport.

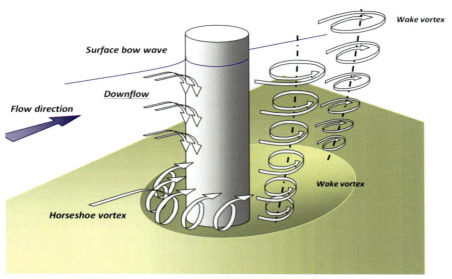

Fig. 1 Depiction of unidirectional flow around a pile on the seabed, penetrating the water surface, and the resultant scour hole. The wake vortices are shown for only one side of the pile, mirror images occurring on the other side.
From BEEMS 118, Cefas, Lowestoft, UK.

Scour around structures has been widely studied, and there are a number of numerical models used by engineers to simulate the effects for widely varying conditions (see references below).

References and further reading

Anandale, G.W., 2006. Scour Technology. Mechanics and Engineering Practice. McGraw-Hill.
Open University Course Team, 1999. Waves, Tides and Shallow-Water Processes. Butterworth-Heineman, Oxford.
Pugh, D., Woodworth, P., 2014. Sea-Level Science: Understanding Tides, Surges, Tsunamis and Mean Sea-Level. Cambridge University Press, p. 395.
Sumer, B., Fredsoe, J., 2002. The Mechanics of Scour in the Marine Environment. Advanced Series on Ocean Engineering, vol. 17 World Scientific, Singapore, p. 536.
Uncles, R.J., Mitchell, S.B. (Eds.), 2017. Estuarine and Coastal Hydrography and Sediment Transport. CUP, Cambridge, p. 351.
Whitehouse, R.J.S., 1998. Scour at Marine Structures. A Manual for Practical Applications. Thomas Telford, London, p. 198.

Linked

Geomorphological terms; Sediment terms

Discharge plumes: 80. Impingement and entrainment (physical)

When an effluent plume enters the receiving body of water, the behavior and characteristics of the discharged effluent in relation to the behavior and characteristics of the waters in the receiving body dictate the eventual fate and effects of the plume. In physical terms, this interaction is regarded as impingement and entrainment—these terms are not to be confused with the same terms used elsewhere in this volume for organisms being taken into coastal industry plants which require a supply of seawater for cooling operation.

Entrainment occurs in stratified fluids when a lower layer is mixed into a more turbulent less dense layer flowing over it. When one water body flows over another of the same density they will interact with each other and exchange mass and momentum when they are similarly turbulent. When the upper layer is less dense, but has a higher velocity than the lower, the interface between the layers can remain stable at low shear rates. As the shear increases, the interface becomes disturbed and small wave-like features are generated as a specific class of internal waves. With further shear increases the waves increase in amplitude and become sharp crested and break intermittently, just like waves on the sea surface. Eventually they break more regularly, each breaking ejecting wisp-like elements and small blobs of the lower fluid into the upper one where they become intimately mixed by the turbulence. This process is called entrainment.

Entrainment is a one-way process that adds fluid (entrained water) into the upper layer and its continuance will gradually alter the properties of the upper layer and lower the elevation of the interface. There is a critical limit above which entrainment starts to occur, involving the ratio of the stabilizing density difference to the destabilizing effect of the shear—this ratio is known as the Richardson number. The relationship between the entrainment rate, the rate at which fluid is transferred between the layers, and the Richardson number has been quantified by experiment, is well understood, and is used in hydraulic modeling to parameterize the exchange processes (Fig. 1).

Fig. 1 Diagrammatic representation of a time sequence of the generation of entrainment from an interfacial wave in a stratified fluid with gradually increasing velocity shear.

References and further reading

Davis, L.R., Shirazi, M.A., 1978. A review of thermal plume modelling. CERL, US EPA, Corvallis, Oregon https://nepis.epa.gov/Exe/ZyPDF.cgi/940032WX.PDF?Dockey=940032WX.PDF. (Accessed 8 April 2023).

Fernando, H.J.S., 1991. Turbulent mixing in stratified fluids. Annu. Rev. Fluid Mech. 23, 455–493.

Lewis, R., 1997. Dispersion in Estuaries and Coastal Waters. John Wiley & Sons, Chichester, p. 312.

Thorpe, S.A., 1969. Experiments on the stability of stratified shear flows. Radio Sci. 4, 1327–1331.

Linked

Entrainment (biological); Plume characteristics and behavior

Discharge plumes: 81. Regulatory mixing zone

The coastal industry manager whose plant discharges materials to the receiving waters will be required to obtain a consent (also termed a license, permit, or authorization) to discharge. Such an authorization will indicate the nature of the material being discharged, its characteristics (temperature, salinity, level of suspended solids), and its constituents as both concentrations and loadings (respectively, amounts per unit volume or weight and total amounts per day or per year). It will also indicate the place of discharge and the means of discharge, for example the structure of the discharge pipe and whether it has single or multiple nozzles; the latter would effect better dilution and dispersion. In some cases, such as the US EPA NDPES (National Pollutant Discharge Elimination Scheme, see https://www.epa.gov/npdes), the size of the mixing zone of the effluent in the receiving area is also delimited—this may be regarded as the regulatory mixing zone.

It is important to note that the concept of a regulatory mixing zone does not always coincide with the use of the term to describe the differing physical process undergone by a discharge into the receiving water. Where a liquid waste is discharged into natural receiving water, the mixing zone is the volume of water into which waste is diluting and is still detectable by measurement. A more narrow definition is used by most UK regulators and in line with latest European Commission guidance is: "an area designated by the Competent Authority (a competent authority approved to regulate European Commission directives) as part of a surface water which is adjacent to the point of discharge and within which the concentrations of one or more contaminants of concern may exceed the relevant Environmental Quality Standard (Chemical or physical) provided that compliance of the rest of the surface water body is not affected." Environmental quality standards can be for chemical concentrations or physical standards, such as temperature, which may be derived from fixed European directives or negotiated locally with the regulator depending on the appropriate legislation.

Most liquid effluents discharged into the sea from coastal industries will tend to rise to the surface because of their lower density resulting from either having a high freshwater content and/or having a temperature higher than ambient. Complications may arise where the receiving water is shallow and the effluent may not rise sufficiently to clear the seabed or where stratification restricts or even arrests the upwards flow of the initial effluent plume. The regulators may define the acceptable mixing zone in terms of meters from the boil or point of discharge, a total area of water, or as a proportion of a water body in 2 or 3 dimensions (Fig. 1).

Mathematical models are used to determine the likely fate and dilution of the effluent in question. They require to be set up with locally collected physical environmental data as well as information regarding the potential effluent and the engineering structure and location of the outfall. No model can hope to cover all situations realistically but can help formulate the worst-case situation (highest concentration or vulnerable receiving environment) and an acceptable mixing zone can then be structured as a 95th percentile or other suitable statistic.

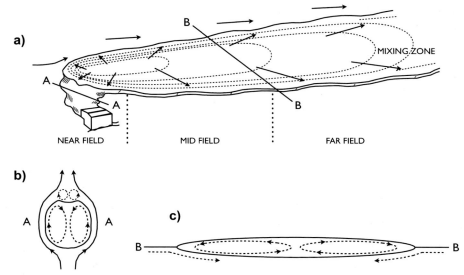

Fig. 1 Representation of a cooling water plume injected into a current. (Not to scale.) (A) Lines show thermal contours at the water surface. (B) Flow around and within the rising plume in the near field on a horizontal section through A-A. (C) Secondary circulation on the vertical plane through B-B in the mid-field.
Based on Dyer, K., Holmes, P., Roast, S., Taylor, C.J.L., Wither, A., 2017. Challenges in the management and regulation of large cooling water discharges. Estuar. Coast. Shelf Sci. 190, 23–30.

Where effluents consist of unusual mixes of substances or some constituents have no existing standards then toxicity tests may be used on the effluent to help determine an acceptable dilution. In this case, whole effluent testing or Direct Toxicity Assessment (DTA, as used in the United States) is used to determine the toxicity of the total effluent, including any synergistic or antagonistic interactions between the constituents. This may be preferable to determining the toxicity of the individual components and making an assumption regarding their synergistic toxicity.

The regulation of the discharge is usually an "end of pipe" standard so the concentration at the edge of the mixing zone is used to back calculate what is acceptable to be discharged. If standards cannot be met, then either the concentration of the substance in question must be reduced or the outfall reengineered until acceptable standards are achieved. The US National Pollutant Discharge Elimination System (NPDES) licensing aims to license the level of such discharges at the edge of the mixing zone.

References and further reading

European Commission, 2010. Technical guidelines for the identification of mixing zones pursuant to Art. 4(4) of the Directive 2008/105/EC C(2010)9369. European Commission, Brussels.

Directive 2000/60/EC of the European Parliament and the Council of 23 October 2000 establishing a framework for the Community action in the field of the water policy, Official Journal of the European Union L 327: 1–73, Brussels.
SEPA, 2009. Modelling coastal and transitional discharges v2, Supporting Guidance (WAT-SG011). Scottish Environmental Protection Agency, Stirling, Scotland.

Linked

Discharge consents, permits licenses, and authorizations; Plume characteristics and behavior; Ecotoxicology assessment

Discharge plumes: 82. Thermal tolerances of organisms

Tolerance levels to each environmental stressor

The potential for any adverse ecological effects of coastal industry discharges is governed by the constituents of the discharge in relation to the tolerances of the organisms in the receiving area. Each organism has a tolerance to a range of natural and usual environmental factors, such as temperature, salinity, pH, dissolved oxygen and hence each species occupies an environment which covers the optimal range of each environmental variable. Outside of those optimal ranges, the organism would be stressed and its health (according to behavioral, physiological, or biochemical aspects, e.g.) would be impaired, what may be termed a reduction in the fitness for survival. With a greater departure from the optimal range, those parameters would then eventually reach sub-lethal levels where the biological functions of the species would be impaired. Even if not immediately, this may or may not eventually lead to the death of an individual or the reduction of the population with resulting effects on the structure of the community and then effects at ecosystem level. Even more severe changes away from optimal conditions would eventually create lethal levels of the environmental variables under which the organism would not survive. For example, an organism encountering waters containing sub-lethal levels of copper would firstly be impaired, even if the species has some capacity to detoxify (sequester or excrete) copper, but with increasing copper then lethal levels would be reached (unless the organism then adopts a behavior to move from the area).

Given the above, less-adapted organisms will be less able to survive specific environmental conditions, whether created naturally or anthropogenically. However, in contrast, if an organism encounters environmental conditions which are closer to its optimum preferred levels then that organism may thrive—this is termed the *stress-subsidy continuum* in which a level of a variable may be a stressor for some organisms but a subsidy (benefit) for others. As an example, an organism adapted to warmer conditions would thrive in a warmer effluent, such as from a coastal power plant, but would suffer under colder conditions, such as an effluent from a desalinization plant.

Hence, it is emphasized that the effects of coastal industry discharges on marine and estuarine plant and animal populations are the net result of the way those populations are adapted to certain environmental conditions, including the tolerable range for temperature and other common environmental parameters (Solan and Whiteley, 2016). Given that each organism has different tolerances to each environmental aspect in a coastal discharge, that each effluent is different, and that changes to the receiving waters caused by coastal and estuarine discharges from industries will have an impact on the biota in the receiving waters, then it is of note that the ecological responses will be site- and species-specific.

Temperature tolerances as an example

Here, temperature is used as an example for the response of an organism to its ambient environmental conditions but it is of note that each factor within a receiving area has to be considered both on its own and cumulatively with all other factors when determining the ecological effects of discharges. In the case of thermal discharges, there is an optimum temperature range within which the organism physiology and development can function effectively; for example, sexual maturation and spawning may be temperature controlled and occur at a specific temperature threshold (Rasmussen, 1973; Solan and Whiteley, 2016).

The temperature of coastal discharges is a ubiquitous stressor, for example, in relation to the warm-water discharges from coastal power plants and the cold-water discharges from desalination plants. As an example of a cumulative interaction, it should be remembered that water temperature has a major impact on the carrying capacity of water for dissolved oxygen with warmer waters carrying less oxygen; hence, warmer waters with lower oxygen levels can create a water quality barrier for migrations, especially of estuarine fishes.

Above and below this optimum temperature range there will be a further range in which the organisms can survive, although at a lower efficiency: within this range enzymatic functions may be inhibited, as may other aspects of development, reproduction, or behavior (e.g., growth, locomotion, feeding, respiration). Generally, organisms will tolerate temperatures within this further range for a limited time, but can recover on return to conditions within their optimum range. Outside this tolerated range, temperatures become too cold or too warm for survival, and the organism will be impaired or may die. For a given species, the warmer and colder temperature limits for survival are termed the upper incipient lethal temperature (UILT) and the lower incipient lethal temperature (LILT).

Predominantly, plant and animal species tolerate the thermal regime in which their populations have evolved. Thus, tropical species tolerate higher temperatures than temperate or polar species, while the latter tolerate lower temperatures, to the point that species below the polar thermocline in polar waters can survive entirely in waters of negative temperature. Within wider zoogeographic zones, populations of a species nearer the poles are more tolerant of lower temperatures, and less tolerant of higher temperatures, than are populations nearer the equator. For example, benthic macrofaunal species off the north-east coast of England suffer deleterious effects at water temperatures above 15°C, while populations of the same species in the English Channel tolerate temperatures above 20°C.

Equally, organisms living in the littoral (intertidal) zone have evolved in a thermal regime which has a greater range than that of sublittoral organisms, and thus they tolerate higher upper temperatures in summer and lower colder temperatures in winter than do sublittoral species. However, there are thermal extremes at which physiological functions break down completely, to the point of protoplasmic coagulation or enzymatic breakdown; these limits are a function of biochemistry and apply equally to populations which have evolved in cold or warm waters.

Further sublethal complications arise where certain thermal conditions are necessary to trigger seasonal behaviors in temperate species. Many marine organisms have thermal thresholds for the onset or maturation and spawning. For example, the littoral barnacle *Semibalanus balanoides* requires a period of water temperature below 10°C to trigger the onset of gametogenesis (the development of reproductive products). Rasmussen (1973) shows the temperature thresholds for maturation, spawning, development, and settlement for many NW European marine and coastal species.

Normally, genetic variation within a population allows a species to adapt to gradual changes in a thermal regime experienced either during migration (extension of the population distribution) or during periods of climatic change. These features have repercussions for the effects of global climate change in which colder-adapted organisms may migrate to higher latitudes. A predominate effect of global change is the alteration of temperature regimes thus allowing species to migrate to higher latitudes.

References and further reading

Rasmussen, E., 1973. Systematics and ecology of the Isefjord marine fauna (Denmark). Ophelia 11, 507.

Solan, M., Whiteley, N. (Eds.), 2016. Stressors in the Marine Environment: Physiological and Ecological Responses: Societal Implications. OUP, Oxford, ISBN: 9780198718826. Hardback.

Wither, A., Bamber, R., Colclough, S., Dyer, K., Elliott, M., Holmes, P., Jenner, H., Turnpenny, A., 2012. Setting new thermal standards for transitional and coastal (TraC) waters. Mar. Pollut. Bull. 64, 1564–1579.

Linked

Phenology; Temperature thresholds which determine spawning times

Discharge plumes: 83. Temperature thresholds which determine spawning times

Fish exhibit water temperature preferences according to their zoogeographic affinities (e.g., colder northerly "Arctic/Boreal" versus warmer southerly "Lusitanian" origins), as well as to their recent temperature experience. In any one season, fish tend to occupy a narrow temperature band within a few degrees of their "optimum" temperature. Whereas mammals are "warm blooded" and regulate their temperature to within a degree or so of their optimum temperature (~37°C in humans), fish are "cold blooded" and with few exceptions can only regulate their body temperature by moving into waters of a suitable temperature. The optimum body temperature for any organism is that at which its many physiological processes perform best and is determined by biochemical reaction rates. Further information on this subject can be found in Wither et al. (2012) and its supporting references.

Certain physiological and developmental processes are optimized for cooler temperatures, and reproduction in some species of fish is an example. Many species of fish spawn during the winter or spring months, which allows the newly hatched fry to take advantage of the full feeding and growth season of the summer-autumn, when plankton and other food resources are most abundant. A detailed review of preferred spawning temperatures and seasons in fish is given in BEEMS (2008). Cold-water species such as salmonid fish and the European smelt (*Osmerus eperlanus*) typically spawn at water temperatures of 6–7°C, while herring (*Clupea harengus*) have been recorded spawning at temperatures between 4.4°C and 15°C. Even some warm-water species prefer cooler water during their spawning period, for example bass (*Dicentrarchus labrax*), which spawn offshore at 10–12°C. Other warm-water species such as the gray mullets *Liza aurata* and *Liza ramada* will spawn at temperatures in excess of 20°C. Fish are able thus able to match their preferred spawning temperature to the appropriate season.

Thermal discharges from power stations can elevate seawater temperatures in the receiving waters and thus have the potential to affect spawning period, timing, and success in some species. This aspect is recognized and allowed for in the planning process for new power stations. Environment Agency (2010) guidance requires that thermal discharges should not impact upon the spawning grounds of thermally sensitive species. Generally, power station siting has the potential only to affect species that spawn close inshore or in estuaries, whereas many species (e.g., cod, *Gadus morhua*, and bass) spawn further offshore or in freshwater (e.g., lampreys and salmonids).

Warm-water species such as bass have been shown to benefit from thermal discharges, which can extend the growing system and result in larger fish (Langford, 1990). Larger fish may produce more eggs and enhance reproductive success. In the case of invertebrates, the changes in local temperature conditions due to thermal discharges can accelerate or delay spawning times; for example, the cold temperature (winter) spawning lugworm, *Arenicola marina*, can have its spawning period delayed and even stopped for one year because of a thermal discharge, whereas the summer spawning intertidal bivalve *Angulus tenuis* has its spawning period extended and commenced earlier when close to a thermal discharge.

Rasmussen (1973) gives the relationship between temperature and spawning and settlement of many estuarine and coastal invertebrates. His extensive data show each species to have thresholds for spawning, larval development in the water column, and settlement to the substrata. These data can be used to indicate the effects of a thermal-discharge-modified seasonal temperature curve on the spawning and development of the species found adjacent to north-west European coastal and estuarine power plants.

References and further reading

BEEMS, 2008. Scientific Advisory Report Series No. 008. CEFAS, Lowestoft, UK.

Environment Agency, 2010. Cooling Water Options for the New Generation of Nuclear Power Stations in the UK. SC070015/SR3, 214 pp., Bristol, UK.

Langford, T.E., 1990. Ecological effects of thermal discharges. Elsevier Applied Science, London, p. 468.

Rasmussen, E., 1973. Systematics and ecology of the Isefjord marine fauna (Denmark). Ophelia 11, 507.

Solan, M., Whiteley, N. (Eds.), 2016. Stressors in the Marine Environment: Physiological and Ecological Responses: Societal Implications. OUP, Oxford, ISBN: 9780198718826.

Wither, A., Bamber, R., Colclough, S., Dyer, K., Elliott, M., Holmes, P., Jenner, H., Taylor, C., Turnpenny, A., 2012. Setting new thermal standards for transitional and coastal (TraC) waters. Mar. Pollut. Bull. 64, 1564–1579.

Linked

Cooling water and direct cooling; Phenology; Source and receiving waters

Discharge plumes: 84. Thermal plume constituents (excluding heat)

Each coastal discharge is a complex of many constituents such that the combined effect of those has to be considered in licensing the discharge by the regulator. While it is not possible to consider all discharges and all constituents, as an example, the plume from a power station may contain:

1. chemical residuals formed in situ from the chlorination of cooling water for biofouling control;
2. waste streams added from on-site processes;
3. miscellaneous discharges on a site-specific basis.

Chemical residuals formed in situ:

(i) in saline waters, added chlorine reacts with bromide to form a free bromine/hypobromous acid mixture which provides the main biocidal action. Many secondary oxidizing agents may be produced, principal among these are bromamines and chloramines particularly if ammonia levels are elevated. These chlorine-produced oxidants (CPO) are reactive and short lived with concentrations declining rapidly on mixing with ambient water. The total of all oxidizing agents in the plume is reported as total residual oxidant (TRO) expressed as a chlorine equivalent;

(ii) the reaction between the CPO and dissolved and particulate carbon in the cooling and receiving waters produces a range of more stable non-oxidizing halogenated organic compounds, generically described as chlorine by-products (CPBs). Bromoform (found typically at concentrations of $5-25\,\mu g\,L^{-1}$) in the cooling water is the most prevalent with other CBPs seldom found at levels $>1\,\mu g\,L^{-1}$. These residuals may be both persistent and toxic and, although they are detected in fish and shellfish, they tend not to bioaccumulate. A list of the principal CBPs found at European sites is given in Table 1.

Chemical containing streams added to the cooling water:

(i) a regular discharge from the water treatment plant containing *inter alia* reactive and non-reactive phosphorus;

(ii) a continuous blowdown from the primary and secondary steam circuits; this blowdown will contain hydrazine used as oxygen scavengers and a variety of nitrogen and phosphorus-based products used as corrosion inhibitors. The volumes and concentrations

Table 1 The principal chlorine by-products formed following chlorination of saline cooling water.

- Bromoform
- Bromodichloromethane (BDCM)
- Chlorodibromomethane (CDBM)
- Monobromoacetic acid (MBAA)
- Dibromoacetic acid (DBAA)
- Dibromoacetonitrile (DBAN)
- 2,4,6-Tribromophenol (2,4,6-TBP)

Table 2 Typical discharges from an EPR nuclear reactor.

Chemical	Expected annual discharge without contingency (kg)	Maximum annual discharge (kg)
Boric acid (boron)	2000 (350)	7000 (1224)
Lithium hydroxide	<1	4.4
Hydrazine	7	14
Morpholine	345	840
Ethanolamine	250	460
Nitrogen compounds (as N) excluding hydrazine, morpholine, and ethanolamine	2350	5060
Phosphate	155	400
Detergents	630	1600
Metals	16	27.5
Suspended solids	655	1400
Chemical oxygen demand	1490	2525

From data on AREVA website.

will be dependent on the reactor/boiler design; typical values for the EPR reactor design are given in Table 2.

(iii) larger volumes of hydrazine are used to protect the high-pressure steam circuit during outages as will occur during maintenance periods. This material after pre-treatment will be discharged in a controlled manner to the main cooling water system.

Other constituents may include

(i) treated sewage;
(ii) site drainage;
(iii) groundwater.

References and further reading

BEEMS, 2011. Scientific Advisory Report Series. CEFAS, Lowestoft. no. 009.
Rajagopal, S., Jenner, H.A., Venugopalan, V.P. (Eds.), 2012. Operational and Environmental Consequences of Large Industrial Cooling Water Systems. Springer, Heidelberg/New York.
Taylor, C.J.L., 2006. The effects of biological fouling at coastal and estuarine power stations. Mar. Pollut. Bull. 53, 30–48.
www.epr-reactor.co.uk/ssmod/liblocal/docs/V3/.

Linked

Non-oxidizing residuals; Plume characteristics and behavior; Water quality considerations

Discharge plumes: 85. Salinity tolerances of organisms

Many coastal industry discharges will potentially change the salinity in the receiving waters. The salinity of any water body is an equilibrium between ionic inputs, dilution, and removals. Changing salinity is a dominant factor in the distribution of both marine and estuarine species and is limiting to freshwater organisms, hence the salinity level and variability are fundamental with far-reaching effects in creating and modifying aquatic ecosystem assemblage structure and functioning.

In estuarine areas, salinity may be changing over very short spatial and temporal scales. The natural salinity pattern from hypersaline areas and the sea to freshwaters (Fig. 1) shows the presence of a continuum of species: marine restricted-salinity tolerance (stenohaline) species, estuarine species tolerant of reduced salinity, and freshwater species with no salinity tolerance. Those species which pass through estuaries, for example diadromous fishes which spend different parts of their life cycle in fresh and seawater, have an ability to change their tolerance between the two extremes of salinity.

Many coastal industries discharge freshwater effluents creating localized brackish conditions, although some industries, such as desalination plants and solute mining areas (the removal of underground salt to create gas storage caverns), will discharge hyper-saline solutions with a salinity greater than the receiving waters (Smyth and Elliott, 2016). The effects of changing salinity on the ecology of different habitats

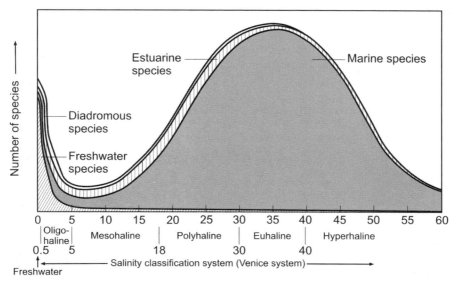

Fig. 1 The species distribution under different salinity regimes.
From Whitfield, A.K., Elliott, M., Basset, A., Blaber, S.J.M., West, R.J., 2012. Paradigms in estuarine ecology—the Remane diagram with a suggested revised model for estuaries: a review. Estuar. Coast. Shelf Sci. 97, 78–90.

are driven ultimately by the underlying physiology and tolerances of organisms and their ability to cope with salinity fluctuations on both long- and short-term scales. Estuarine species are often euryhaline, adapted to tolerate fluctuating salinity, whereas many marine species are stenohaline and limited by their narrow range of physiological tolerance (Smyth et al., 2014). Importantly, salinity is one of the main drivers of the distribution of fish in estuaries (Whitfield et al., 2022) such that a large freshwater discharge could create a barrier to fish migrations.

For some estuarine species, lowered salinities may be a subsidy, that is, benefiting the estuarine-adapted organisms by reducing competition, whereas for non-tolerant (marine or wholly freshwater) species they are a stressor. Salinities at the margins or outside the tolerance range of particular species will prevent their occurrence, change their behavior, or limit reproduction and germination, reducing their fitness for survival in that environment. Salinity can act synergistically or antagonistically with other environmental stressors. For example, factors such as nutrient availability, temperature, pH, and dissolved oxygen have all been shown potentially to interact, producing a dynamic and changing environment where one factor can exacerbate or negate the detrimental effects of another. Anthropogenic activities and effluents can modify local salinity regimes and their effects depend on the assimilative capacity both of the area and the species.

Often as part of the discharge license negotiations, coastal industries will need to be aware of the effects of salinity changes on species individuals and populations and on communities and the whole ecosystem. Indeed, it is emphasized that for many marine organisms, freshwater should be regarded as a toxicant.

References and further reading

Smyth, K., Elliott, M., 2016. Chapter 9: Effects of changing salinity on the ecology of the marine environment. In: Solan, M., Whiteley, N. (Eds.), Stressors in the Marine Environment: Physiological and Ecological Responses: Societal Implications. OUP, Oxford, ISBN: 9780198718826, pp. 161–174. , Hardback.

Smyth, K., Mazik, K., Elliott, M., 2014. Behavioural effects of hypersaline exposure on the lobster *Homarus gammarus* (L) and the crab *Cancer pagurus* (L). J. Exp. Mar. Biol. Ecol. 457, 208–214.

Solan, M., Whiteley, N. (Eds.), 2016. Stressors in the marine environment: physiological and ecological responses: societal implications. OUP, Oxford, ISBN: 9780198718826.

Whitfield, A.K., Elliott, M., Basset, A., Blaber, S.J.M., West, R.J., 2012. Paradigms in estuarine ecology—the Remane diagram with a suggested revised model for estuaries: a review. Estuar. Coast. Shelf Sci. 97, 78–90.

Whitfield, A.K., Able, K.W., Blaber, S.J.M., Elliott, M. (Eds.), 2022. Fish and Fisheries in Estuaries—A Global Perspective. John Wiley & Sons, Oxford, ISBN: 9781444336672, pp. 458–552.

Linked

Ecosystem resilience, resistance, recovery

Section 9

Monitoring

Introduction

Monitoring is the rigorous collection of environmental and operational information which demonstrates the spatial and/or temporal changes in a set of parameters. By definition, it is required to determine the status of an area, the need for environmental management, or the efficacy of environmental management. It should be accompanied by a similarly rigorous assessment and reporting framework. At its simplest, it involves routine measurement of a parameter to determine whether it is within relevant and acceptable/legally agreed limits; this is defined as compliance monitoring, that is, the ability to comply with a regulation or license condition. Monitoring is also required to detect any changes in a system or process which may instigate a responsive action; for example, as is the case with biofouling monitoring in which the detection of fouling populations will enable the plant operator to take remedial action. At its most complex, monitoring is undertaken in order to determine whether the parameter being monitored is changing or not, and if so whether this change is significant, and what (if any) predictable outcome may result in the longer term (or to a period required by the project).

Whereas the term monitoring is used loosely, there are several types of monitoring mentioned in legislative and management guidance. All of these types are required to accommodate natural systems which are inherently variable, and especially coastal, marine, and estuarine systems which vary more than others. With particular reference to coastal industry discharges, the environmental manager may be required by the regulator to detect a change due to the industry activities (the so-called signal) among the natural variability in a parameter (the "noise"), hence the need to derive the signal:noise ratio.

Variation in the physical and physicochemical environmental conditions, such as temperature, salinity, depth, and extent, occur from short term to very long term:

- hourly in response to weather;
- twice daily in response to semi-diurnal tides;
- fortnightly in response to the spring-neap tidal cycle;
- monthly in response to seasons in temperate zones or to monsoons in response to rainfall and river flow in tropical systems;
- annually and longer in response to climatic changes, to the North Atlantic Oscillation, to El Niño effects, or to the sunspot cycle.

On top of this extensive background of habitat variation, animals and plants also show marked variation with time in terms of their population sizes, ranges, densities, and condition (including presence or absence). This is particularly the case in marine systems, where the communities vary apparently stochastically with time, responding to differential recruitment, to range extensions through migration, to genetic variation, and to competition with other organisms in the community, these other organisms also showing variation.

When potential impacts exist from coastal industries, these may also impart some change to the habitat conditions. Thus, monitoring is required to identify the presence, degree, and significance of any such changes, against the very complex background of natural variation described before. Rigorous monitoring procedures are therefore not only necessary, but must be planned carefully in terms of frequency, duration, and detail in order to encompass all those naturally varying conditions of the background environment which may have a significant influence on the parameter being monitored. Most importantly, monitoring is required to determine whether any management response has achieved the desired effect in controlling adverse effects of coastal industries.

Monitoring: 86. Management framework monitoring types and definitions

It is a management axiom that you cannot manage any activity unless you can measure it. Hence the management of environmental activities, impacts, and consequences require monitoring, the detection of spatial and temporal patterns, and the ability to meet any regulatory requirements (see accompanying entries). Where the processes are understood or based on underlying theory (termed deterministic) or at least can be explained according to other events (termed empirical and previous experience), if there is the need to interpolate, extrapolate, or predict the outcomes or even the impacts of scenarios, then modeling will be required (see accompanying entries). As long as the limitations of the monitoring and modeling are known then their outputs can be used in management.

Monitoring is required to be rigorous and defendable with aspects measured in a structured way and thus related to the nature of the system and the nature of the activity. The analyses required to have extensive AQC/QA (analytical quality control/quality assurance) using standard operating procedure methods and protocols. This is especially important given the possibility that the results of monitoring would be used in legal proceedings if an industry exceeds its license conditions.

The prevailing governance framework has produced many types of monitoring: for example, surveillance monitoring, condition monitoring, operational monitoring (Table 1). The measurement of change can be related to and incorporate indicators of change and those indicators are required to have specific characteristics. Again, it is axiomatic that the pre-determined definition of acceptable change, that is, what adverse change in the ecosystem is acceptable in an ecological, social, and legal

Table 1 Examples of types of monitoring and relevance to determination of environmental status (e.g., good environmental status (GES) for the EU Marine Strategy Framework Directive).

Type of monitoring	Relevance to GES determination
Surveillance monitoring—a "look-see" approach which begins without deciding what are the endpoints followed by a post hoc detection (a posteriori) of trends and suggested management action	Assessment of ecological components within an area, very often of the structural rather than the functional aspects, and often in cases where the threshold, reference, or baseline conditions have not been defined in advance
Condition monitoring—used by nature conservation bodies and other relevant bodies to determine the present status of an ecosystem, area, or a species; it could be linked to biological valuation (e.g., Derous et al., 2007a,b)	The status of a designated nature conservation area (e.g. SAC, SPA, MPA), including its Favorable Conservation Status linked to its conservation objectives (for a given feature, species, or habitat)

Continued

Table 1 Continued

Type of monitoring	Relevance to GES determination
Operational monitoring—used by industry for business reasons (e.g., for a dredging scheme linked to aims for management and to determine if an area requires further dredging) Compliance monitoring—used by industry and linked to license setting for effluent discharge, disposal at sea, etc. Self-monitoring—being carried out by the developer/industry under the "polluter pays principle" or the "damager debt principle" but often sub-contracted to an independent and quality-assured/controlled laboratory Check monitoring—where an Environmental Protection Agency checks self-monitoring to ensure that a developer is performing appropriate monitoring	Not relevant for GES especially as it is related to a small activity area which would be subjected to an operating license; however, the produced data will be of value if the regulatory authorities can build it into their assessments and it has been quality controlled
Toxicity testing—as a predictive approach needed for license setting, used by regulators to determine. compliance of the license conditions with required standards Investigative monitoring—applied research on cause and effect, to explain any deviation from perceived or required quality Diagnostic monitoring—determining effects but link to cause; synonymous with investigative monitoring	Only relevant for GES determination if used as cause-and-effect determination and therefore linking the status of a species or habitat, or reduction in it, to a given pressure (emanating from one or more activities)
Feedback monitoring—real-time analysis, linked to predetermined action; for example, monitoring during dredging on condition that the activity is controlled/prevented/stopped if a deleterious change is observed; this relies on acceptance that any early warning signal will be related to an ultimate affect	Linked to the Programme of Measures, as long as a measure is the responsibility of an activity or developer

Modified from Elliott, M., Houde, E.D., Lamberth, S.J., Lonsdale, J.-A., Tweedley, J.R., 2022. Chapter 12. Management of fishes and fisheries in estuaries. In: Whitfield, A.K., Able, K.W., Blaber, S.J.M., Elliott, M. (Eds.), Fish and Fisheries in Estuaries—A Global Perspective. John Wiley & Sons, Oxford, ISBN: 9781444336672, pp. 706–797.

framework, should be accompanied by a decision of what action will be required to mitigate or compensate if the change is not acceptable.

Monitoring was previously defined merely as the repeated taking of measurements and assessments whether at spatial and/or temporal levels (Gray and Elliott, 2009). In order to detect change therefore requires monitoring the system by determining when to assess and what to assess. Despite this, monitoring *sensu stricto* implies having a pre-defined (*a priori*) baseline (or standard, threshold, trigger point) against which the monitoring results are judged and where ideally an action is pre-agreed. For example, the receiving area for a cooling water discharge is monitored for temperature against a standard and if that standard is breached then the managers should have agreed on the action to be taken. This contrasted with "surveillance" in which an area is assessed and perhaps an *a posteriori* change detected again followed by a subsequent discussion of what action to take.

With the advent of the different EU Directives, for example, the term monitoring has been expanded and often this has led to greater confusion as well as an expansion of terms (see Table 1). For example, compliance monitoring will be included when discussing whether a developer has met the terms of their license/authorization/permit for the Integrated Pollution Prevention and Control (IPPC) regulations. In contrast, condition monitoring is required when, for a designated nature conservation site, there is the need to determine if the conservation features for which the site was designated fall within Favorable Conservation Status and meet the Conservation Objectives. As an example, the UK Defra and JNCC marine monitoring strategies for UK waters are indicated in the websites given in the "References and further reading" section.

References and further reading

De Jonge, V.N., Elliott, M., Brauer, V.S., 2006. Marine monitoring: its shortcomings and mismatch with the EU Water Framework Directive's objectives. Mar. Pollut. Bull. 53 (1-4), 5–19.

Derous, S., Agardy, T., Hillewaert, H., Hostens, K., Jamieson, G., Lieberknecht, L., Mees, J., Moulaert, I., Olenin, S., Paelinckx, D., Rabaut, M., Rachor, E., Roff, J., Stienen, E.W.M., van der Wal, J.T., Van Lancker, V., Verfaillie, E., Vincx, M., Weslawski, J.M., Degraer, S., 2007a. A concept for biological valuation in the marine environment. Oceanologia 49 (1), 99–128.

Derous, S., Austen, M., Claus, S., Daan, N., Dauvin, J.-C., Deneudt, K., Depestele, J., Desroy, N., Heessen, H., Hostens, K., Marboe, A.H., Lescrauwaet, A.-K., Moreno, M.P., Moulaert, I., Paelinckx, D., Rabaut, M., Rees, H., Ressureicao, A., Roff, J., Santos, P.T., Speybroeck, J., Stienen, E.W.M., Tatarek, A., Hofstede, R.T., Vincx, M., Zarzycki, T., Degraer, S., 2007b. Building on the concept of marine biological valuation with respect to translating it to a practical protocol: viewpoints derived from a joint ENCORA-MARBEF initiative. Oceanologia 49 (4), 579–586.

Ducrotoy, J.P., Mazik, K., Elliott, M., 2011. Bio-sedimentary indicators for estuaries: a critical review. Union des océanographes, de France, Paris, ISBN: 978-2-9510625-2-8, pp. 1–77.

Elliott, M., 2011. Marine science and management means tackling exogenic unmanaged pressures and endogenic managed pressures—a numbered guide. Mar. Pollut. Bull. 62, 651–655.

Gray, J.S., Elliott, M., 2009. Ecology of Marine Sediments: Science to Management. OUP, Oxford, p. 260.

JNCC Marine Monitoring Handbook 2001 – Joint Nature Conservation Committee - https://hub.jncc.gov.uk/assets/ed51e7cc-3ef2-4d4f-bd3c-3d82ba87ad95.

McLusky, D.S., Elliott, M., 2004. The Estuarine Ecosystem; Ecology, Threats and Management, third ed. OUP, Oxford, 216 pp.

UK Clean Sea Monitoring Strategy. http://www.cefas.defra.gov.uk/our-science/observing-and-modelling/monitoring-programmes/clean-seas-environment-monitoring-programme.aspx.

UKMMAS, UK Marine Monitoring and Assessment Strategy (UKMMAS). http://www.defra.gov.uk/environment/marine/science/ukmmas/.

Linked

Compliance monitoring; Coastal plant environmental modeling

Monitoring: 87. Survey, experimental and modeling approaches

In order to obtain data, information, and understanding regarding the effects of the natural system on the coastal industry or the effects of the coastal industry plant on the natural system, several approaches can be adopted. A field survey approach will provide information on the ambient, natural features (the hydrographic patterns, the ecological populations in the receiving waters, the sediment and bathymetry, etc.) which, if targeted specifically for the impact assessment, will be precisely focused on the two types of effects indicated before. An initial survey will also produce a baseline against which future changes can be judged. A survey will provide spatial and temporal data and information and may indicate correlations between components and processes but will leave uncertainty regarding cause-effect chains (causality).

Hence a more investigative approach will be needed; as an example, the carrying out of field or laboratory experiments (e.g., the use of the entrainment mimic unit (EMU) to determine settlement patterns inside cooling water systems) both to control for the large field variability and to determine causality in environmental responses. These experiments should be specified with a level of sampling, replication, and analyses to give statistical rigor (e.g., using power analysis). The experimental approach should follow the rigorous and highly defendable cyclical scientific method of first setting a statistical hypothesis (or better still, a null hypothesis) based on scientific theory, previous observations (and published information). This is then followed by designing an experiment and setting analytical and statistical criteria to allow the hypothesis to be accepted or rejected; eventually this may lead to further hypotheses being set.

In order for the coastal industry manager and their regulators to understand the prevailing environment and the effects of the coastal industry and any discharge on that environment, a rigorous information base is required (Table 1). A modeling approach may involve deterministic models (in which there is an underlying theory being used/tested); an empirical approach (where the underlying theory may not necessarily be known but where empirical data are obtained by field and/or laboratory studies and summarized/interpreted); and/or stochastic, predictive models, etc. The modeling will allow scenarios to be generated and tested, extrapolations or interpolations to be made to supplement field data, and the linking of experimental results of field consequences and vice versa.

Table 1 The 9 stages in the provision of data, information, and understanding needed for management (in Elliott, 2011).

Stage	Topic	Information produced
1	Behavior/characteristics of the system	Of the intertidal, subtidal, lagoon, estuarine, open coastal areas, etc.
2	Physical/chemical nature of system	Its hydrography, topography, bathymetry, salinity regime, nutrient status, etc.
3	Physical and chemical behavior of additives to system	Their dispersion in a solid or liquid phase, solubility, transport, sequestration, etc.
4	Behavior/characteristics of an activity in the environment	E.g., whether there is a barrier to the flow of materials and biota, or the disruption of processes
5	Habitat at risk from modification or materials addition	E.g., whether there is a surface feature (monolayer), or effects in the water column, water-substratum interface, sediment, supralittoral, intertidal, circalittoral, infralittoral, shelf
6	Inert or biologically effective action	Whether there is a direct toxic nature, secondary toxic nature (after modification in or of habitat)
7	Biotic and non-biotic component(s) at risk	E.g., which of the phytoplankton, zooplankton, pelagic nekton, demersal nekton, hyperbenthos, epifauna, infauna, microphytobenthos, macroalgae, saltmarsh, reedbeds, wading birds, wildfowl are at risk
8	Behavior of contaminants within organisms	E.g., their uptake, sequestration, storage, excretion, passage to progeny, passage to prey
9	Structure and functioning of biological system	The response at any of the levels of biological organization

Modified from McLusky, D.S., Elliott, M., 2004. The Estuarine Ecosystem; Ecology, Threats and Management, third ed. OUP, Oxford, 216 pp.

References and further reading

Baker, J.M., Wolff, W.J. (Eds.), 1987. Biological Surveys of Estuaries and Coasts. Estuarine and Brackish Water Sciences Association Handbook. Cambridge University Press, Cambridge.

Bayne, B.L., Clarke, K.R., Gray, J.S. (Eds.), 1988. Biological effects of pollutants: results of a practical workshop. Mar. Ecol. Prog. Ser. 46 (1–3), 1–278.

Eleftheriou, A., McIntyre, A.D. (Eds.), 2005. Methods for the Study of Marine Benthos, third ed. Blackwell Science, Oxford. IBP Handbook No. 16.

Elliott, M., 2011. Marine science and management means tackling exogenic unmanaged pressures and endogenic managed pressures – a numbered guide. Mar. Pollut. Bull. 62, 651–655. http://dx.doi.org/10.1016/j.marpolbul.2010.11.033.

Elliott, M., Houde, E.D., Lamberth, S.J., Lonsdale, J.-A., Tweedley, J.R., 2022a. Chapter 12. Management of fishes and fisheries in estuaries. In: Whitfield, A.K., Able, K.W., Blaber, S.J.M., Elliott, M. (Eds.), Fish and Fisheries in Estuaries—A Global Perspective. John Wiley & Sons, Oxford, ISBN: 9781444336672, pp. 706–797.

Elliott, M., Franco, A., Ramos, S., Hemingway, K.L., Marshall, S., 2022b. Appendix B study methods: data processing, analysis and interpretation. In: Whitfield, A.K., Able, K.W., Blaber, S.J.M., Elliott, M. (Eds.), Fish and Fisheries in Estuaries—A Global Perspective. John Wiley & Sons, Oxford, ISBN: 9781444336672, pp. 941–1005.

Gray, J.S., Elliott, M., 2009. Ecology of Marine Sediments: Science to Management. OUP, Oxford, p. 260.

Krebs, C.J., 1998. Ecological Methodology, second ed. Addison Wesley Longman, Harlow.

Schmidt, R.J., Osenberg, C.W. (Eds.), 1996. Detecting Ecological Impacts: Concepts and Applications in Coastal Habitats. Academic Press, San Diego, CA.

Southwood, T.R.E., Henderson, P.A., 2000. Ecological Methods, third ed. Blackwell, Oxford.

Linked

Cause-consequence-response frameworks; Compliance monitoring; Management framework monitoring types and definitions; Monitoring types and definitions

Monitoring: 88. Coastal plant environmental monitoring

Appropriate monitoring of the environment surrounding a coastal industry site is essential for long-term sustainability of operation. Monitoring provides a context for current operation and provides the basis for assessing possible changes which may affect ecosystem health and operational security. Using a coastal power station as an example, the following basic monitoring will be required:

1. physical changes including hydromorphology, sea level, storminess;
2. water quality at intake, outfall, and within the mixing zone to protect both the environment and plant;
3. biological and microbiological processes which influence biofouling;
4. the impact of the intake, specifically on fish and other biota;
5. the impact of the discharge on the surrounding ecosystem.

The choice of parameters and features monitored is critical, but of equal importance is the choice of method and frequency of sampling to optimize resources. Certain water quality parameters can effectively be measured continuously to provide a surrogate system for wider system changes. Routine sampling programs may be adopted, but for many physical and biological components comprehensive sampling programs at relatively infrequent intervals may provide the most useful data. The most appropriate sampling regimes are inevitably site specific, but Table 1 presents the suggested summary monitoring program with typical sampling regimes that are shown after the various items.

Routine (R) can range from anything between daily and seasonal or annually. Where a comprehensive survey (I) is indicated, this would normally be undertaken before commissioning but thereafter intervals could be as long as 5 years. A combination of sampling strategies may be appropriate. As an example, in the case of a cooling water discharge in which antifouling measures have been used, analysis of the full range of chlorine by-products is expensive. However, once a comprehensive survey has characterized the CPBs around a station, regular monitoring of the dominant species (usually bromoform) will provide adequate information and protection with full analysis needed only at extended intervals.

Since an operating cooling water system (CWS) does not allow visual inspection, a special device is needed for the monitoring of biofouling. Those devices (see Fig. 1) have settlement plates where an optimal environment for settlement and growth is enabled. In general, they are located in front of the main condensers (where optimal antifouling is needed). These devices will then give information on the rate of settlement and growth of organisms, the performance of the applied water treatment program/strategy, and eventually the occurrence of unknown problems/species in the CWS.

Table 1 Examples of monitoring at a coastal power plant.

Component	Monitoring aspect	C = continuous; R = routine; I = comprehensive surveys at lesser intervals
1. Physical environment	Changes around intake and outfall	I
	Behavior of plume and mixing zone	I
	Changes to local sea level	I
2. Water quality	Salinity and temperature	C
	Dissolved oxygen in mixing zone	C/R
	(Unionized) ammonia	R
	pH	R
	Residual oxidants and chlorine by-products (CBPs) in discharge	R/I
3. Biofouling	Biofouling monitor[a]	C
	Inspection of CWS screens and outside structures	R
	Opportunistic inspection within CWS during outages	R
4. Intake	Fish impingement and entrainment Species and biomass on inlet screens	R/I
	Other biota	I
5. Discharge	Intertidal biota	I
	Local fish populations	I
	Monitoring of designated features	R

[a] Biofouling monitors.

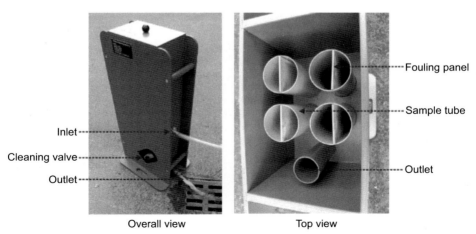

Fig. 1 KEMA biofouling monitor. Flow is bottom top along the fouling panels which can be inspected visually. Sediment is trapped in the bottom side and is released by opening the cleaning valve.
Ack. H.A. Jenner.

References and further reading

Claudi, R., Jenner, H.A., Mackie, G.L., 2011. Monitoring: the underestimated need in macrofouling control. In: Rajagopal, S., Jenner, H.A., Venugopalan, V.P. (Eds.), Operational and Environmental Consequences of Large Industrial Cooling Water Systems. Springer, New York, p. 522.

Bamber, R.N., Turnpenny, A.W.H., 2011. Entrainment of organisms through power station cooling water systems. In: Rajagopal, S., Jenner, H.A., Venugopalan, V.P. (Eds.), Operational and Environmental Consequences of Large Industrial Cooling Water Systems. Springer, New York, p. 522.

Linked

Biofouling control by chlorination; Thermal plume constituents; Source and receiving waters

Monitoring: 89. Compliance monitoring

Compliance monitoring is the assessment of the implementation of environmental requirements placed on an industry or operator by a statutory regulator, for example, an environmental protection agency, fisheries authority, harbor manager. In almost all cases, an industry will be given a license or planning permission to operate for either the whole plant or specific aspects of the activity. In modern environmental management, all activities and potential harm to the environment, including land, air, and waters, will be licensed and assessed together through an Integrated Pollution Prevention and Control (IPPC) program from the regulatory body.

Compliance is achieved when all requirements are met and any desired changes achieved. Monitoring may be of the compliance with environmental legal requirements (i.e., the license or planning requirements) or the monitoring of environmental quality. Compliance monitoring will vary according to the regulation issued by the competent authority, the site involved, and the characteristics of the receiving environment. It can be performed by the operator or the regulator depending on regulatory requirements, technical competences, and resources but under the polluter pays/developer debt principles, it should be funded by the industry.

Different regulations stipulate different requirements so monitoring may be continuous, periodic, or remote but would usually be subject to audit by the regulator and be required to be valid, reliable, accurate, and relevant. Assessment and analysis should be undertaken by properly qualified scientific staff operating under an approved quality audit program. For example, all elements of an AQC/QA (analytical quality control/quality assurance) program should be followed which may even involve getting environmental and discharge sample results verified by an independent laboratory. For example, the UK Good Laboratory Practice (GLP) Compliance Monitoring Programme is the mechanism whereby the GLP compliance of test facilities that conduct regulatory studies is assessed and monitored. Test facilities are subject to inspections by the Good Laboratory Practice Monitoring Authority (GLPMA), the government body charged with enforcing the GLP regulations. The frequency of GLP compliance monitoring inspections is determined by the outcome of a risk assessment that is conducted in accordance with current GLPMA risk assessment procedures.

The licenses are likely to contain both the concentrations (the level per unit volume or weight, e.g., $\mu g\,L^{-1}$) and the content (loadings, the total amount of material discharged in a given period, e.g., $kg\,day^{-1}$ or $tonnes\,year^{-1}$) of each constituent discharged to the environment. Statistical limits, required for environmental discharges, are usually set on concentration maxima or ranges and their frequency. As a general trend, governments are encouraging regulators (the competent authority) to simplify and streamline regulations (in the UK termed the Better Regulation initiative) and this inevitably requires more self-monitoring, more targeted monitoring but perhaps with less monitoring overall. How much self-monitoring an industry is allowed is highly contentious and in the end is partly a political decision designed to encourage industrial development. One of the aims is to ensure monitoring is more comparable across sectors and in this connection the Scottish Environmental Protection Agency, for instance, has produced a Compliance Assessment Scheme Manual to meet this aim.

As a further example, the Environment Agency for England has extensive guidance on monitoring requirements and has instigated a monitoring Certification Scheme (MCERTS) which covers a wide range of services and monitoring equipment. This covers continuous or portable monitoring of stacks or aquatic emissions, flow measurements, soils testing, Direct Toxicity assessment, and specific associated software programs. Other UK agencies are adopting the same scheme and a similar approach.

References and further reading

http://www.betterregulation.gov.uk/ideas/viewidea.cfm?proposalid=ea85b82054ac45c3aa2cbb57eb2ca13d.
http://www.environment-agency.gov.uk/business/regulation/31821.aspx.
http://www.environment-agency.gov.uk/static/documents/Business/MCERTS_introduction_letter_v2.pdf.
http://www.iema.net/download/legal/Practitioner%20Vol%206%20Legal%20Compliance.pdf.
http://www.sepa.org.uk/about_us/publications/better_regulation.aspx.

Linked

Coastal plant environmental monitoring; Management framework monitoring types and definitions

Section 10

Management background

Introduction

Sustainable development will rely on a complex interplay of science, technology, and management skills. This section includes some of the fundamental management philosophies which underpin the more holistic approach required to achieve sustainable development and management of coastal industries. Currently, the main underlying philosophy is *Ecosystem Based Management* and the *Ecosystem Based Approach*—as first developed by the UN Convention for Biological Diversity, this is defined as a strategy for the integrated management of land, water, and living resources that promotes conservation and sustainable use in an equitable way. As such, its application will help to reach a balance of these three objectives of the convention. It is based on applying appropriate scientific methodologies focused on levels of biological organization which encompass the essential processes, functions, and interactions among organisms and their environment. Furthermore, it recognizes that humans, with their cultural diversity, are an integral component of ecosystems.

In managing the environment and looking for successful and sustainable solutions, we can no longer just be concerned with single sciences. We should also look more broadly, for example at relevant business models such as the PEST/PESTLE[a] model used by many industries covering many sectors. This has been expanded into sustainable environmental management which has recently been codified into the 10-tenets approach which can be applied to new coastal industrial developments (see the Introductory chapter to this volume).

Industries are required to comply with management regimes from inside the company; these ensure clear operational and environmental management. These aim to determine the effect of the external environment on the industry, and they have to respond to external management initiatives to determine and control the effect of the industry on the receiving environment. The latter includes both the natural aspects and the societal environment. Hence, these require a clear understanding of the endogenic pressures (those emanating inside the management area) and exogenic pressures (those emanating from outside the management area) that apply in the marine environment and which may affect the industry and what controls may be available or possible. This centers around a cause-consequence-response model which provides a valuable and necessary tool for describing the complex relationship between our needs from marine, coastal, and estuarine systems; the consequences of those needs; and the

---

[a] From business, the Political Economic Social Technological Legal Environment of a business.

means of tackling any problems (i.e., the management responses) arising from those needs and consequences.

Until recently, economic analyses associated with industrial development have concentrated exclusively on human-oriented factors. Today, a clear understanding of what may be termed the socio-ecological system is required. Hence, this includes the ecosystem services associated with the marine environment and the societal goods and benefits associated with them. These aspects are a vital part of the broader perspective required for sustainable development and management solutions.

In particular, any coastal industry is subject to management and governance regimes imposed on it from regulatory bodies. In turn, those bodies are responding to local, national, regional, and international agreements and instruments. Hence, governance is an integral part of the management of coastal industries and the way in which their role in the environment is controlled for the benefit of nature and society. In this respect, governance is defined as the sum of the policies, politics, administration, and legislation required on managing the environment.

Management background: 90. Sustainable environmental management

In managing the environment and looking for successful and sustainable solutions we can no longer just be concerned with single sciences—for example, it is valuable for us to take ideas from the business literature which suggests that the environment of an organization is summarized by the four categories of PEST (political, economic, social, and technological constraints) (Palmer and Hartley, 2008). This has been expanded to the PESTLE analysis which includes the fifth, legal aspect. We can then juxtapose this to reinforce the idea that the organization and management of an environment is subjected to the same constraints.

This approach recognizes that while natural scientists may want to emphasize the natural science, we have to be aware of (and work with) wider disciplines. These features are important in not only preventing the deterioration in ecosystem health (see entry 38) but also in restoring and allowing a degraded ecosystem to recover (Elliott et al., 2007). The four aspects of PEST and the five of PESTLE have been separately expanded through several iterations to emphasize that successful and sustainable environmental management requires a set of 10-tenets, that is, a "soft law."

The three basic principles of sustainable development, that uses of the environment have to encompass society, economics, and the natural environment, have long been adopted by governments and regulatory bodies. Linked to this, any coastal industry would aim that their business should be economically viable, technologically feasible, and environmentally sustainable. All of these aspects have been embedded within national and international strategies for many years.

In order to address all features of solutions to environmental problems, and to give as much guidance as possible, these initial three principles have been augmented by a further seven considerations to give rise to the 10-tenets of sustainable and successful environmental management (Elliott, 2013; Barnard and Elliott, 2015) (see Table 1). In cases where environmental management measures are proposed or introduced to address the adverse impacts of development, it has been postulated that the measures should meet these 10-tenets in order to be both successful and sustainable.

For successful and sustainable environmental management, these tenets have to be applied to the particular activity (e.g., all aspects of constructing, operating, and decommissioning the coastal plant and industry), the wider area (e.g., all related uses and users of a water body), and/or the mitigation and/or compensation measures aimed at minimizing impacts (e.g., habitat creation to restore lost wetlands). Hence, the responses to reducing the cause and consequences of risks and hazards have to cover the many aspects, the 10-tenets of sustainable management (Table 1). To date, these 10-tenets have been applied to various industries, such as ports, to the management of habitats such as beaches, and to particular problems such as marine litter and microplastics.

Table 1 The 10-tenets as applied to coastal industries to lead to sustainable and successful environmental management as related to coastal hazards and risk.

Tenet	Meaning	Examples for hazards and risk prevention and response
Environmentally/ecologically sustainable	That the measures will ensure that the ecosystem features are safeguarded	That the natural ecology is maintained where possible
Technologically feasible	That the methods and equipment for ecosystem and society/infrastructure protection are available	Flood barriers, shore protection, treatment plants of chemical pollutants, mechanisms to prevent the inflow of biological organisms
Economically viable	That a cost-benefit assessment of the environmental management indicates sustainability but that adaptation to hazards is within financial budgets	Compensation schemes for those people and areas affected; that industry in the national interest and large urban areas are protected; that measures for pollution reduction are funded
Socially desirable/tolerable	That the environmental management measures are as required or at least are understood by society as being required; that society regards the protection as necessary	The society is educated regarding the effects and implications of coastal hazards and thus has a high level of preparedness; that the societal "memory" of disasters is accommodated
Ethically defensible (morally correct)	That the wishes and practices of current and future individuals are respected in decision-making	Dealings with individuals are at the highest level and that no single sector is favored unduly; that the costs of present action to be borne by future generations is considered (e.g., economic discounting)
Culturally inclusive	That local customs and practices are protected and respected	That indigenous peoples, habits and customs are incorporated into decision-making; aboriginal (first-nation) rights are defended
Legally permissible	That there are regional, national, governance bloc (e.g., European), or international agreements and/or statutes which will enable the management measures to be performed; that either under regular or emergency statutes the hazard protection can be achieved	International agreements for aid and minimizing hazards or the consequences of it; national laws and agreement allowing regional and national bodies to act even in emergencies; governance mechanisms are adequate

Table 1 Continued

Tenet	Meaning	Examples for hazards and risk prevention and response
Administratively achievable	That the statutory bodies such as governmental departments, environmental protection and conservation bodies are in place and functional to enable the successful and sustainable management	Flood management schemes, erosion protection schemes, shoreline management plans, etc. have been created; that there are contingency plans showing the command structure to respond to hazards and disasters; that there are bodies to carry out these actions within the governance framework
Effectively communicable	That all horizontal links and vertical hierarchies of governance are accommodated and decision-making is inclusive	That all sectors are aware of the important issues and involved decision-making; that all stakeholders have the opportunity to participate in decision-making
Politically expedient	That the management approaches and philosophies are consistent with the prevailing political climate	That there is pressure on politicians to carry out measures; that politicians are aware of the risks and the consequences of either not being prepared nor having suitable responses for the hazards occurring

Based on Elliott, M., Day, J.W., Ramachandran, R., Wolanski, E., 2019. Chapter 1—A synthesis: what future for coasts, estuaries, deltas, and other transitional habitats in 2050 and beyond? In: Wolanski, E., Day, J.W., Elliott, M., Ramachandran, R. (Eds.), Coasts and Estuaries: The Future. Elsevier, Amsterdam, pp. 1–28, ISBN 978-0-12-814003-1. Adapted and expanded from Elliott, M., 2013. The 10-tenets for integrated, successful and sustainable marine management. Mar. Pollut. Bull. 74(1), 1–5.

References and further reading

Barnard, S., Elliott, M., 2015. The 10-tenets of adaptive management and sustainability—applying an holistic framework for understanding and managing the socio-ecological system. Environ. Sci. Policy 51, 181–191.

Borja, A., Elliott, M., 2019. Editorial—So when will we have enough papers on microplastics and ocean litter? Mar. Pollut. Bull. 146, 312–316. https://doi.org/10.1016/j.marpolbul.2019.05.069.

Elliott, M., 2013. The *10-tenets* for integrated, successful and sustainable marine management. Mar. Pollut. Bull. 74 (1), 1–5.

Elliott, M., Burdon, D., Hemingway, K.L., Apitz, S., 2007. Estuarine, coastal and marine ecosystem restoration: confusing management and science - a revision of concepts. Estuar. Coast. Shelf Sci. 74, 349–366. https://doi.org/10.1016/j.ecss.2007.05.034.

Elliott, M., Day, J.W., Ramachandran, R., Wolanski, E., 2019. Chapter 1—A synthesis: what future for coasts, estuaries, deltas, and other transitional habitats in 2050 and beyond? In: Wolanski, E., Day, J.W., Elliott, M., Ramachandran, R. (Eds.), Coasts and Estuaries: The Future. Elsevier, Amsterdam, ISBN: 978-0-12-814003-1, pp. 1–28.

Palmer, A., Hartley, B., 2008. The Business Environment, sixth ed. McGraw-Hill Higher Education, London.

Linked

Causes of and solutions to estuarine, coastal, and marine degradation; Ecosystem restoration; Ecohydrology and ecoengineering; Ecosystem resilience, resistance, recovery; Precautionary principle and approach

Management background: 91. Marine, coastal, and estuarine activities

Estuarine, coastal, and marine areas are the sites of multiple human activities, many of which are industries in their own right or are the results of these activities. All of these need managing and all lead to pressures (the mechanism of change) which in turn may change the natural and the human systems, especially the ecosystem services provided by the natural environment from which society derives goods and benefits after inputting human capital. While an environmental manager of an activity has to manage the effects on that activity on the environment, they also have to manage the effects of the environment on their activity. This includes the repercussions of other activities on the industry in question. Hence there is the need to determine the cumulative, synergistic, and antagonistic effects of all the activities taken together. Table 1 presents a generic list of these activities.

Each activity may be regarded as having a footprint which may be relatively small and the size of the area subject to an Environmental Impact Assessment. In turn, that activity footprint will create pressures-footprints which in turn again create effects-footprints on the natural and social systems. For example, while a shore-based pipeline may have a small footprint as the area in which it was constructed, the pressures emanating from that pipeline, such as the input of contaminating materials, may be transported further than the activity-footprint. In turn, the effects-footprints on the natural and social systems may be even greater given that those contaminants may be taken up by mobile animals such as fish or sea mammals which then migrate distances from the activity. Similarly, the activity-footprint may ultimately affect human users of the system away from the area, such as stopping local fishermen from using their gear. These three types of footprint then require management measures, what may be termed management response-footprints.

Table 1 Generic list of marine activities.

Sector	Examples of activities
Aquaculture	Culture of finfish, macroalgae, predator control, shellfisheries
Extraction of living resources	Benthic trawling (e.g., scallop dredging); discharging fishery wastes; netting (e.g., fixed nets); pelagic trawls; potting/creeling; suction (hydraulic) dredging; bait digging; seaweed and saltmarsh vegetation harvesting; bird egg and shellfish hand collecting, curio collecting
Transport and shipping	Ejecting litter and debris; mooring/beaching/launching; shipping; producing shipping wastes; operating ferries; noises; nearshore roads
Renewable energy	Building and operating devices for renewable (tide/wave/wind) power generation
Non-renewable (fossil fuel) energy	Building and operating oil and gas installations, power stations; discharging thermal wastes (cooling water), marine fracking

Continued

Table 1 Continued

Sector	Examples of activities
Non-renewable (nuclear) energy	Nuclear effluent discharge; nuclear power construction and operation; thermal discharge (cooling water)
Extraction of non-living resources	Water abstraction and operating desalination plants; mining for inorganic and particulate materials; non-living maerl, rock/minerals extraction by coastal quarrying; sand/gravel (aggregates) extraction; water for salt extraction
Navigational dredging	Capital and maintenance dredging; removal of substratum; dredged material disposal
Coastal infrastructure	Artificial reefs and barrage building; beach replenishment; communication infrastructure (cables); culverting lagoons; building dock/port facilities, groynes; land claim, marinas, pipelines; removal of space and substrata; constructing seawalls/breakwaters, urban dwellings, that is, housing and other buildings
Land-based industry	Industrial effluent treatment and discharge; industrial/urban emissions (air); discharging particulate waste; desalination effluent; sewage and thermal discharge
Agriculture	Agricultural waste production; coastal farming; coastal forestry; operating land/waterfront drainage
Tourism/recreation	Angling; boating/yachting, diving/dive site operation; litter and debris production; public beach use; tourist resort and water sports operation
Military	Disposal areas operation, infrastructure building, munitions testing and use; warfare
Research and education	Animal sanctuaries; marine archaeology; marine research; engaging in field education and training
Carbon capture and storage	Exploration, construction, operation of carbon capture and storage

Modified and expanded from Elliott, M., Burdon, D., Atkins, J.P., Borja, A., Cormier, R., de Jonge, V.N., Turner, R.K., 2017. "*And DPSIR begat DAPSI(W)R(M)!*"—a unifying framework for marine environmental management. Mar. Pollut. Bull. 118 (1–2), 27–40. https://doi.org/10.1016/j.marpolbul.2017.03.049, Lepage, M., Capderrey, C., Meire, P., Elliott, M., 2022. Chapter 8. Estuarine degradation and rehabilitation. In: Whitfield, A.K., Able, K.W., Blaber, S.J.M., Elliott, M. (Eds.), Fish and Fisheries in Estuaries—A Global Perspective. John Wiley & Sons, Oxford, pp. 458–552, ISBN 9781444336672.

References and further reading

Cormier, R., Elliott, M., Borja, Á., 2022. Managing marine resources sustainably—the 'management response-footprint pyramid' covering policy, plans and technical measures. Front. Mar. Sci. 9, 869992. https://doi.org/10.3389/fmars.2022.869992.

Elliott, M., Burdon, D., Atkins, J.P., Borja, A., Cormier, R., de Jonge, V.N., Turner, R.K., 2017. "*And DPSIR begat DAPSI(W)R(M)!*"—a unifying framework for marine environmental management. Mar. Pollut. Bull. 118 (1–2), 27–40. https://doi.org/10.1016/j.marpolbul.2017.03.049.

Elliott, M., Borja, A., Cormier, R., 2020. Activity-footprints, pressures-footprints and effects-footprints—walking the pathway to determining and managing human impacts in the sea. Mar. Pollut. Bull. 155, 111201. https://doi.org/10.1016/j.marpolbul.2020.111201.

Lepage, M., Capderrey, C., Meire, P., Elliott, M., 2022. Chapter 8. Estuarine degradation and rehabilitation. In: Whitfield, A.K., Able, K.W., Blaber, S.J.M., Elliott, M. (Eds.), Fish and Fisheries in Estuaries—A Global Perspective. John Wiley & Sons, Oxford, ISBN: 9781444336672, pp. 458–552.

Linked

Marine, coastal and estuarine activities; Sustainable environmental management

Management background: 92. Endogenic managed pressures and exogenic unmanaged pressures

The pressures likely to produce change in the marine and estuarine environment, and for which we need good science, can be separated into two sets: those emanating from activities (including industries, see entry 91) within the system under management (a sea area, a part of the coast, an estuary) and which we can control and those emanating from outside the system (globally or from the catchment) which are not under our control when managing a particular system. Each of these requires an ability to detect, understand, and manage change in the marine, coastal, and estuarine environment; therefore, environmental change is caused by the combined impacts of these two: respectively, *endogenic managed pressures* and *exogenic unmanaged pressures*.

In the case of the endogenic managed pressures, management of a particular area has to respond to the causes and consequences of the pressures whereas it only responds to the consequences of the exogenic unmanaged pressures. For example, endogenic managed pressures will include the effects of an industry, such as a power plant in an estuary or an offshore wind farm, that we can control through design and licensing the causes and the consequences of those pressures. In the case of exogenic unmanaged pressures, for example, relative sea-level change rise through global warming or isostatic rebound, we do not control the causes of this when managing a particular area but we do have to respond to the consequences, for example, by building higher dykes or creating more wetland to absorb rising water levels.

Each of the pressures may be regarded as having a footprint which may be larger than the activity-footprint but, because of the dynamic nature of the marine environment, smaller than the footprint of effect on the natural and social systems. While regulators aim to control the pressures- and effects-footprints, in reality, it is the activity-footprint which is licensed, not least through an Environmental Impact Assessment. The footprint of the endogenic pressures may extend beyond the management areas whereas the footprint of an exogenic pressure may extend to international and perhaps global scales.

The separation between these two types of pressure depends on the definition of the management area. If this is an estuary, then any pressures derived from outside the estuary are exogenic—for example, nutrient inputs from agriculture in a catchment may be an exogenic unmanaged pressure when we are attempting to manage an estuary but they become an endogenic managed pressure when we are managing the whole catchment from freshwaters to the sea.

The endogenic managed pressures can in turn be divided simply into two types—those things which we put into the system and those which we take out. For example, pollutants and infrastructure such as buildings and bridges are placed into the system (think of a coastal power station intake pipe or industrial plant on the banks of an estuary as a big particle!), and we take out physical resources such as aggregates and biological resources such as fisheries. These aspects, however, merge when we remove marine space by putting in land claim on which is built a coastal industry. Most importantly, this separation of the pressures affecting marine systems allows us to know and appreciate what, why, and how we can and cannot manage human activities (Tables 1 and 2).

Table 1 Examples of endogenic managed pressures (EnMP) (where the causes and consequences are managed within the management area).

Pressure	Description
Smothering	By man-made structures/disposal at sea
Substratum loss	Sealing by permanent construction (coastal defenses/wind turbines), change in substratum due to loss of key physical/biological features, replacement of natural substratum by another type (e.g., sand/gravel to mud)
Changes in siltation	Change in concentration of suspended solids in the water column (dredging/runoff)
Abrasion	Physical interaction of human activities with the seafloor/seabed flora and fauna causing physical damage (e.g., trawling, dredging for shellfish)
Selective extraction of nonliving resources (habitat removal)	Aggregate extraction/removal of surface substrata
Underwater noise	Shipping/acoustic surveys/coastal subsea construction
Litter	Waste products disposed of inappropriately into the marine environment
Thermal regime change	Temperature change (average, range, variability) due to thermal discharge (local)
Salinity regime change	Salinity and freshwater runoff change (average, range, variability)
Introduction of synthetic compounds	Pesticides, anti-foulants, pharmaceuticals
Introduction of non-synthetic compounds	Heavy metals, hydrocarbons
Introduction of radionuclides	Radionuclides
Introduction of other substances	Solids, liquids, or gases not classed as synthetic/non-synthetic compounds or radionuclides
Nitrogen and phosphorus enrichment	Input of nitrogen and phosphorus (e.g., fertilizer, sewage)
Input of organic matter	Input of organic matter (industrial/sewage effluent, agricultural runoff, aquaculture, discards, etc.)
Introduction of microbial pathogens	Sewage, certain food industry wastes, temperature promotion of microbial pathogen growth
Introduction of non-indigenous species and translocations	Through fishing activity/netting/aquaculture/shipping
Selective extraction of species	Removal and mortality of target (e.g., fishing) and non-target (e.g., bycatch, cooling water intake) species
Death or injury by collision	Caused by impact with moving parts of a human activity (ships, propellers, tidal turbines)
Barrier to species movement	Obstructions preventing natural movement of mobile species by barrages, causeways, tidal and wind turbines, along migration routes

Continued

Table 1 Continued

Pressure	Description
Emergence regime change	Change in natural sea level (mean, variation, range) due to man-made structures (local)
Water flow rate changes	Change in currents (speed, direction, variability) due to man-made structures (local)
pH changes	Change in pH (mean, variation, range) due to runoff/change in freshwater flow, etc. (local)
Electromagnetic changes	Change in the amount and/or distribution and/or periodicity of electromagnetic energy from electrical sources (e.g., underwater cables)
Change in wave exposure	Change in size, number, distribution, and/or periodicity of waves along a coast due to man-made structures (local) or climate change (large scale)

Modified and expanded from Elliott, M., Burdon, D., Atkins, J.P., Borja, A., Cormier, R., de Jonge, V.N., Turner, R.K., 2017. "*And DPSIR begat DAPSI(W)R(M)!*"—a unifying framework for marine environmental management. Mar. Pollut. Bull. 118 (1–2), 27–40. https://doi.org/10.1016/j.marpolbul.2017.03.049, Lepage, M., Capderrey, C., Meire, P., Elliott, M., 2022. Chapter 8 Estuarine degradation and rehabilitation. In: Whitfield, A.K., Able, K.W., Blaber, S.J.M., Elliott, M. (Eds.), Fish and Fisheries in Estuaries—A Global Perspective. John Wiley & Sons, Oxford, pp. 458–552, ISBN 9781444336672.

Table 2 Examples of physicochemical and biological exogenic unmanaged pressures (ExUP) (where the consequences are managed within the management area but the causes require global action).

Pressure	Description
Thermal regime change	Temperature change (average, range, variability) due to climate change (large scale)
Salinity regime change	Salinity and freshwater runoff change (average, range, variability) due to climate change (large scale)
Emergence regime change	Change in natural sea level (mean, variation, range) due to climate change (large scale) and isostatic rebound
Water flow rate changes	Change in currents (speed, direction, variability) due to climate change (large scale)
pH changes	Change in pH (mean, variation, range) due to climate change (large scale); volcanic activity (local)
Change in wave exposure	Change in size, number, distribution, and/or periodicity of waves along a coast due to climate change (large scale)
Introduction of non-indigenous species	Entry of species from other areas that may be alien, non-indigenous, or invasive and so which may or may not cause adverse consequences in an area

Modified and expanded from Elliott, M., Burdon, D., Atkins, J.P., Borja, A., Cormier, R., de Jonge, V.N., Turner, R.K., 2017. "*And DPSIR begat DAPSI(W)R(M)!*"—a unifying framework for marine environmental management. Mar. Pollut. Bull. 118 (1–2), 27–40. https://doi.org/10.1016/j.marpolbul.2017.03.049, Lepage, M., Capderrey, C., Meire, P., Elliott, M., 2022. Chapter 8. Estuarine degradation and rehabilitation. In: Whitfield, A.K., Able, K.W., Blaber, S.J.M., Elliott, M. (Eds.), Fish and Fisheries in Estuaries—A Global Perspective. John Wiley & Sons, Oxford, pp. 458–552, ISBN 9781444336672.

References and further reading

Cormier, R., Elliott, M., Borja, Á., 2022. Managing marine resources sustainably—the 'management response-footprint pyramid' covering policy, plans and technical measures. Front. Mar. Sci. 9, 869992. https://doi.org/10.3389/fmars.2022.869992.

Cowie, J., 2013. Climate Change: Biological and Human Aspects, second ed. Cambridge University Press, Cambridge.

Elliott, M., 2011. Marine science and management means tackling exogenic unmanaged pressures and endogenic managed pressures—a numbered guide. Mar. Pollut. Bull. 62, 651–655.

Elliott, M., Burdon, D., Atkins, J.P., Borja, A., Cormier, R., de Jonge, V.N., Turner, R.K., 2017. "*And DPSIR begat DAPSI(W)R(M)!*"—a unifying framework for marine environmental management. Mar. Pollut. Bull. 118 (1–2), 27–40. https://doi.org/10.1016/j.marpolbul.2017.03.049.

Elliott, M., Borja, A., Cormier, R., 2020. Activity-footprints, pressures-footprints and effects-footprints—walking the pathway to determining and managing human impacts in the sea. Mar. Pollut. Bull. 155, 111201. https://doi.org/10.1016/j.marpolbul.2020.111201.

Lepage, M., Capderrey, C., Meire, P., Elliott, M., 2022. Chapter 8. Estuarine degradation and rehabilitation. In: Whitfield, A.K., Able, K.W., Blaber, S.J.M., Elliott, M. (Eds.), Fish and Fisheries in Estuaries—A Global Perspective. John Wiley & Sons, Oxford, ISBN: 9781444336672, pp. 458–552.

Wolanski, E., McLusky, D.S. (Eds.), 2011. Treatise on Estuarine & Coastal Science. In: Kennish, M.J., Elliott, M. (Eds.), Volume 8. Human-Induced Problems (Uses and Abuses) in Estuaries and Coasts, Elsevier, Amsterdam, p. 315.

Linked

Activity-, pressures-, effects-, and management response-footprints; Ecosystem resilience, resistance, recovery; Mechanical, thermal, and chemical stressors

Management background: 93. Cause-consequence-response frameworks (DPSIR, DAPSI(W)R(M) approaches)

Marine management, as with all environmental management, is implicitly or explicitly based on a cause-consequence-response framework whereby human activities such as those emanating from coastal industries then lead to consequences. In turn, the consequences need to be managed, often through what are termed a Programme of Measures. Those consequences are regarded as effects on both the natural system and the way society uses the natural system, which then need management actions to alleviate, reduce, or remove those consequences. As a manifestation of this approach, the driver-pressure-state-impact-response (DPSIR) framework has long existed (since the early 1990s) to integrate the relationships between development and their pressures and impacts to the environment (Patricio et al., 2016).

By combining ideas on our needs for the marine and estuarine systems, the consequences of those needs, and the means of tackling any problems resulting from those needs and consequences, the five elements of DPSIR framework give us a valuable philosophy for tackling and communicating our methods of marine management (McLusky and Elliott, 2004; Patricio et al., 2016; Elliott, 2011). This cyclical framework considers the *Drivers* (human activities and economic sectors responsible for the pressures), *Pressures* (particular stressors on the environment), *State* (the characteristics and conditions of the environment), *Impacts* (changes in the natural and human system and the way in which we use the marine area), and *Responses* (the creation of different policy options and economic instruments to overcome the state changes and impacts).

To the DPSIR we may also add Recovery (a reduction in the state changes as the result of these actions), this giving a sixth element in the DPSIRR framework (Elliott et al., 2007; Borja et al., 2010). The EU project KNOWSEAS has replaced the I for Impact with W for Welfare (hence DPSWR), both to avoid the long-standing confusion and separate impacts on the natural system (the state changes) from impacts on the human system and also to reinforce the societal-ecosystem links.

There will be a single DPSIR cycle for each major driver (e.g., power generation is one cycle) but this interacts with cycles for wild capture fisheries, recreational fisheries, tourism, other industry, etc. These ideas were expanded to require a set of 15 DPSIR-ES&SB (Ecosystem Services & Societal Benefits) postulates (see Atkins et al., 2011; Elliott, 2011). As an example, a power station development fulfills the need for power by society (D) which in turn will lead to loss of space, requirement for cooling water, and aggregates (P) which could change the ecological health of the benthos and the fish community (S). If not checked, these changes on the natural system would lead to a loss of amenity and fisheries (I). To prevent the latter then requires economic and legal instruments (R) (Fig. 1).

Over time, various areas of confusion have developed in the use of DPSIR; for example, a driver could be a human need, an activity, a sector of activities, or a stressor; a pressure could be an activity, a sector, or a mechanism of change; state

Management background

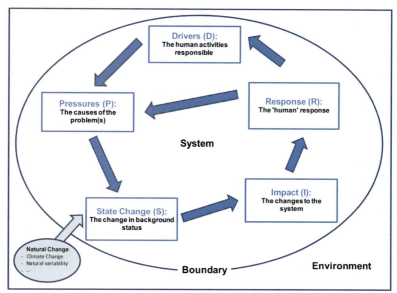

Fig. 1 The DPSIR cycle.
Based on Elliott, M., Burdon, D., Atkins, J.P., Borja, A., Cormier, R., de Jonge, V.N., Turner, R.K., 2017. "*And DPSIR begat DAPSI(W)R(M)!*"—a unifying framework for marine environmental management. Mar. Pollut. Bull. 118 (1–2), 27–40. https://doi.org/10.1016/j.marpolbul.2017.03.049; Patrício, J., Elliott, M., Mazik, K., Papadopoulou, K.-N., Smith, C.J., 2016. DPSIR—Two decades of trying to develop a unifying framework for marine environmental management? Front. Mar. Sci. 3, 177. https://doi.org/10.3389/fmars.2016.00177.

could be the characteristics of a system or the changes to those; impacts could be on the natural and/or social system; and responses were poorly defined. Hence DPSIR has been modified and refined into the most recent, and arguably a more complete, approach such as the DAPSI(W)R(M) (pronounced *dap-see-worm*) framework (Figs. 2 and 3) (Patrício et al., 2016; Elliott et al., 2017). In this, *Drivers* of basic

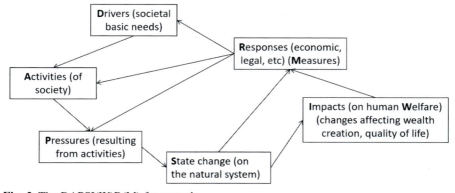

Fig. 2 The DAPSI(W)R(M) framework.

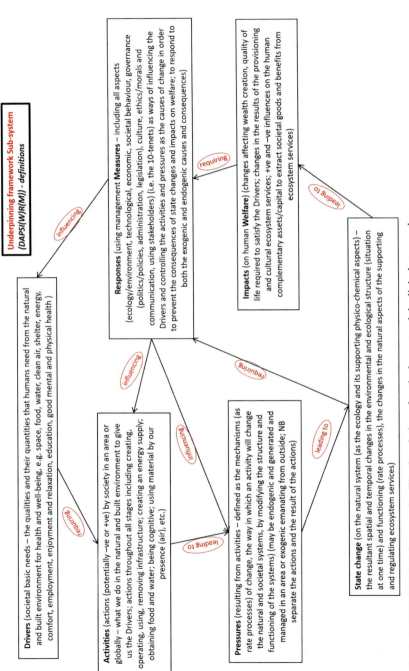

Fig. 3 The DAPSI(W)R(M) framework—The description of each element and the links between them.

human needs and values (such as the need for food and recreation) need to be fulfilled by *Activities* (e.g., fishing, tourism) that create *Pressures* (e.g., seabed abrasion, pollution); in turn, those pressures, as the mechanisms of change, lead to *State changes* on the natural system (e.g., turbidity increase, oxygen depletion) and *Impacts (on human Welfare)* for the human system (e.g., biodiversity loss, ecosystem services provision depletion). The *Response (using management Measures)*, that is, a policy response, then implies that society responds to those environmental and societal consequences, not least using a Programme of Measures, as defined in the EU Water Framework Directive and the EU Marine Strategy Framework Directive (Elliott et al., 2017). Those responses then include fulfilling the 10-tenets (see entry 90) to cover all aspects of sustainable management of coastal industries and their relationship with the natural and human environments.

References and further reading

Atkins, J.P., Burdon, D., Elliott, M., Gregory, A.J., 2011. Management of the marine environment: integrating ecosystem services and societal benefits with the DPSIR framework in a systems approach. Mar. Pollut. Bull. 62 (2), 215–226.

Borja, Á., Dauer, D.M., Elliott, M., Simenstad, C.A., 2010. Medium- and long-term recovery of estuarine and coastal ecosystems: patterns, rates and restoration effectiveness. Estuar. Coast. 33, 1249–1260.

Elliott, M., 2011. Marine science and management means tackling exogenic unmanaged pressures and endogenic managed pressures—a numbered guide. Mar. Pollut. Bull. 62, 651–655.

Elliott, M., Burdon, D., Hemingway, K.L., Apitz, S., 2007. Estuarine, coastal and marine ecosystem restoration: confusing management and science—a revision of concepts. Estuar. Coast. Shelf Sci. 74, 349–366.

Elliott, M., Burdon, D., Atkins, J.P., Borja, A., Cormier, R., de Jonge, V.N., Turner, R.K., 2017. "And DPSIR begat DAPSI(W)R(M)!"—a unifying framework for marine environmental management. Mar. Pollut. Bull. 118 (1-2), 27–40. https://doi.org/10.1016/j.marpolbul.2017.03.049.

McLusky, D.S., Elliott, M., 2004. The Estuarine Ecosystem; Ecology, Threats and Management, third ed. OUP, Oxford, p. 216.

Patrício, J., Elliott, M., Mazik, K., Papadopoulou, K.-N., Smith, C.J., 2016. DPSIR—Two decades of trying to develop a unifying framework for marine environmental management? Front. Mar. Sci. 3, 177. https://doi.org/10.3389/fmars.2016.00177.

Sekovski, I., Newton, A., Dennison, W.C., 2012. Megacities in the coastal zone: Using a driver-pressure-state-impact-response framework to address complex environmental problems. Estuar. Coast. Shelf Sci. 96, 48–59.

Linked

Sustainable environmental management

Management background: 94. Socioecological system—Ecosystem services and societal goods and benefits

The links between the human use of the coastal, estuarine, and marine systems by coastal industries and the resulting health and status of the natural environment may be regarded as the socioecological system (O'Higgins et al., 2020). Many aspects of ecosystem management now center around ensuring that the natural system can produce ecosystem services from which society can obtain goods and benefits. Hence the importance of the ability of coastal industries to contribute to providing such goods and benefits while protecting ecosystem functioning and services.

Coastal industry managers are now familiar with the term ecosystem services in that the prevailing legislation under which they operate has these as a central pillar. Furthermore, their regulators will emphasize that the industry and its activities have to protect ecosystem services and deliver societal goods and benefits. Despite this, there is still confusion in the terminology.

Beaumont et al. (2007) refer to ecosystem goods and services as "the direct and indirect benefits people obtain from ecosystems" (Fig. 1). They view ecosystem goods as distinguished from services in representing the "materials produced" that are obtained from natural systems for human use. In the context of identifying, defining, and quantifying goods and services provided by marine biodiversity alone, Beaumont

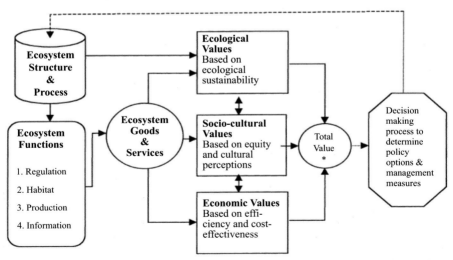

Fig. 1 Framework for the integrated assessment and valuation of ecosystem functions, goods, and services. * refers to aggregation and weighting of different aspects.
From De Groot, R.S., Wilson, M.A., Boumans, R.M.J., 2002. A typology for the classification, description and valuation of ecosystem functions, goods and services. Ecol. Econ. 41, 393–408.

et al. (2007) introduce a further category to those of the UN 2005 Ecosystem Assessment (https://www.millenniumassessment.org/en/index.html). Thus, their assessment framework comprises:

- Production services which involve products and services obtained from the ecosystem;
- Regulating services which are the benefits obtained from the regulation of ecosystem processes;
- Cultural services which are the non-material benefits people obtain from ecosystems;
- Option use values which are associated with safeguarding the option to use the ecosystem in an uncertain future, and
- Supporting services which are those that are necessary for the production of all other ecosystem services, but do not yield direct benefits to humans.

The generic term "goods and services" was then modified to indicate that a fully functioning ecosystem maintains a set of ecosystem services and that these are separated into fundamental services or characteristics (the physicochemical environment) and final services (the biological elements and processes resulting from the fundamental services which will lead to the benefits for society) (Fig. 2). That fundamental structure (the natural capital and the ecosystem structure and functioning) and final ecosystem services then produce societal benefits although these require the introduction of human capital to be obtained. Human capital can be regarded as inputting energy, time, skills, knowledge, and money or the ability for humans to be sentient beings, thereby allowing them to appreciate aesthetic values.

For example, the natural system can maintain the hydrographic processes which create the conditions for invertebrates as food for fishes and then harvesting the fishes requires boats, harbors, and the skills to use those fish. As another example, the natural processes can deliver marine sands and gravels but these become marine aggregates for construction when the vessels and infrastructure are created to exploit them. As a further example, the natural system can produce a blue whale but human capital is required for society to confer a greater value to that animal than just if it was yet another animal.

Hence coastal industries have to operate within the ecological system in which physicochemical functioning delivers ecological structure and functioning (Fig. 3, upper part) which then produce ecosystem services (central part). With the addition of human capital (such as energy, money, time, and skills and the fact that humans are sentient beings who can enjoy nature for its own sake), the ecosystem services then lead toward societal goods and benefits (Fig. 3, lower section). The societal goods and benefits can then be valued both as TEV (total economic value) and TSV (total system value) in which the latter may include components for which it is difficult to derive a monetary value (use/non-use, tangible/non-tangible, material/non-material, and "feel-good" values).

As the most recent iteration of this model (Fig. 4), it is emphasized that the term ecosystem services only refers to the central part of the model and should always be distinguished from societal goods and benefits. The model suggests that supporting services are no different from ecosystem structure and functioning, and that cultural services are a misnomer in that the natural environment does not recognize "culture"

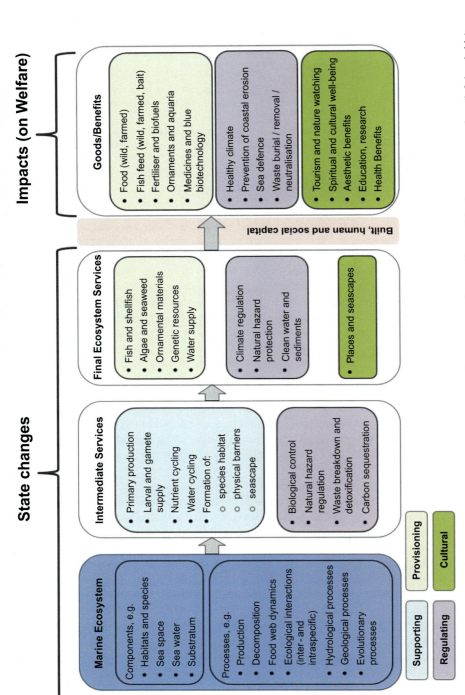

Fig. 2 State changes to the natural system reflected by changes in the marine ecosystem, intermediate and final ecosystem services (left-hand side), and impacts (on human welfare) reflected by changes to the provision of societal goods and benefits (right-hand side). Modified and expanded from In: Turner, R.K., Schaafsma, M., (Eds.), 2015. Coastal Zones Ecosystem Services: From Science to Values and Decision Making. Springer Ecological Economic Series. Springer Internat. Publ. Switzerland, ISBN 978-3-319-17213-2.

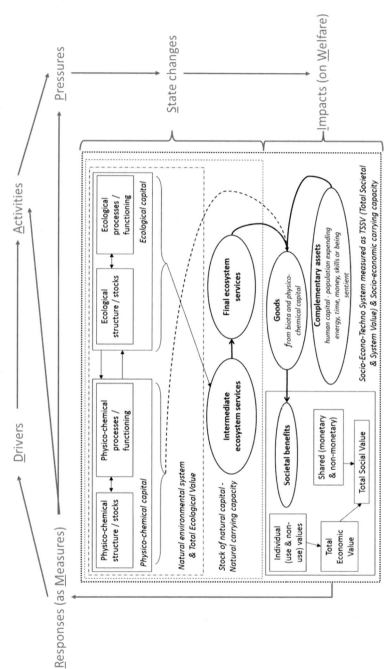

Fig. 3 The socioecological system with the DAPSI(W)R(M) cause-consequence-response system superimposed. From Elliott, M., Burdon, D., Atkins, J.P., Borja, A., Cormier, R., de Jonge, V.N., Turner, R.K._, 2017. "_And DPSIR begat DAPSI(W)R(M)!_"—a unifying framework for marine environmental management. Mar. Pollut. Bull. 118 (1–2), 27–40. https://doi.org/10.1016/j.marpolbul.2017.03.049.

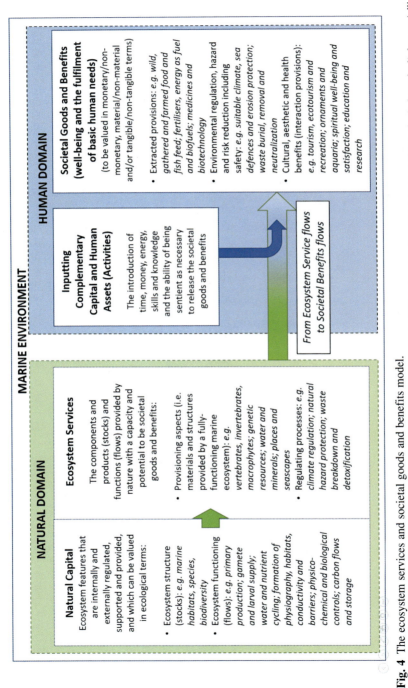

Fig. 4 The ecosystem services and societal goods and benefits model. From Elliott, M., 2023. Marine ecosystem services and integrated management: "There's a crack, a crack in everything, that's how the light gets in"! Mar. Pollut. Bull. 193. https://doi.org/10.1016/j.marpolbul.2023.115177.

which only appears after the addition of human capital. Finally, the model further indicates that ecosystem services are an intermediate step giving flows from ecosystem structure and functioning (natural capital) to societal goods and benefits.

References and further reading

Atkins, J.P., Burdon, D., Elliott, M., Gregory, A.J., 2011. Management of the marine environment: integrating ecosystem services and societal benefits with the DPSIR framework in a systems approach. Mar. Pollut. Bull. 62 (2), 215–226.

Beaumont, N.J., Austen, M.C., Atkins, J.P., Burdon, D., Degraer, S., Dentinho, T.P., Derous, S., Holm, P., Horton, T., Van Ierland, E., Marboe, A.H., Starkey, D.J., Townsend, M., Zarzycki, T., 2007. Identification, definition and quantification of goods and services provided by marine biodiversity: Implications for the ecosystem approach. Mar. Pollut. Bull. 54 (3), 253–265.

De Groot, R.S., Wilson, M.A., Boumans, R.M.J., 2002. A typology for the classification, description and valuation of ecosystem functions, goods and services. Ecol. Econ. 41, 393–408.

Elliott, M., 2023. Marine ecosystem services and integrated management: *"There's a crack, a crack in everything, that's how the light gets in"*! Mar. Pollut. Bull. 193. https://doi.org/10.1016/j.marpolbul.2023.115177.

Elliott, M., Burdon, D., Atkins, J.P., Borja, A., Cormier, R., de Jonge, V.N., Turner, R.K., 2017. *"And DPSIR begat DAPSI(W)R(M)!"*—a unifying framework for marine environmental management. Mar. Pollut. Bull. 118 (1–2), 27–40. https://doi.org/10.1016/j.marpolbul.2017.03.049.

Elliott, M., Borja, A., Cormier, R., 2020. Managing marine resources sustainably: a proposed integrated systems analysis approach. Ocean Coast. Manag. 197, 105315. https://doi.org/10.1016/j.ocecoaman.2020.105315.

Fisher, B., Turner, R.K., Morling, P., 2009. Defining and classifying ecosystem services for decision making. Ecol. Econ. 68 (3).

O'Higgins, T.G., Lago, M., DeWitt, T.H. (Eds.), 2020. Ecosystem Based Management and Ecosystem Services: Theory, Tools and Practice. Springer, Amsterdam, ISBN: 978-3-030-45842-3, https://doi.org/10.1007/978-3-030-45843-0. ISBN 978-3-030-45843-0 (eBook).

Pascual, M., Borja, A., Franco, J., Burdon, D., Atkins, J.P., Elliott, M., 2012. What are the costs and benefits of biodiversity recovery in a highly polluted estuary? Water Res. 46, 205–217.

Turner, R.K., Schaafsma, M. (Eds.), 2015. Coastal Zones Ecosystem Services: From Science to Values and Decision Making. Springer Ecological Economic Series, Springer Internat. Publ, Switzerland, ISBN: 978-3-319-17213-2.

Linked

Endogenic managed pressures, and exogenic unmanaged pressures; Marine, coastal, and estuarine activities; Sustainable environmental management

Section 11

Governance and management

Introduction

The activities of coastal industries are regulated by laws, guidelines, and agreements, many of which emanate from local, national, regional, or global governance. Hence coastal industry managers require some knowledge of the practices and principles of marine and environmental governance. If a country is a member of a larger bloc, such as the European Union, then the country is responsible for enacting EU law but the country would ensure that its citizens and enterprises follow the regulations that allow a country to stay within the wider (EU) law. In the case of other countries, the state may still be part of larger blocs, such as Regional Seas Conventions (OSPAR in the case of the United Kingdom) in which agreements are non-binding. However, again, the country would ensure that an individual, organization, or company is also following guidelines which ensure that the country does not jeopardize its membership of the bloc.

Environmental governance may be defined as the sum of the policies, politics, administration, and legislation, and coastal industries need to be aware of each of these elements which create their business environment (as embedded in the PESTLE and 10-tenets frameworks—see entry 90). Successful environmental outcomes rely heavily on good governance and good governance stems from the cross disciplinary integration of evidence gathering, analysis, and considered regulatory and planning control.

Governance has to be integrated across two dimensions—firstly, horizontal integration across all the sectors in marine, coastal, and estuarine environments (e.g., fisheries, navigation, urbanization, power generation). Secondly, there has to be vertical integration from local management to global environmental management and where the sequence can operate from bottom-up (starting at the local level) to top-down (dependent on diktat from bodies as high as the United Nations). Taken together, these management and governance measures constitute what may be regarded as management response-footprints.

Most if not all countries have similar legal instruments, environmental visions, and administrative bodies much of which is to encourage better networking and closer cooperation across environmental sectors with input from government departments, agencies, industry, and academia. This general shift to better governance is reflected in more coordinated marine environmental legislation such as that produced by the European Commission with the Habitats and Species Directive, the Water Framework Directive, and the Marine Strategy Framework Directive all having a wider vision and more stakeholder involvement than reflected in previous national or regional regulations. In turn, many of the regulations may be combined to give integrated planning systems, again typified by the EU Maritime Spatial Planning Directive.

Governance in most countries centers on incorporating internationally recognized policies, politics, legislation, and administration by horizontal and vertical integration of the management organogram to accomplish the vision of The Ecosystem Approach. This involves the following universally accepted principles: ecologically sustainable development, inter-generational equity, the precautionary principle, conservation of biological diversity and ecological integrity, ecological valuation, economic valuation of environmental factors, the "damager debt"/"polluter pays" principle, waste minimization, and public participation—the role of individuals and ethics.

Governance and management: 95. Governance of the coastal and marine environment

Governance is defined as the sum of policies, politics, administration, and legislation. This can be given in more detail in several ways as, firstly, the method or system of government or management—the act of directing government and monitoring (through policy design, implementation, and evaluation) the long-term strategy and direction; planning, influencing, and conducting policy and affairs.

Secondly, it covers the traditions (set of laws, procedures, and common practice), policies, institutions, and processes that determine how power is exercised (accountability/openness), how citizens (stakeholders) are given a voice (democratization/participation), and how decisions are made.

Thirdly, governance is the processes and systems by which government, society, or an organization operates—the functions of elected representatives, their committees, and officers. Fourthly, as required for sustainable development, it encompasses political accountability, freedom of association and participation, a sound judicial system, bureaucratic accountability, freedom of information and expression, and capacity building.

Lastly, governance, is "the exercise of political, economic and administrative authority in the management of a country's affairs at all levels. It is a neutral concept comprising the complex mechanisms, processes, relationships and institutions through which citizens and groups articulate their interests, exercise their rights and obligations and mediate their differences" (UNDP definition, www.emro.who.int/mei/mep/Healthsystemsglossary.htm).

Governance therefore includes the administrative bodies, the policy-making and implementing bodies and framework, and the legislative framework and legislation required to achieve, in this case, sustainable environmental management of coastal industries (See Table 1) (Boyes and Elliott, 2015; Cormier et al., 2022). It is not synonymous with "government." By definition it has a hierarchical organization involving vertical coordination (creating cooperation among various hierarchical levels of government—central, regional, and local), horizontal coordination—creating cooperation within a specific level of hierarchy (e.g., at the local level among local government, all sectors, and various stakeholders), and temporal coordination with the aim to achieve the optimal phasing of management actions (Fig. 1).

International law is a set of rules that regulates the conduct of States (e.g., the agreements in the inner circle of Fig. 1). Legally binding forms of international law comprise treaties conventions, custom, general principles and, as a subsidiary means of determining the law, decisions of international tribunals. These sources are increasingly supplemented by non-binding guidelines, codes, and practices which can influence the conduct of States. Treaties are written agreements, binding on States when they enter into force. Individual treaties specify when they enter into force, and normally this is once a minimum number of States have ratified the agreement. Much marine law is to be found in treaty law. Customary law is a body of law derived from

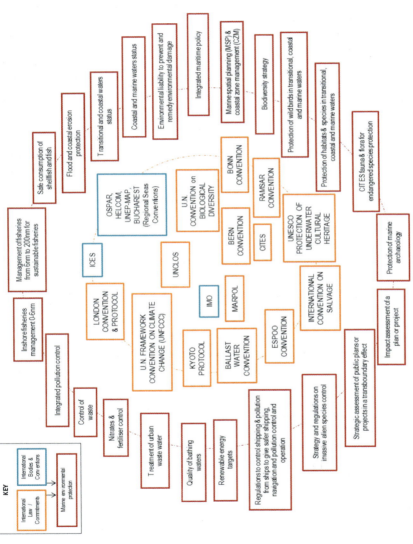

Fig. 1 Categories of environmental legal instruments—An example for the marine environment showing the vertical (from local to global) and horizontal (across all sectors) integration.
From Boyes, S.J., Elliott, M., 2014. Marine legislation—the ultimate 'horrendogram': international law, European directives & national implementation. Mar. Pollut. Bull. 86 (1–2), 39–47. https://doi.org/10.1016/j.marpolbul.2014.06.055.

Table 1 Governance: Governance to achieve marine environmental management.

What are the "legitimate" uses of the system? (What are the accepted uses?)
What are the human demands on the system?
What/where are the conflicts? (Do activities conflict spatially/temporally?)
Can the physical/biological system cope? (Can adverse effects be detected?)
Should/can the activity be stopped? (Can the conflicts be tolerated?)
Would zoning resolve potential/actual conflicts?
Are mitigation/compensation measures feasible? (Should mitigation/compensation measures be carried out?)
What governance aspects (policies, politics, administration, and legislation) are required to address these issues?

the practice of States that is considered to be legally required or permitted. Practice is generally taken to be official government conduct, as reflected in public statements, legislation, judicial decisions, and diplomatic measures. Practice must be general and consistent and customary law is generally binding on all States. General principles are principles of law common to most legal systems, or rules of general application, such as freedom of the high seas.

For a more detailed interpretation, Boyes and Elliott (2014) show the European and UK dimension of these marine governance instruments and recognizes that the management of the marine area is intimately linked with the management of coasts, estuaries, freshwaters, and terrestrial areas. This gives the relevant EU Directives and the way in which they are to be implemented as spatial (and occasional temporal) designations. Next it gives the UK enabling legislation from which these aspects will translate to corresponding instruments from the devolved administrations. Finally, the figure indicates the means of implementing the national instruments, and thus, by extension, the implementation of the EU Directives.

The legislative instruments are considered to be in four interlinked groups: firstly, those aimed at controlling, licensing, and authorizing existing activities such as shellfisheries and complex industries. Secondly, there are those instruments aimed at ensuring that new developments (plans and projects) have either no or acceptable environmental consequences or that those consequences are mitigated or compensated. This includes the instruments related to carrying out Environmental Impact Assessments and producing Environmental Statements. This also cross-refers to the need for Appropriate Assessments and Habitat Regulations Assessments for designated conservation features. The latter in turn links through the third group which relates to conservation designations, the relevant directives, regulations, and objectives but for predefined species and habitats. The final group relates to wide-ranging directives which define and aim for holistic assessments of quality and are related to the Ecosystem Approach. They aim to manage an area on a multi-sectoral basis by focusing on the net result of all the activity-pressures-impacts-response chains within an area.

The aim of many of these instruments is both simple and similar—to determine: what is the normal system, how does it vary, has there or will there be any changes due to natural or anthropogenic causes, can these be detected, are they acceptable or unacceptable, and what can we do about them. The figure emphasizes the main aim of marine management—to maintain and protect the natural functioning of the system while at the same time deliver society's needs—what may be described as ensuring the fundamental and final ecosystem services are maintained and delivered, while providing societal goods and benefits.

Some rules of international law aim to regulate the conduct of individuals but normally such rules have to be transposed into domestic law for them to take effect. Some States make international law automatically part of domestic law, while others only recognize its validity once it has been incorporated into domestic law, through, for example, legislation.

Policy can be defined as a set of decisions which are oriented toward a long-term goal or to a particular problem. A policy does not generally denote what is actually done to achieve or address such goals/problems. Policy can include proposals, initiatives, and legislation which are intended to achieve the environmental aims in specific fields or sectors of activity (e.g., fisheries, navigation, industrial discharges, the red boxes in the outer part of Fig. 1). Such decisions by governments are often embodied in legislation and usually apply to a Member State as a whole rather than to one part of it.

As an example, for the European Union, Directives are decisions which are binding on Member States (Boyes and Elliott, 2014). Each Member State is required to achieve a particular result although the European Commission does not dictate the means of achieving that result. As such, Directives are well suited to environmental measures, since the decisions on how to implement them is left to the individual Member States, which each will have different methods for setting environmental laws through enabling legislation.

Statutory Instruments, also referred to as secondary, delegated, or subordinate legislation, are the most important form of delegated legislation. They allow the provisions of an Act of Parliament to be subsequently brought into force or altered without Parliament having to pass a new act. They generally deal with matters which are too detailed to be included in an Act of Parliament and can easily be amended or repealed thus enabling governments to respond quickly to changing circumstances. Statutory instruments can be used to amend, update, or enforce existing primary legislation.

Regulations are legislative acts of general application, and as long as they are sufficiently precise they are normally effective. In the case of the European Union, they are the most rigorous form of legislation as they provide instructions which are applicable throughout the EU and are binding upon the Member States. Regulations become law in all Member States the moment they come into force, without the requirement for any implementing measure, and automatically override conflicting domestic provisions. National governments do not have to take action themselves to implement EU regulations as they are automatically part of national law. Finally,

EU countries are subject to the jurisdiction of the European Court of Justice and any transgressions may lead to infraction proceedings with the ability to levy fines to the country involved. If private companies are responsible for the failure to comply with EU law, then action would be taken against the company under national law.

In the case of countries not within such a larger bloc such as the EU, including the United Kingdom post-Brexit, environmental laws are made at the national level. For example, the United Kingdom recently passed the Environment Act 2012 and the Fisheries Act 2020 in order to transpose some previous EU regulations into UK law. In this case, any infringement of those national laws by a company is then subject to nationally decided penalties and legal proceedings.

References and further reading

Bell, S., McGillivray, D., Pedersen, O.W., 2013. Environmental Law, eighth ed. OUP, Oxford. 788 pp.

Boyes, S.J., Elliott, M., 2014. Marine legislation—the ultimate 'horrendogram': international law, European directives & national implementation. Mar. Pollut. Bull. 86 (1–2), 39–47. https://doi.org/10.1016/j.marpolbul.2014.06.055.

Boyes, S.J., Elliott, M., 2015. The excessive complexity of national marine governance systems—has this decreased in England since the introduction of the Marine and Coastal Access Act 2009? Mar. Policy 51, 57–65. https://doi.org/10.1016/j.marpol.2014.07.019.

Cormier, R., Elliott, M., Borja, Á., 2022. Managing marine resources sustainably—the 'management response-footprint pyramid' covering policy, plans and technical measures. Front. Mar. Sci. 9, 869992. https://doi.org/10.3389/fmars.2022.869992.

Davies, K., 2016. Understanding European Union Law. Routledge, London.

Linked

Estuaries; Marine, coastal, and estuarine activities; Sustainable environmental management

Governance and management: 96. Integrated Coastal Zone Management (ICZM)

Coastal Zone Management has been defined as a dynamic process in which a coordinated strategy is developed and implemented for the allocation of environmental, sociocultural, and institutional resources to achieve the conservation and multiple use of the coastal zone. Coastal Zone Management typically is concerned with resolving conflicts among many coastal uses and determining the most appropriate use of coastal resources (Sorensen, 1997).

In this case it involves a set of underlying legal principles to support ICAM (Integrated Coastal Area Management): the precautionary principle, the principle of preventative action, the polluter pays principle, the responsibility not to cause transboundary environmental damage, the rational and equitable use of natural resources, and the involvement of the public.

Coastal and marine management involves many interlinked pieces of international, regional (European), and national statutory instruments (e.g., see entries 95, 107–109) and, within each country, statutory bodies.

References and further reading

Cummins, V., O Mahony, C., Connolly, N., n.d. Review of Integrated Coastal Zone Management & Principals of Best Practice Prepared for the Heritage Council by the Coastal and Marine Resources Centre, Environmental Research Institute University, College Cork Ireland, pp. 84. http://cmrc.ucc.iehttp://www.globalislands.net/greenislands/docs/ireland_coastal_zone_review.pdf.

Green, D.R. (Ed.), 2010. Coastal Zone Management. Thomas Telford, London, ISBN: 978-0-7277-3516-4, p. 454.

McLusky, D.S., Elliott, M., 2004. The Estuarine Ecosystem; ecology, threats and management, third ed. OUP, Oxford, p. 216.

Sorensen, J., 1997. National and international efforts at integrated coastal management: definitions, achievements and lessons. Coast. Manag. 25 (1), 3–41. www.fao.org/docrep/w8440e00.htm.

Linked

Governance of the coastal and marine environment; Marine, coastal, and estuarine activities

Governance and management: 97. Administrative and regulatory aspects

The final stage of the cause-consequence-response chain implies that we have sufficient legal instruments which will allow or require marine users to control and limit the adverse effects of their activities. The legislative instruments then require implementation by administrative bodies, often regarded and listed in the statutory instruments as competent authorities. These instruments are implemented in all countries through a series of government departments and ministries and through agencies and other executive bodies with statutory powers. It is of note than many government departments have a marine, coastal, or estuarine remit (Table 1).

It is axiomatic in environmental management that the use of a code of conduct or a guideline is not taken as seriously by users as would be a statutory instrument. Such instruments reach the users through a hierarchy from global and international agreements, to regional (e.g., European in the case of EU states), and then in turn to national and devolved legislation. The national and devolved legislation is then required as the enabling legislation (acts, regulations, etc.) for implementing the wider agreements, directives, treaties, etc.

The competent authorities shown in Fig. 1 are mostly governmental bodies or their statutory agencies but they can also include private companies such as port authorities and other coastal industries which have been given a legal competence. Fig. 1 indicates examples of some of the main approaches used by those competent authorities to achieve environmental management of estuaries and coastal and marine areas.

All countries have pollution control bodies and agencies which have broadly similar regulations and which are the local competent authorities for enacting the environmental instruments, for example Directives in the case of EU Member States; in turn the agencies may come under the purview of a government department, for example the Department for Food and Rural Affairs in the United Kingdom. In cases where there are separate bodies for the control of pipeline discharges, diffuse pollution, vessel discharges waste, fisheries, nature conservation, flood defenses, etc. then

Table 1 Types of government departments with a marine competency (with their agencies).

- Environment
- Food rural affairs
- Fisheries and conservation
- Rural affairs
- Business, skills, innovation, energy, and climate change
- Foreign office
- International development office
- Defense
- Transport
- Communities and local government
- Culture, media, and sport
- Home office
- Cabinet office

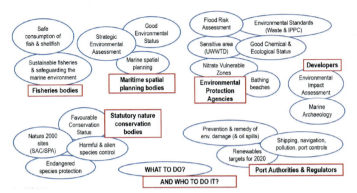

Fig. 1 The dominant organizations with a statutory role in marine and coastal management and the main management measures employed.
From Elliott, M., Houde, E.D., Lamberth, S.J., Lonsdale, J.-A., Tweedley, J.R., 2022. Chapter 12. Management of fishes and fisheries in estuaries. In: Whitfield, A.K., Able, K.W., Blaber, S. J.M., Elliott, M. (Eds.), Fish and Fisheries in Estuaries – A Global Perspective. John Wiley & Sons, Oxford, UK, pp. 706–797. ISBN 9781444336672.

the agencies cooperate to try and ensure that regulations are applied in a similar way with consistent standards wherever possible.

The details of responsibilities of environmental protection agencies are complex but in principle they are responsible to a greater or lesser degree for protecting biodiversity, improvement of contaminated land, licensing radioactive emissions, the disposal and management of waste and the point source emission to atmosphere, surface freshwaters, groundwaters, and marine waters.

References and further reading

Bell, S., McGillivray, D., Pedersen, O.W., 2013. Environmental Law, eighth ed. OUP, Oxford. 788 pp.
Boyes, S.J., Elliott, M., 2014. Marine legislation—the ultimate 'horrendogram': international law, European directives & national implementation. Mar. Pollut. Bull. 86 (1–2), 39–47. https://doi.org/10.1016/j.marpolbul.2014.06.055.
Boyes, S.J., Elliott, M., 2015. The excessive complexity of national marine governance systems—has this decreased in England since the introduction of the Marine and Coastal Access Act 2009? Mar. Policy 51, 57–65. https://doi.org/10.1016/j.marpol.2014.07.019.
Cormier, R., Elliott, M., Borja, Á., 2022. Managing marine resources sustainably—the 'management response-footprint pyramid' covering policy, plans and technical measures. Front. Mar. Sci. 9, 869992. https://doi.org/10.3389/fmars.2022.869992.

Linked

Discharge consent, permit, license, authorizations; Governance to achieve marine environmental management; Integrated Coastal Zone Management

Governance and management: 98. Nature conservation designations

Coastal industry managers will be charged with ensuring that their activities do not affect designated nature conservation sites irrespective of whether the industry is in, adjacent to or further away from the site. Therefore, in addition to discussions with the local environmental protection agency and the marine licensing agency, they will require discussions with the local statutory nature conservation body.

As indicated in Table 1, there is a plethora of nature conservation designations, each emanating from a particular piece of legislation (in the case of a country or European designation) or an agreement (in the case of local, regional, and global

Table 1 Examples of marine nature conservation designations.

Acronym	Title	Originator
PSSA	Particularly Sensitive Sea Areas	Global, International Maritime Organisation
SAC	Special Areas of Conservation	EU Habitats and Species Directive
SPA	Special Protected Areas	EU Wild Birds Directive
MPA	Marine Protected Areas	EU Maritime Spatial Planning Directive, global
SSSI	Sites of Special Scientific Interest	United Kingdom
OECM	Other Effective Conservation Measures	Global
EBSA	Ecologically and/or Biologically Sensitive Areas	Global
HPMA	Highly Protected Marine Areas	United Kingdom
MCZ	Marine Conservation Zones	United Kingdom
NTZ	No-Take Zones	Global
EFH	Essential Fish Habitat	United States, United Kingdom, etc.
BSH	Broad Scale Habitats	United Kingdom
HSCI	Habitats and Species of Conservation Importance	United Kingdom
EMS	European Marine Sites	EU Natura 2000 Directives
FOCI	Feature of Conservation Importance	United Kingdom
VMEs	Vulnerable Marine Ecosystems	FAO
Ramsar	Sites under the Ramsar Wetlands Convention	Global
Examples of Predominantly Terrestrial (United Kingdom)		

Site of Nature Conservation Interest (SNCI), Site of Importance for Nature Conservation (SINC), County Wildlife Site; Site of Metropolitan Importance for Nature Conservation; Regionally Important Geological Sites (RIGS); Areas of Outstanding Natural Beauty (AONBs), National Parks, National Nature Reserves (NNR)

designations); countries may implement local regulations to implement the EU Directives. The sites will be designated to protect specific and designated features (species and habitats, these may be termed the conservation objectives) from plans or projects (the industrial activities).

The regulatory body will then require an assessment of the potential effects of the activity; this may be an Appropriate Assessment in the case of the EU Natura 2000 Directives (the Habitats & Species and Wild Birds Directives), a Habitats Regulations Assessment (HRA), or an Environmental Impact Assessment (EIA). It is emphasized that while the statutory body is not required to demonstrate that there will be an adverse environmental effect by the activity, the developer will be asked to demonstrate that there will not be an adverse environmental effect. Demonstrating a negative effect is challenging and may not always be possible. An adverse environmental effect although demonstrated may still be allowed if it is decided by the competent authority that there are good reasons for this and the effects cannot be mitigated, the so-called IROPI—imperative reasons of overriding public interest.

The coastal industry manager will have to be particularly aware of the spatial and temporal scale of their activities in relation to the nature conservation site. Given the dynamic nature of coastal, marine, and estuarine areas, it is possible that an activity in one area can produce pressures and effects that have an effect far away from the industry, hence the importance of determining the areas affected (the so-called activity-, pressures-, and effects-footprints), and noting that the adverse effects are not only on the natural features but also on the societal features of an area.

Some nature designated sites will allow activities as long as they are shown not to adversely affect the designated features or where mitigation measures have been employed, for example by creating habitats. Under some designations, activities are not allowed, for example no-take zones or no-trawl zones in which fishing will be prohibited. Under some of the designations, commercial activities may be prohibited but recreational ones allowed.

The legislation and agreements are likely to dictate that the designated areas are maintained or restored to the given status and hence activities will be controlled to restrict the pressures and effects. Any causes of actual or potential degradation will then have to be removed, reduced, or mitigated or, failing that, compensated. Compensation will be of three types—to compensate the users of an area (e.g., economic compensation for fishermen affected), the resource affected (e.g., by restocking with fish or replanting seagrasses), or the habitat affected (e.g., by re-creating habitats elsewhere, such as by wetland creation).

Once a site is degraded then restoration measures will need to be implemented in order to return the site to an acceptable nature conservation status. Restoration may include geoengineering, that is, changing the physical shape and structure of the area, and ecoengineering. Ecoengineering is of two types—Type A in which the natural habitats are restored on the basis that organisms will then recolonize the area with natural recruitment patterns; failing that, Type B ecoengineering will be used to artificially replace organisms if natural recruitment is not successful, for example by restocking, reseeding, or replanting.

The 2022 Convention on Biological Diversity agreed that countries would aim for 30% of their areas to be protected for nature and biodiversity by 2030 with a third of that being strictly protected, that is, where activities are greatly (strictly or strongly) controlled; this is described as the "$30 \times 30 + 10$" approach. Hence it is likely that in the coming years the designated areas will increase in size.

It is also emphasized that some areas will have more than one designation. For example, many European Marine Sites (EMS) will be designated both for their bird populations and other species and habitats; hence they may be an EMS, SAC, SPA, and Ramsar Site. As such, the protected areas may range in size from very localized areas to large areas as in the case of EBSAs (Ecologically and/or Biologically Sensitive Areas) covering large ocean areas.

Each country will have its own nature protection designations, many of which may be for terrestrial areas which will include terrestrial coastal areas, possibly up to high water tide mark or even including intertidal areas.

Examples of UK local designations

A Site of Nature Conservation Interest (SNCI) is a designation used in many parts of the United Kingdom to protect areas of importance for wildlife and geology at a regional scale. In other parts of the country the same designation is known by various other names, including Site of Importance for Nature Conservation (SINC), County Wildlife Site, and Site of Metropolitan Importance for Nature Conservation. Overall, the designation is referred to as a "non-statutory wildlife site" or a "Local Wildlife Site" as part of Local Site systems, which include Local Geological Sites and Regionally Important Geological Sites (RIGS). Designated sites are protected by local authorities from most development. Site selection is usually done by the County Wildlife Trust, RIGS Groups or Geology Trusts, or their equivalents. Selection is objective and is normally based upon a recent survey specifically designed for SNCI, but selection on the basis of existing, published information may also occur. The approach is similar to that used for the selection of biological or geological Sites of Special Scientific Interest (SSSI), but the thresholds are lower.

Selection is primarily for habitats of inherent wildlife interest, but some sites may be selected for supporting rare or scarce species of plants or animals outside such habitats. The areas concerned may be areas of "natural" habitats, or they may be man-made. Once identified, designation and protection of the areas are carried out by local authorities through planning policies in their development plans. National government guidance requires all development plans to include such policies. The variation in names for the designation reflects its separate existence in the different development plans for different areas. In the guidance, the designation is referred to as a "Local Site," which may be divided into Local Geological Sites and Local Wildlife Sites.

In some areas, the designation is subdivided, or additional and more local designations may also be used. For example, in the London Metropolitan area, the following three designations are used: Site of Metropolitan Importance for Nature Conservation (equivalent to SINCs elsewhere), Site of Borough Importance for

Nature Conservation, and Site of Local Importance for Nature Conservation. These designations are restricted to sites of importance for wildlife.

References and further reading

Dunstan, P., Calumpong, H., Celliers, L., Cummins, V., Elliott, M., Evans, K., Firth, A., Guichard, F., Hanich, Q., de Jesus, A.C., Hildago, M., Mohammed, E., Lozano-Montes, H.-M., Meek, C.L., Polette, M., Purandare, J., Smith, A., Strati, A., Vu, C.T., 2021. Developments in management approaches. In: World Oceans Assessment II. vol. II. United Nations, New York, pp. 441–469. (Chapter 27). Downloaded from https://www.un.org/regularprocess/woa2launch.

Elliott, M., Mander, L., Mazik, K., Simenstad, C., Valesini, F., Whitfield, A., Wolanski, E., 2016. Ecoengineering with ecohydrology: successes and failures in estuarine restoration. Estuar. Coast. Shelf Sci. 176, 12–35. https://doi.org/10.1016/j.ecss.2016.04.003.

Ramsar Convention—see https://rsis.ramsar.org/.

Wolanski, E., Elliott, M., 2015. Estuarine Ecohydrology: An Introduction. Elsevier, Amsterdam, ISBN: 978-0-444-63398-9, p. 322.

www.naturalengland.org.uk/ourwork/conservation/designatedareas/localsites/default.aspx.

www.naturalengland.org.uk/ourwork/conservation/designatedareas/sac/conservationobjectives.aspx.

Linked

Nature conservation bodies; Integrated Coastal Zone Management; Integrated management of marine areas

Governance and management: 99. Environmental and operational managers

Operational managers are the senior staff within coastal industries who traditionally have been responsible for running a site or production unit to ensure that the quality of product or service is achieved at a high level and best cost. Their role is to plan and control all the subfunctions to ensure that the operation as a whole operates efficiently and smoothly.

Originally, environmental managers began to be recruited by some employers simply to make sure their organization understood and met developing environmental legislation which, up to the early 1990s, was largely in connection with the need to control aquatic or atmospheric emissions. They would be concerned with determining the Best Practical Environmental Option (BPEO) and the Best Available Technologies (BAT), occasionally extended to BATNEEC—the Best Available Technologies Not Entailing Excessive Costs. At present, this is no longer sufficient and existing and forthcoming regulations require the balancing of the organization carbon footprint, energy consumption, sustainability, and waste disposal, as well as traditional pollution control, and to ensure that environmental issues are considered all the way through an operational chain from conception and site surveying, through construction and operation, to decommissioning. Environmental managers should therefore be brought into all key decision-making discussions from purchasing all the way through to production.

Industry environmental managers will be charged with ensuring that best available procedures and Standard Operating Procedures are used, possibly and often including ISO standards (which may be CEN standards in Europe or British Standards in the United Kingdom). These may be combined into an environmental management system agreed between the industry and the regulators. For example, ISO 14001 and 14031 are concerned with Environmental Performance Evaluation (EPE) which *provides a robust and repeatable process to compare past and present environmental performance using Key Performance Indicators.*

Where external contractors are involved, the industrial environmental management system needs to ensure that environmental management protocols and procedures of contractors and subcontractors align with their own. Customers and stakeholders expect high levels of environmental performance and these demands may influence their trade or investment on ethical or practical grounds. For large industrial and agricultural activities with a high pollution potential, Integrated Pollution Prevention and Control (IPPC) legislation (a directive in the case of the European Union) already requires a wide-scale assessment of potential environmental effects even to the point of returning the site to its original condition at the completion of the activity.

Operational and environmental managers must now work in a harmonized way from the planning stage onwards to make sure that environmental issues are no longer an afterthought but are core to the business function of the enterprise. Environmental consequences planned from the beginning should be less of a financial burden on the business. Waste minimization and reuse can, under certain circumstances, reduce overall costs. There are examples with combustion plant of waste heat previously discharged being used to reduce process heating by sophisticated heat exchangers with net environmental improvement.

Both operational and environmental divisions can be subject to audit by appropriate quality management systems (e.g., ISO 14001 Environmental Management Systems) and all organizations can benefit from having that outside independent review of their approach. Not only does the use of a management auditing system drive up standards and awareness within an organization but the certification achieved demonstrates to others that the management has high levels of ambition for their environmental and operational performance.

There is now a need by environmental managers to address the vertical integration and coordination of development policies and sustainability policies, from local through national to international levels and vice versa (Fig. 1). Hence there is a joint need for a horizontal integration of operational controls and conservation measures across sectors. This especially involves the environmental operations manager in the use of risk assessment and risk management techniques (for example using the ISO standard Bow-tie analysis) such as IEC/ISO 31010 (IEC/ISO, 2009). The environmental manager will be responsible for carrying out operational controls to reach

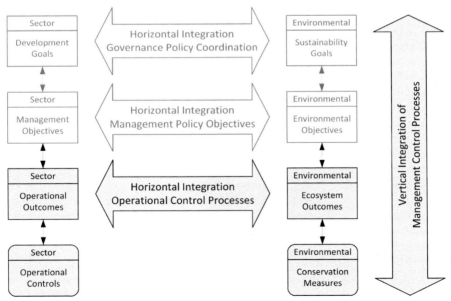

Fig. 1 Roles and responsibilities of environmental managers—Horizontal and vertical integration.
From Cormier, R., Elliott, M., Rice, J., 2019. Putting on a Bow-tie to sort out who does what and why in the complex arena of marine policy and management. Sci. Total Environ. 648, 293–305. https://doi.org/10.1016/j.scitotenv.2018.08.168.

operational outcomes which in turn are required for fulfilling the management objectives and the higher national and international development goals (Fig. 1). The risk assessment technique of the ISO 31000 risk management standard (ISO, 2018) is an effective method well suited to this role. The value to coastal industry managers of the risk management process of ISO 31000 is emphasized here given that an analysis of the measures and actions is needed both to reduce the risks and horizontally to integrate operational controls and conservation measures.

References and further reading

Campos, L.M.S., Aparecida de Melo Heizen, D., Verdinelli, M.A., Miguel, P.A.C., 2015. Environmental performance indicators: a study on ISO 14001 certified companies. J. Clean. Prod. 99, 286–296. https://doi.org/10.1016/j.jclepro.2015.03.019.

Cormier, R., Kannen, A., Elliott, M., Hall, P., Davies, I.M. (Eds.), 2013. Marine and Coastal Ecosystem-Based Risk Management Handbook. International Council for the Exploration of the Sea, Copenhagen. ICES Cooperative Research Report, No. 317, March 2013. 60 pp. ISBN 978-87-7472-115-1.

Cormier, R., Kannen, A., Elliott, M., Hall, P., 2015. Marine Spatial Planning Quality Management System. ICES Cooperative Research Report No. 327. 106 pp, https://doi.org/10.17895/ices.pub.5495.

IEC/ISO, 2009. Risk Management—Risk Assessment Techniques. International Organization for Standardization IEC/ISO 31010:2009.

ISO, 2018. Risk Management Principles and Guidelines. International Organization for Standardization ISO 31000:2018(E).

www.decisionsciences.org/decisionline/vol30/30_3/pom30_3.pd.

www.iso14000-iso14001-environmental-management.com/.

Linked

BAT, BATNEEC, best practice, and Integrated Pollution Prevention and Control (IPPC); Environmental assessment

Governance and management: 100. Nature conservation bodies

The coastal industry manager will have to ensure that their activities do not cause harm to the local biodiversity, species, and habitats and those ecological components further afield that may be affected. If this is not possible, then they will have to demonstrate that any adverse effects are acceptable and tolerated by society under the prevailing regulations (e.g., using the IROPI concept, Imperative Reasons of Overriding Public Interest). This requires extensive discussion with statutory and local nature conservation bodies at all stage of the development from its inception to the decommissioning. While some of those bodies may have a statutory remit, that is, their competence is outlined in legal instruments, some of the bodies may be informal or voluntary, the environmental Non-Governmental Organizations (eNGOs, see later). In particular, these organizations will be statutory consultees in Planning Applications and in carrying out the scoping during Environmental Impact Assessments.

The titles of nature conservation bodies will be particular to each country but, using the United Kingdom as an example, general features can be presented. Firstly, public nature conservation bodies include both statutory ones (governmental or quasi-governmental) and non-statutory, often voluntary. There are also a number of non-departmental government bodies that have a role to play in the processes of national government and which may have regional remits and competencies. In the United Kingdom, the Joint Nature Conservation Committee (JNCC) is the government agency which coordinates both national and international work of the three country agencies for nature conservation: Natural England, NatureScot, and Natural Resources Wales, and also links to the Department for Agriculture, Environment and Rural Affairs (DAERA) in Northern Ireland. Again, as an example, JNCC has Thematic Programs: Agricultural Biodiversity, Forest Biodiversity, Marine and Coastal Biodiversity, Inland Waters Biodiversity. Crosscutting Issues are 2010 Biodiversity Target, Alien Species, Climate Change, Conference of the Parties, Global Strategy for Plant Conservation, Global Taxonomy Initiative, Governance Law and Policy, Identification, Monitoring and Indicators, Protected Areas and In Situ conservation, Economics Trade and Incentive Measures, Scientific Assessment, Sustainable use of Biodiversity.

At a country level, Natural England (NE) brings together the former English Nature, the Countryside Agency, and the Rural Development Service. Natural England aims to conserve and enhance the natural environment, for its intrinsic value, the well-being and enjoyment of people, and the economic prosperity that it brings. Again, it has Thematic Programs: Agricultural Biodiversity, Forest Biodiversity, Inland Waters Biodiversity, Marine and Coastal Biodiversity. Crosscutting Issues are Alien Species, Identification, Monitoring and Indicators, Protected Areas and In Situ conservation, Public Awareness and Education, Sustainable use of Biodiversity, Governance Law and Policy.

In the case of Scotland, NatureScot (Scottish Natural Heritage until 2020) is the government agency that advises on and promotes the conservation of Scotland's wildlife and natural features. As expected, its work revolves around Thematic Programs: Agricultural Biodiversity, Forest Biodiversity, Inland Waters Biodiversity, Marine and Coastal Biodiversity, Island Biodiversity. Crosscutting Issues are Alien Species, Climate Change, 2010 Biodiversity Target, Identification, Monitoring and Indicators, Protected Areas and In Situ conservation, Public Awareness and Education, Governance Law and Policy.

Similarly, for Wales, Cyfoeth Naturiol Cymru/Natural Resources Wales was formed in 2013 with responsibility for functions formerly undertaken by the Countryside Council for Wales, Environment Agency (Wales), and the Forestry Commission (Wales). This is an example of a coordinated and integrated environmental body which combines nature conservation, environmental protection and pollution control, flooding and erosion, and sustainable development. Unusually among many countries, the Welsh Senedd (devolved parliament) has specifically passed the Well-being of Future Generations (Wales) Act 2015 which has recently (2022) encompassed seven goals to improve the social, economic, environmental, and cultural well-being of Wales. The seven well-being goals are a prosperous Wales, a resilient Wales, a healthier Wales, a more equal Wales, a Wales of cohesive communities, a Wales of vibrant culture and thriving Welsh language, a globally responsible Wales.

The principal responsibilities of the three bodies mentioned before include the coordination of all relevant nature conservation designations (see entry 98) and including the designation of Sites of Special Scientific Interest—areas of particular value for their wildlife or geology, Areas of Outstanding Natural Beauty (AONBs), and National Parks, as well as declaring National Nature Reserves. They also exercise regulation in those functions impinging on nature conservation in accordance with national and international legislation. As indicated, within the United Kingdom this links to the functions of the Department of Agriculture Environment and Rural Affairs (DAERA) in Northern Ireland. DAERA is a division of the Northern Ireland Executive (the devolved government given that environment is a devolved competency in the United Kingdom) and have responsibility for inter alia biodiversity, species and habitats, wildlife licensing and management, and marine conservation.

Non-governmental bodies, eNGO

All countries have a very broad range of non-governmental nature bodies, shortened to eNGO, engaged in nature conservation. These may be funded by voluntary contributions and subscriptions with some public funding, and they may be regarded as non-statutory consultees in planning applications and Environmental Impact Assessments. For example, in the United Kingdom these include the Royal Society for Protection of Birds (RSPB), the National Trust, County Wildlife Trusts, the Wildfowl and Wetlands Trust, the Woodland Trust, and the Marine Conservation Society. Internationally these include the WWF (Worldwide Fund for Nature), the IUCN (International Union for Conservation of Nature), Greenpeace, Fauna and Flora International, and Wetlands International.

References and further reading

Earll, R., 2018. Marine Conservation: People, Ideas, Action. Pelagic Publishing, Exeter, ISBN: 978-1-78427-176-3, p. 303.
Hiscock, K., 2014. Marine Biodiversity Conservation: A Practical Approach. Earthscan from Routledge, Abingdon, ISBN: 978-0-415-72355-8, p. 289.
JNCC—http://jncc.defra.gov.uk/page-3339-theme=textonly.

Linked

Habitats and species legislation; Nature conservation designations

Governance and management: 101. Discharge consent, permit, license, authorizations

It is axiomatic that any activity by a coastal industry which has the potential to damage the environment requires permission. Such a permission may be termed a discharge consent, a permit, a license, or an authorization. In the case of contaminants, it will specify that under a particular set of defined and documented circumstances, a discharge may be made as long as it does not exceed agreed chemical and/or physical limits. These limits might be peak values; a concentration; a percentile of concentration over a specified period; or a total load of material discharged over a given time, often a day or year.

By inference this is a special permission taking into consideration the ability of the receiving environment to absorb the discharge without any permanent ecological harm. The amount of detail needed and recorded is highly variable depending on the legislation involved and the importance of the control required. Major economic and social issues may need to be included in weighing up the evidence before any permissions are granted.

The license may use either or both of two approaches—the Environmental Quality Objectives and Standards (EQO, EQS) approach or the Uniform Emission Standards (UES) approach.

In the EQO/EQS approach, the amount of material permitted for discharge is set according to a set of agreed objectives (e.g., Table 1) and the assimilative capacity of the receiving body of water. A highly natural dispersing water body (such as an open coastal area or a high flushing estuary) is likely to have a larger capacity to degrade, disperse, and assimilate contaminants thereby reducing the likelihood of an environmental effect, whereas an industry discharging to a body of water which is poorly diluting or which has already received contaminants may only be allowed to discharge lower levels of materials. This could disadvantage an industry at the mouth of an estuary or river which receives many discharges. This is particularly

Table 1 Examples of Environmental Quality Objectives, such as those adopted for estuaries in the United Kingdom, in that the quality of the estuary should allow.

1. The protection of all existing defined uses of the estuary system.
2. The ability to support on the bottom the biota necessary for sustaining sea fisheries.
3. The ability to allow the passage of migratory fish at all stages of the tide.
4. The estuary's benthos and resident fish community and populations are consistent with the hydrophysical conditions.
5. The levels of persistent toxic and tainting substances and microbial contamination in the biota should be insignificant and should not affect it being taken by predators including humans.

important when the estuary has a catchment covering several countries, each with their own pollution and discharge control approaches and regulations.

The UES approach ensures that a particular industry or type of effluent is given the same discharge limits irrespective of the receiving environment. In this case, an industry is not penalized according to its location. Under both approaches, the levels for discharge are determined according to the toxic levels of the materials discharged; their half-life; bioaccumulation potential; and their ability to disperse, accumulate, or be degraded. Examples of the license conditions for land-based discharges are given in Table 2, and Table 3 gives an example of the US NDPES (National Pollutant Discharge Elimination Scheme) system. The latter also stipulates the nature of monitoring near a discharge and these permits may also stipulate the size of the mixing zone, the area of allowable environmental change defined in order to protect the wider environment.

The permitted level of a chemical in a discharge is often determined according to laboratory or field toxicity information based on the behavior of the substance in isolation. Other substances and their degradation products discharged from the same or other industries within the catchment may exhibit antagonistic and synergistic effects within a matrix of variable salinity, pH, temperature, etc. An alternative approach

Table 2 Examples of the conditions imposed within a consent (license/authorization/permit) for a discharge into coastal and estuarine waters.

1. Types of outlet and number of diffuser ports;
2. Source of material discharged (type of industry);
3. pH, BOD_5, suspended solids, temperature;
4. Emission levels as concentrations based on toxicity information + ability of receiving areas to degrade/disperse/assimilate;
5. Maximum loadings of key parameters (amount per day, month);
6. Volume of discharge per day and per hour;
7. Position of sampling points within plant;
8. Information on record keeping and good practice;
9. Toxicity-based element (DTA).

Table 3 US system—National Pollutant Discharge Elimination Scheme (NPDES) (Clean Water Act—Water Pollution Control Act, issued by EPA).

Permits given stipulate:
1. Effluent characteristics and type of treatment to effluent;
2. Type of discharge system (diffuser type);
3. Dimensions of mixing zone and characteristics and concentrations at edge of mixing zone;
4. Type of testing of physical dispersion of effluent;
5. Oceanographic/hydrographic monitoring;
6. Other physical parameter monitoring (temp., weather, salinity);
7. Benthic communities in receiving areas;
8. Fate and effects of pollutants, especially persistent contaminants.

intended to overcome this complexity is whole effluent testing often termed Direct Toxicity Assessment (DTA). An example of a DTA requirement based on UK guidance is: "the effluent shall be conclusively deemed to comply with the terms of this consent when a sample thereof taken at the sampling point and diluted 125 times with seawater and tested according to the procedure set out in the document headed 'Toxicity test for effluent discharges to saline waters' attached to this consent, exhibits a cumulative percentage mortality as hereinafter defined as not greater than 50 per cent."

In the United States, State and federal licenses are needed and that the latter can be more strict than the former. Similarly, in a European context, a Member State can set more strict conditions but not less strict than those laid down in EU legislation controlling polluting discharges. The term permit or authorization is also the term used under the European Integrated Pollution Prevention and Control Directive derived regulations where industries and agricultural activities with a high pollution potential are permitted to operate with agreed controls and conditions. Such an IPPC Authorization has to consider discharges to the whole environment (air, land, and water) and may use a database system such as the USEPA EpiSuite or the UK "H" system.

References and further reading

Elliott, M., Houde, E.D., Lamberth, S.J., Lonsdale, J.-A., Tweedley, J.R., 2022. Management of fishes and fisheries in estuaries. In: Whitfield, A.K., Able, K.W., Blaber, S.J.M., Elliott, M. (Eds.), Fish and Fisheries in Estuaries—A Global Perspective. John Wiley & Sons, Oxford, ISBN: 9781444336672, pp. 706–797 (Chapter 12).
McLusky, D.S., Elliott, M., 2004. The Estuarine Ecosystem: Ecology, Threats and Management, third ed. OUP, Oxford. 216 pp.
http://www.defra.gov.uk/environment/quality/permitting/.
http://www.doeni.gov.uk/niea/water-home/regulation_of_discharges_industrial.htm.
https://www.sepa.org.uk/regulations/water/.
www.netregs.gov.uk/.

Linked

BAT, BATNEEC, best practice, and Integrated Pollution Prevention and Control (IPPC); Compliance monitoring; Ecotoxicology

Governance and management: 102. Breaching regulations

Once a permit (or license, consent, authorization) has been issued it is accompanied by operational and compliance monitoring (entry 89). The consequences of breaching environmental regulations are complex because of the point at which action is taken by the regulatory authority and the potential consequences that follow. A permit to discharge may allow the occasional failure of a standard as an agreed statistic but a very high degree failure or repeated failure would lead to enforcement action. Although the consequences of breaching a permit vary with country and type of Environmental Protection Agency (EPA), as an example the Environment Agency (EA) for England is given but with generic features being identified. The EPA would normally issue a warning and enter into discussion after a breach in the first instance (Table 1). The severity of the response then increases with successive breaches.

The EA have a compliance classification scheme which characterizes breaches and the agency response (Table 2). The Compliance Classification Scheme (CCS) provides consistency across different regulatory regimes in the reporting of non-compliance with permit conditions. The information from their scheme contributes to an activity's Operational Risk Assessment (OpRA) risk rating or profile, through the compliance rating attribute. The scheme categorizes non-compliance based on the potential to cause environmental damage.

If breaches are considered sufficiently serious to demand action, then a range of actions or sanctions can be triggered (Table 3). The agency may decide when offenders are repeat offenders to invoke a civil sanction instead of traditional prosecution. It is emphasized that different EPA in different countries may use differing terms for compliance policies but with the same outcome.

Conservation legislation is very different but prosecution or warning still forms part of the armory. One option under the (English) Conservation of Habitats and Species Regulations 2010 is the ability to compulsorily obtain a controlling interest over land if the owner has breached a previously arranged management agreement. The regulations also make it an offense (subject to exceptions) to deliberately capture, kill,

Table 1 Formal options following one or more breaches of an environmental regulation.

- Issue a warning
- Statutory enforcement notices or works notices
- Prohibition notices
- Suspension or revocation of permit
- Injunction
- Carry out remedial works (claim back cost)
- Civil sanctions
- Other civil or financial sanctions
- Formal caution
- Prosecution
- Sanctions in combination

Table 2 The Compliance Classification Scheme (derived from the English Environment Agency).

Category	Is a non-compliance with a condition at an Environment Agency-permitted site that if it were classified as an incident ...
CCS category 1	would have the potential to have a major environmental impact that would be classified as a CICS category 1.
CCS category 2	would have the potential to have a significant environmental impact that would be classified as a CICS category 2.
CCS category 3	would have the potential to have a minor environmental impact that would be classified as a CICS category 3.
CCS category 4	has no potential to have an environmental impact.

Table 3 Civil sanctions can be of six types.

- Compliance Notice—a regulator's written notice requiring actions to comply with the law, or to return to compliance, within a specified period;
- Restoration Notice—a regulator's written notice requiring steps to be taken, within a stated period, to restore harm caused by non-compliance, so far as possible;
- Fixed Monetary Penalty—a low-level fine, fixed by legislation, that the regulator may impose for a specified minor offense;
- Enforcement Undertaking—an offer, formally accepted by the regulator, to take steps that would make amends for non-compliance and its effects;
- Variable Monetary Penalty—a proportionate monetary penalty, which the regulator may impose for a more serious offense;
- Stop Notice—a written notice which requires an immediate stop to an activity that is causing serious harm or presents a significant risk of causing serious harm.

disturb, or trade in the animals listed in Schedule 2 of the regulations or pick, collect, cut, uproot, destroy, or trade in the plants listed in Schedule 4.

References and further reading

Bell, S., McGillivray, D., Pedersen, O.W., 2013. Environmental Law, eighth ed. OUP, Oxford. 788 pp.
http://www.environment-agency.gov.uk/business/regulation/31823.aspx.
www.ucl.ac.uk/cserge/Ogus_and_Abbot.pdf.

Linked

Discharge consents, permits, licenses, and authorizations

Governance and management: 103. UN Conference on Environment and Development (UNCED); Convention for Biological Diversity (CBD)

While coastal industry managers will not be responsible for a country meeting or failing to meet its international obligations on nature protection, any breaches of national law and any enforced (or accidental) degradation of the natural environment would lead to such international regulations and agreements being breached. As such, it is emphasized that coastal industry environmental managers need to be aware of these wider considerations.

The global UN Convention on Biological Diversity, initiated in 1992, laid down basic and wide-ranging principles for the protection of biodiversity. Its nation signatories then agree to implement its principles through national or regional enabling legislation. In the case of Europe, the CBD was transposed into EU law through the Habitats and Species Directive and then into the regulations and acts of each Member State which serves as enabling legislation. The CBD defined and highlighted the Ecosystem Approach in which a national state's aims, actions, and outcomes have to maintain and protect biodiversity while at the same time having regard to societal needs.

The Ecosystem Approach as defined by the UN Convention for Biological Diversity is based on 12 principles (see Table 1). It is notable that the first 4 of these relate to societal desires, economics, and management and, in the order in which they were written, it is number 5 before ecology is mentioned. Perhaps this reinforces the fact that the economic and social aspects of marine, estuarine, and coastal management may have equal or perhaps even greater weight than the ecological aspects, especially in financially difficult times. Because of this, it is increasingly emphasized to stakeholders and policy-makers that there is the need to consider the ability of the marine environment to deliver a set of fundamental structural elements and functioning processes creating final ecosystem services leading to societal benefits (Turner and Schaafsma, 2015; Potschin et al., 2016).

As these 12 principles then map on to the 10-tenets of successful and sustainable environmental management and on to the Programme of Measures (see entry 90) this shows the complexity of the system. However, in particular, it shows the need for a multidisciplinary approach linking natural and social sciences, especially the ability to protect Ecosystem Services and deliver Societal Benefits.

The UN Conference on Environment and Development (UNCED) first met in Rio de Janeiro in 1992 (as the "Earth Summit") and then reconvened in New York in 1995,

Table 1 Twelve principles of the Ecosystem Approach (CBD, 2007).

1. The objectives of management of land, water, and living resources are a matter of societal choices.
2. Management should be decentralized to the lowest appropriate level.
3. Ecosystem managers should consider the effects (actual or potential) of their activities on adjacent and other ecosystems.
4. Recognizing potential gains from management, there is usually a need to understand and manage the ecosystem in an economic context. Any such ecosystem-management program should (a) reduce those market distortions that adversely affect biological diversity, (b) align incentives to promote biodiversity conservation and sustainable use, (c) internalize costs and benefits in the given ecosystem to the extent feasible.
5. Conservation of ecosystem structure and functioning, in order to maintain ecosystem services, should be a priority target of the ecosystem approach.
6. Ecosystem must be managed within the limits of their functioning.
7. The ecosystem approach should be undertaken at the appropriate spatial and temporal scales.
8. Recognizing the varying temporal scales and lag effects that characterize ecosystem processes, objectives for ecosystem management should be set for the long term.
9. Management must recognize that change is inevitable.
10. The ecosystem approach should seek the appropriate balance between, and integration of, conservation and use of biological diversity.
11. The ecosystem approach should consider all forms of relevant information, including scientific and indigenous and local knowledge, innovations and practices.
12. The ecosystem approach should involve all relevant sectors of society and scientific disciplines.

Johannesburg in 2002, and then again in Rio de Janeiro in 2012 ("Rio+20"). The overriding principle of the first Earth Summit was "think global, act local" and it aimed for environmental sustainability; as such it produced the blueprint for the 21st Century ("Agenda 21"). It also aimed for sustainable development which is defined as "that which meets the needs of the present without compromising the ability of future generations to meet their own needs." It is seen as the guiding principle for long-term global development and so sustainable development consists of three pillars: economic development, social development, and environmental protection. Sustainability therefore becomes embedded in each nation-state environmental management philosophy (see Fig. 1 as an example).

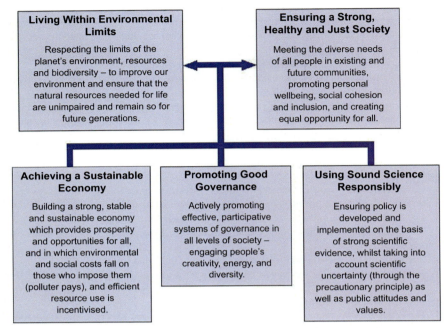

Fig. 1 An Environmental Sustainability Framework (a UK example from the Department for Environment, Food and Rural Affairs, Defra).

References and further reading

CBD, 2007. Principles for the Ecosystem Approach. Available from: https://www.cbd.int/ecosystem/principles.shtml. (Accessed 24 September 2023).

Potschin, M., Haines-Young, R., Fish, R., Turner, R.K., 2016. Routledge Handbook of Ecosystem Services. Routledge, Abingdon, ISBN: 978-1-138-02508-0, p. 629.

Turner, R.K., Schaafsma, M., 2015. Coastal Zones Ecosystem Services: From Science to Values and Decision Making. Springer Link, ISBN: 978-3-319-17214-9, p. 240.

See also the websites
http://www.un.org/esa/dsd/agenda21/.
http://www.uncsd2012.org/about.html.
http://www.cbd.int/.

Linked

Nature conservation bodies; Nature conservation designations

Governance and management: 104. Mitigation, amelioration, enhancement, compensation

After carrying out a rigorous risk assessment of the activities of a coastal industry, the environmental manager then has to manage or prevent that risk by implementing actions which may include restoring and enhancing degraded areas or bringing in compensation measures. This applies to all environmental management in all countries and these actions may be required by legal instruments and wider agreements. For example, although this will differ with country, these measures are used in the United Kingdom largely within the context of the Town and Country Planning Environmental Impact Regulations (as revised, 2011) and the Habitats and Species Regulations, 2010. The Institute of Ecology and Environmental Management (IEEM) has produced Guidelines for Ecological Impact Assessment in the United Kingdom, which have been endorsed by all of the major regulators and nature conservation bodies; these guidelines are used here as general guidance to show how these terms are used in environmental management and regulation.

Mitigation measures seek to reduce the impact of a development or operation. Compensation may be required where no mitigation is possible or all mitigation measures have been exhausted and a significant residual impact remains. Amelioration (or enhancement) measures seek to improve the existing situation. In the early years of Environmental Impact Assessment, mitigation and compensation were the only measures considered. In recent years, a growing body of opinion has evolved that considers that new developments should deliver net ecological gain rather than simply be designed to achieve damage limitation. The phrase "creative conservation" has been used recently in this context.

Therefore, coastal industry managers and indeed proponents of any scheme should incorporate, as part of the proposals for scheme design and implementation, measures that are required to deliver ecological enhancements as well as measures to:

- avoid negative ecological impacts—especially those that could be significant;
- reduce negative impacts that cannot be avoided; and
- compensate for any remaining significant negative ecological impacts.

Wherever possible, enhancement and mitigation measures should be developed and incorporated into a project as part of the design process, as measures that are fully integrated into a project from the start have a greater guarantee of delivery. The objective should always be to agree the identified measures between the proponent of a project, the regulatory bodies, and the wider stakeholders so that they become part of the scheme that is subject to detailed assessment.

Priority should be given to the avoidance of impacts at source, whether through the redesign of a project or by regulating the timing or location of activities. If it is not possible to avoid significant negative impacts, opportunities should be sought to reduce the impacts, ideally to the point that they are no longer significant. If this is not possible, but the scheme is permitted, compensation may be appropriate. These measures should be designed to meet specific ecological objectives that will deliver meaningful compensation for the negative impacts that are predicted.

Compensation can of three types: (i) to compensate the users of an area, for example by the economic compensation for fishermen affected by a development or landowners whose value is affected by the development; (ii) the resource affected, for example by supplanting the resource used by others and effected by the development such as by restocking with fish or replanting seagrasses; or (iii) the habitat affected, for example by re-creating habitats elsewhere, such as by wetland creation.

In this regard, the environmental, ecological, and operational objectives may include:

- primary, high-level objectives related to the reasons for compensation, gross area of compensation, and overall quality of compensation, and
- detailed objectives comprising specific ecological requirements for the habitat compensation scheme in terms of, for example, the number of birds and the habitats that they require.

Resource and habitat compensation, included within the activity of ecoengineering or Nature-Based Solutions, often carries a degree of uncertainty in that it may not be effective nor replace the ecological resources damaged by the development. Furthermore, even if it is effective, there may be a temporary or permanent loss of ecological value due to a time lag between damage occurring and the new habitat becoming fully functional. Evidence should be provided of the effectiveness of recommended mitigation, compensation, and enhancement measures and to what extent their success can be guaranteed. If possible, information from similar projects should be used to support statements about the level of success that can be reasonably expected.

Mitigation should be presented in terms of the integrity or conservation status of the resources or features to which it applies. For example, mitigation may be designed to ensure that the status of a species population can be maintained following development.

Following from the EU Habitats and Species Directive, the UK Habitats and Species Regulations 2010 provide for the control of potentially damaging operations, whereby consent from the competent statutory agency may only be granted once it has been shown through appropriate assessment that the proposed operation will not adversely affect the integrity of the site and especially the conservation objectives for which the site was designated. When considering potentially damaging operations, competent authorities apply the "precautionary principle," that is, consent cannot be given unless it is ascertained that there will be no adverse effect on the integrity of the site.

Again, it is emphasized that it is not the duty of the regulator to demonstrate that the coastal industry activity will have an adverse environmental effect but rather that the developer has to demonstrate that their activities will not have an adverse environmental effect. In instances where damage could occur, an operation may only proceed where it forms part of a plan or project with no alternative solutions, which must be carried out for reasons of overriding public interest. In such instances, for example, this requires government approval, in the case of the United Kingdom, the Secretary of State must secure compensation to ensure the overall integrity of the Natura 2000 system.

As a further example here, in England, in alignment with the 25 Year Environment Plan, the Environment Act 2021 made provision for environmental "net gain" by way of amendments to the Town and Country Planning Act 1990 (TCPA) and the Planning Act 2008 (in relation to Nationally Significant Infrastructure Projects (NSIPs)). As such, development on land and in intertidal locations will be required to deliver mandatory Biodiversity Net Gain. In June 2022, Defra consulted on how marine environmental gain will be applied as a complementary mandatory measure below the Low Water Mark. The aims of marine net gain relate primarily to nature recovery and biodiversity in the marine environment, taking a "nature first" approach while recognizing wider environmental benefits. It is expected that eventually marine net gain should cover nearly all new marine developments in English waters.

The enhancement of sites includes aspects of the restoration and re-creation of habitats, usually using ecoengineering (Nature-Based solutions) which may be designed to replace habitats lost due to the development or replace in an increased amount the habitats affected. Current practice is that at least "like-for-like" should be achieved although more commonly there is the need to create more habitat than was lost, that is, net gain. This then will also compensate for historical losses.

References and further reading

Defra Net Gain—https://www.gov.uk/government/collections/biodiversity-net-gain.

Elliott, M., Mander, L., Mazik, K., Simenstad, C., Valesini, F., Whitfield, A., Wolanski, E., 2016. Ecoengineering with ecohydrology: successes and failures in estuarine restoration. Estuar. Coast. Shelf Sci. 176, 12–35. https://doi.org/10.1016/j.ecss.2016.04.003.

Institute of Ecology and Environmental Management Guidelines for Ecological Impact Assessment in the UK. www.ieem.net/ecia.

Livingston, R.J., 2006. Restoration of Aquatic Systems. CRC Press, Taylor & Francis, Boca Raton, FL, p. 423.

Perrow, M.R., Davy, A.J., 2002a. Handbook of Ecological Restoration. Principles of Restoration, vol. 1 Cambridge University Press, Cambridge, p. 444.

Perrow, M.R., Davy, A.J., 2002b. Handbook of Ecological Restoration. Restoration in Practice, vol. 2 Cambridge University Press, Cambridge, p. 599.

The Conservation of Habitats and Species Regulations 2010, no 490. http://www.legislation.gov.uk/uksi/2010/490/contents/made.

Wolanski, E., Elliott, M., 2015. Estuarine Ecohydrology: An Introduction. Elsevier, Amsterdam, ISBN: 978-0-444-63398-9, p. 322.

Linked

Appropriate assessment; Habitats and Species directive; Nature conservation designations; Sustainable environmental management; Sustainable solutions; Ecohydrology with ecoengineering

Governance and management: 105. Sustainable solutions

As emphasized throughout this volume, the essence of successful and sustainable environmental management is to protect and maintain the natural system, its ecological structure and functioning, while at the same time allowing the environment to produce ecosystem services from which society, after putting in human capital, can obtain goods and benefits. The successful operation of a coastal industry therefore has to achieve both of these aims and the coastal industry manager should be aware of where their activities fit within this overall scheme. In particular, the coastal industry manager should aim for environmentally sustainable solutions to any actual or potential problems created by their industry.

A sustainable solution can be defined as a product or outcome which meets the principles of sustainable development. The concept of sustainable development can be interpreted in many different ways, but at its core is an approach to development that seeks to balance different, and often competing, needs against an awareness of the environmental, social, and economic limitations we face as a society.

Living within our environmental limits is one of the central principles of sustainable development, but the focus of sustainable development is far broader than just the environment. It is also about ensuring a strong, healthy, and just society. This means meeting the diverse needs of all people in existing and future communities; promoting personal well-being, social cohesion and inclusion; and creating equal opportunity. Sustainable development is about finding better ways of doing things, both for the present and the future. We might need to change the way we work and live now, but this does not mean our quality of life will be reduced.

The way we approach development affects everyone and the impacts of our decisions as a society have very real consequences for people's lives. Poor planning of communities, for example, reduces the quality of life for the people who live in them. Relying on imports rather than growing food locally puts a country at risk of food shortages. Sustainable development provides an approach to making better decisions on the issues that affect all of our lives.

The concept of sustainable development first received major international recognition in 1972 at the UN Conference on the Human Environment held in Stockholm. The term was not referred to specifically, but the international community agreed to the concept—now fundamental to sustainable development—that both development and the environment, until then addressed as separate issues, could be managed in a mutually beneficial way. Hence, sustainable development was the solution to the problems of environmental degradation discussed by the Brundtland Commission in the 1987 report "Our Common Future."

The remit of the Brundtland Report was to investigate the numerous concerns that had been raised in previous decades, that human activity was having severe and negative impacts on the planet, and that patterns of growth and development would be unsustainable if they continued unchecked. Sustainable development formed the basis of the United Nations Conference on Environment and Development held in Rio de Janeiro in 1992. The summit marked the first international attempt to draw up action plans and strategies for moving toward a more sustainable pattern of development. More recently, the World Summit on Sustainable Development was held in Johannesburg in 2002. The summit delivered three key outcomes: a political declaration, the Johannesburg Plan of Implementation, and a range of partnership initiatives. Key commitments included those on sustainable consumption and production, water and sanitation, and energy.

Our Common Future included what is deemed the classic definition of sustainable development: "development which meets the needs of the present without compromising the ability of future generations to meet their own needs." In the context of coastal industries, sustainable design and management solutions can be defined through best practice guidance, such as that developed for the Environment Agency in 2010.

The concepts of sustainable development were further put into operation with creation of the UN 2030 Agenda for Sustainable Development which as adopted by all UN Member States in 2015. This was aimed to create "a shared blueprint for peace and prosperity for people and the planet, now and into the future." This agenda adopted the 17 Sustainable Development Goals (SDGs) which aim to cover all aspects of development, including the need to protect natural and built environment, biodiversity, and the quality of life. Furthermore, they recognize the urgent need to address climate change and preserve river catchments, seas, and oceans. Of particular interest to coastal environmental managers is SGD 14, Life Below Water which addresses coastal and marine areas. The SDGs have a set of targets and associated actions although these targets and any available indicators have been criticized for not being SMART (specific, measurable, achievable, realistic, and time bounded); consequently, it will not be clear if and when such targets have been achieved.

References and further reading

Cormier, R., Elliott, M., 2017. SMART marine goals, targets and management—is SDG 14 operational or aspirational, is '*Life Below Water*' sinking or swimming? Mar. Pollut. Bull. 123, 28–33. https://doi.org/10.1016/j.marpolbul.2017.07.060.
Cormier, R., Elliott, M., Borja, Á., 2021. Measuring success: indicators and targets for SDG 14. In: Leal, F.W., Azul, A.M., Brandli, L., Lange, S.A., Wall, T. (Eds.), Life Below Water. Encyclopedia of the UN Sustainable Development Goals. Springer, Cham, https://doi.org/10.1007/978-3-319-71064-8_113-1.
Sustainable Development Commission—www.sd-commission.org.uk.
Turnpenny, A.W.H., Coughlan, J., Ng, B., Crews, P., Rowles, P., 2010. Cooling Water Options for the New Generation of Nuclear Power Stations in the UK. Science Report SC070015/SR, Environment Agency, Bristol.
UN Sustainable Development Goals—https://sdgs.un.org/goals.

Linked

AT, BATNEEC, best practice; Sustainable environmental management

Governance and management: 106. Ecological, socioecological, and socioeconomic valuation

It is accepted that a good and healthy environment is required for a good and healthy economy and although coastal industries could damage their local environment and still be a successful short-term business, this would neither be sustainable nor successful in the long term. Despite this, the economic well-being of a business is central to its survival and so the coastal industry manager will have to help enable both the economic success of their business and fulfilling the environmental obligations on their business imposed by regulators and the wider society. All of these aspects have to be included in environmental planning not just for the immediate area adjacent to a coastal industry but also to the wider marine area, as encompassed in Integrated Coastal Zone Management (ICZM) and Maritime Spatial Planning (MSP).

Environmental economics applies the concepts and analytical approaches of economics to better understand environmental issues and to the design of policies to address those issues. A basic argument underpinning environmental economics is that environmental amenities (or environmental goods) have an economic value and there are environmental costs of economic growth that may go unaccounted in current market models. Environmental amenities and goods include clean water, clean air, a healthy biodiversity, a tolerable climate, etc. The socioecological model shown in the Introduction to this volume emphasizes that a healthy natural physicochemical and ecological structure and functioning will give a flow of ecosystem services from which society gains goods and benefits after inputting human capital.

Ecological Economics is a transdisciplinary field, which attempts to bridge the human domain (Economics) and the natural environmental domain (Ecology). It recognizes the importance of the efficient allocation of resources but seeks a much deeper understanding of the relationship between economic development and resource use. As well as the valuation of the natural and social environment, ecological economics also encompasses issues of sustainability, ecological-economic system accounting, ecological-economic modeling at various spatial scales, and exploring innovative tools for environmental management. Hence the coastal environmental manager will be required by environmental regulators to use such tools.

The Ecosystem Approach by definition links natural scientific evidence with economic valuation to contribute to better policy decision-making and it systematically covers a much wider range of potential impacts on the natural environment than just natural environmental concerns. This is a developing field given the need for both scientific evidence and robust economic valuation techniques, especially in the marine environment. In turn this becomes part of an integrated management and assessment framework (Table 1).

The estuarine, coastal, and marine environments can be valued in a continuum covering three terms: ecological valuation, socioecological valuation, and socioeconomic valuation. "Ecological value" can be defined as "the intrinsic value of biodiversity, without reference to anthropogenic use; it is measured in non-monetary terms"

Table 1 The inclusion of ecological, socioecological, and socioeconomic valuation in estuarine, coastal, and marine decision-making.

- Integrated Coastal Zone Management (ICZM) and Marine Spatial Planning (MSP) involve an integrated planning system for managing the seas, coasts, and estuaries;
- Marine and coastal activity licensing is needed to ensure the sustainable use of the natural resources, where regulation has been streamlined to improve the overall efficiency;
- With the inclusion of the different types of valuation, monetary incentives may be used to change the behavior of industry and society;
- Marine Protected Areas may be required to be designated to provide better protection for the marine environment but also where socioeconomic impacts may have to be taken into account for marine conservation zones;
- Fisheries, marine management, and conservation bodies seek to balance the social and economic benefits of exploiting the marine and coastal resources, such as the use of the coastal area and waters, with the need to protect the local habitats, or even to help the marine environment recover from past exploitation.

Table 2 Ecological valuation assesses the value of key habitats and species with reference to their characteristics—Questions being addressed by nature conservation regulators.

Characteristic	Explanation
Fragility	Are there any delicate habitats and species?
Irreplaceability	Is the component a fundamental part of the system with major consequences in ecological terms[a] if it is lost?
Vulnerability/sensitivity	How vulnerable or sensitive is the component to specific natural hazards and pressures[b]?
Pristineness/naturalness	How untouched are they?
Uniqueness	Are they unique in local, national, regional, or global terms?
Rarity	How rare are they in local, national, regional, or global terms?

[a] If this is in societal terms then this refers to anthropogenic valuation.
[b] If this relates to activities or anthropogenic pressures then it is societal valuation.

(Table 2) (see Derous et al., 2007a,b). Ecological valuation relates to inherent ecological properties such as fragility, connectivity, uniqueness, irreplaceability, redundancy, and vulnerability of selected genes, species, communities, habitats, and ecosystems.

Socioecological valuation relates to the development of ecosystem services from those biological levels of organization and based on supporting and regulating aspects of ecosystem structure and functioning. In turn these give the provisioning services which in turn can deliver societal goods and benefits, including cultural benefits. That socioecological valuation then ensures that individual and societal welfare and well-being can be measured and safeguarded.

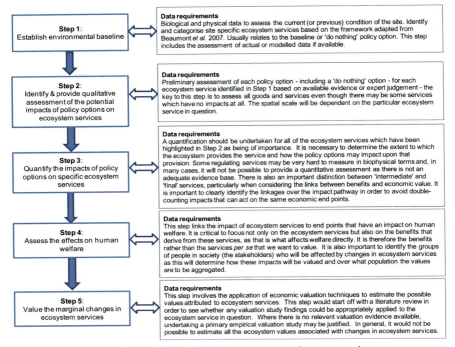

Fig. 1 Evaluation of policy options using an ecosystem services approach. Modified from Defra, 2007. An Introductory Guide to Valuing Ecosystem Services. Defra, London. http://www.defra.gov.uk/environment/policy/natural-environ/using/value.htm.

These aspects are increasingly justified against both an economic ecosystem services approach and a societal goods and benefits approach—these constitute the socio-ecological system (see Fig. 1 and references). While regulators are increasingly required to include an economic appraisal of options for environmental management, industry has always regarded the economic imperative as paramount. Hence any actions, whether for the benefit of plant operation or in order to meet environmental concerns, have an economic appraisal. Of course, superimposed over this are the plant and human safety aspects. The costs of operations are thus balanced against benefits even though the former may be economically calculated while the latter may include both economic and also less tangible "feel-good" aspects such as environmental protection, for example the protection of a wetland for its bird populations adjacent to the industrial plant and discharges. The less tangible aspects can also be valued economically (see Table 3 giving economic valuation methods) as well as by ecological valuation (see Pascual et al., 2012). Ecological valuation integrates the fragility and rarity of ecological elements, each of which may also have societal importance.

Socioeconomical valuation relates to the societal and national macroeconomics, including the employment related to the environment, the economic aspects related to public health and public safety, and the protection of people and coastal assets from

Table 3 Economic valuation methods and examples of their application in the marine environment.

Economic valuation method	Description	Marine example
Market Analysis (MA)	Where market prices of outputs (and inputs) are available. Marginal productivity net of human effort/cost. Could approximate with market price of close substitute. May require shadow pricing where prices do not reflect social valuations.	Deriving the social and economic value of shellfish, such as oysters, from market prices.
Productivity Gains and Losses (PGL)	Change in net return from marketed goods: a form of (dose-response) market analysis.	Improvements in water quality leading to reduced purification requirements following shellfish harvesting which would be reflected in higher net returns.
Production Function Analysis (PFA)	An ecosystem good or service treated as one input into the production of other goods: based on ecological linkages and market analysis.	The use of wetlands as fish nursery areas for species which eventually become commercial catches.
Hedonic Pricing (HP)	Derive an implicit price for an environmental good from analysis of goods for which markets exist and which incorporate particular environmental characteristics.	House prices are determined by the characteristics of the houses, including environmental features such as their proximity to marine leisure facilities.
Travel Cost Method (TCM)	Cost incurred in reaching a recreation site as a proxy for the value of recreation. Expenses differ between sites (or for the same site over time) with different environmental attributes.	The costs borne by visitors to bird watching sites may be interpreted as the minimum value they attached to that site.
Contingent Valuation Method (CVM)	Construction of a hypothetical market by direct surveying of a sample of individuals and aggregation to encompass the relevant population. Problems of potential bias.	The public might be asked to value a hypothetical environmental improvement, such as increased biodiversity.
Cost-of-Illness (COI)	The benefits of pollution reduction are measured by estimating the possible savings in direct out-of-pocket expenses resulting from illness and opportunity costs.	Loss of earnings due to illness caused by poor water quality.

Continued

Table 3 Continued

Economic valuation method	Description	Marine example
Choice Experiment Method (CEM)	Discrete choice model which assumes the respondent has perfect discrimination capability. Uses experiments to reveal factors that influence choice.	Can be used to investigate preference trade-offs involving security of water supply and biodiversity.
Damage Avoidance Costs (DAC)	The costs that would be incurred if the ecosystem good or service was not present.	A saltmarsh provides a natural form of flood prevention.
Defensive Expenditure Costs (DEC)	Costs incurred in mitigating the effects of reduced environmental quality. Represents a minimum value for the environmental function.	The cost of cooling water ponds to mitigate cooling water discharge effects.
Net Factor Income (NFI)	Estimates changes in producer surplus by subtracting the costs of other inputs in production from total revenue and ascribes the remaining surplus as the value of the environmental input.	The economic benefits of improved water quality can be measured by the increased revenues from greater aquaculture productivity when water quality is improved.
Relocation Costs (RLC)	Expenditures involved in the relocation of affected agents or facilities: a particular form of defensive expenditure.	The costs of relocating activity following managed realignment or marine-based wind farms.
Replacement/ Substitution costs (R/SC)	Potential expenditures incurred in replacing the function that is lost, for example, by the use of substitute facilities or "shadow projects."	The costs associated with the creation of intertidal habitat to compensate for habitat lost following industrial development.
Restoration costs (RC)	Costs of returning the degraded ecosystem to its original state. A total value approach; important ecological, temporal, and cultural dimensions.	The costs of rehabilitating an affected/degraded wetland.

From Cooper, K., Burdon, D., Atkins, J., Weiss, L., Somerfield, P., Elliott, M., Turner, K., Ware, S., Vivian, C., 2010. Seabed Restoration Following Marine Aggregate Dredging: Do the Benefits Justify the Costs? MEPF/09/P115 Marine Aggregate Levy Sustainability Fund (MALSF).

both locally (endogenic) managed pressures and externally (exogenic) managed pressures such as climate change. The types of valuation can be made in monetary/non-monetary, material/non-material, and tangible/non-tangible terms and constitute both natural and human capital.

Finally, it is emphasized that the viability of the construction, operation, and decommissioning of coastal industries and their ability to meet environmental and operational regulations and imperatives rely on finding technological and economic solutions to problems. For example, the costs and benefits of installing a fish return and recovery system or antifouling measures in a coastal power plant cooling system are related to balancing technological, economic, and environmental considerations against risks to the plant and the natural system.

References and further reading

Beaumont, N.J., Austen, M.C., Atkins, J.P., Burdon, D., Degraer, S., Dentinho, T.P., Derous, S., Holm, P., Horton, T., Van Ierland, E., Marboe, A.H., Starkey, D.J., Townsend, M., Zarzycki, T., 2007. Identification, definition and quantification of goods and services provided by marine biodiversity: implications for the ecosystem approach. Mar. Pollut. Bull. 54 (3), 253–265.

Beaumont, N.J., Austen, M.C., Mangi, S.C., Townsend, M., 2008. Economic valuation for the conservation of marine biodiversity. Mar. Pollut. Bull. 56, 386–396.

Birol, E., Karousakis, K., Koundouri, P., 2006. Using economic valuation techniques to inform water resources management: a survey and critical appraisal of available techniques and an application. Sci. Total Environ. 365, 105–122.

Cooper, K., Burdon, D., Atkins, J., Weiss, L., Somerfield, P., Elliott, M., Turner, K., Ware, S., Vivian, C., 2010. Seabed Restoration Following Marine Aggregate Dredging: Do the Benefits Justify the Costs? MEPF/09/P115 Marine Aggregate Levy Sustainability Fund (MALSF).

de Groot, R.S., Wilson, M.A., Boumans, R.M.J., 2002. A typology for the classification, description and valuation of ecosystem functions, goods and services. Ecol. Econ. 41, 393–408.

Defra, 2007. An Introductory Guide to Valuing Ecosystem Services. Defra, London. http://www.defra.gov.uk/environment/policy/natural-environ/using/value.htm.

Derous, S., Agardy, T., Hillewaert, H., Hostens, K., Jamieson, G., Lieberknecht, L., Mees, J., Moulaert, I., Olenin, S., Paelinckx, D., Rabaut, M., Rachor, E., Roff, J., Stienen, E.W.M., van der Wal, J.T., Van Lancker, V., Verfaillie, E., Vincx, M., Weslawski, J.M., Degraer, S., 2007a. A concept for biological valuation in the marine environment. Oceanologia 49 (1), 99–128.

Derous, S., Austen, M., Claus, S., Daan, N., Dauvin, J.-C., Deneudt, K., Depestele, J., Desroy, N., Heessen, H., Hostens, K., Marboe, A.H., Lescrauwaet, A.-K., Moreno, M.P., Moulaert, I., Paelinckx, D., Rabaut, M., Rees, H., Ressurreicao, A., Roff, J., Santos, P.T., Speybroeck, J., Stienen, E.W.M., Tatarek, A., Hofstede, R.T., Vincx, M., Zarzycki, T., Degraer, S., 2007b. Building on the concept of marine biological valuation with respect to translating it to a practical protocol: viewpoints derived from a joint ENCORA-MARBEF initiative. Oceanologia 49 (4), 579–586. https://bibliotekanauki.pl/articles/48603.

Pascual, M., Borja, A., Franco, J., Burdon, D., Atkins, J.P., Elliott, M., 2012. What are the costs and benefits of biodiversity recovery in a highly polluted estuary? Water Res. 46, 205–217.

Pol, P.O., 2019. Introduction to Environmental Economics. The James Hutton Institute, Aberdeen.

Potschin, M., Haines-Young, R., Fish, R., Turner, R.K., 2015. Routledge Handbook of Ecosystem Services. Routledge, Abingdon, ISBN: 978-1-138-02508-0, p. 629.

Turner, R.K., Schaafsma, M., 2015. Coastal Zones Ecosystem Services: From Science to Values and Decision Making. Springer Link, ISBN: 978-3-319-17214-9, p. 240.

Websites

Value Transfer Guidelines: http://www.defra.gov.uk/environment/policy/natural-environ/using/valuation/index.htm.

HM Treasury Green Book: http://www.hm-treasury.gov.uk/green_book_guidance_environment.htm.

What nature can do for you: a practical introduction to making the most of natural services, assets and resources in policy and decision making: http://www.defra.gov.uk/environment/policy/natural-environ/using/index.htm.

Linked

Sustainable environmental management; Ecosystem services and societal benefits; Environmental and operational managers; Sustainable solutions; Integrated management of catchment, transitional and coastal waters

Governance and management: 107. Habitats and species legislation—Example of the EU Directive

The coastal industry environmental manager will be challenged with implementing several major pieces of legislation (typified by the Integrated Pollution Prevention and Control legislation) relating to the adverse effects of operational and production methods including the consequences of occupying habitats and discharging materials to those habitats and ecosystems (environmental protection legislation), and the protection of local and regional biodiversity from adverse industrial consequences (habitats and species legislation). It is illustrative to give more details of these legal environmental instruments. Although European legislation is given here, the environmental governance in other parts of the world shows many similarities (see Elliott et al., 2022).

The EC adopted the Council Directive 92/43/EEC on the Conservation of natural habitats and of wild fauna and flora (commonly known as the "Habitats & Species Directive" or, potentially misleadingly, as the "Habitats Directive") in 1992. The directive is the means by which the European Union meets its obligations under the Bern Convention. The main aim of the Habitats Directive is to promote the maintenance of biodiversity by requiring Member States to take measures to maintain or restore natural habitats and wild species listed in the Annexes to the Directive at a favorable conservation status, introducing robust protection for those habitats and species of European importance. This then becomes the central part of the European Union Biodiversity Strategy and then this aims to fulfill the global (UN) Biodiversity Strategy. In applying these measures, Member States are required to take account of economic, social, and cultural requirements, as well as regional and local characteristics.

The provisions of the directive require Member States to introduce a range of measures, including:

- Maintain or restore European protected habitats and species listed in the Annexes at a favorable conservation status as defined in Articles 1 and 2;
- Contribute to a coherent European ecological network of protected sites by designating Special Areas of Conservation (SACs) for habitats listed on Annex I and for species listed on Annex II. These measures are also to be applied to Special Protection Areas (SPAs) classified under Article 4 of the Birds Directive. Together SACs and SPAs make up the Natura 2000 network (Article 3);
- Ensure conservation measures are in place to appropriately manage SACs and ensure appropriate assessment of plans and projects likely to have a significant effect on the integrity of an SAC. Projects may still be permitted if there are no alternatives, and there are imperative reasons of overriding public interest (IROPI). In such cases, compensatory measures are necessary to ensure the overall coherence of the Natura 2000 network (Article 6);
- Member States shall also endeavor to encourage the management of features of the landscape that support the Natura 2000 network (Articles 3 and 10);

- Undertake surveillance of habitats and species (Article 11),
- Ensure strict protection of species listed on Annex IV (Article 12 for animals and Article 13 for plants).
- Report on the implementation of the directive every 6 years (Article 17), including assessment of the conservation status of species and habitats listed in the Annexes to the Directive.

The directive was amended in 1997 by a technical adaptation directive. The annexes were further amended by the Environment Chapter of the Treaty of Accession 2003 and in 2007 when Bulgaria and Romania joined the EU. Transposition to UK legislation was enacted by a series of regulations between 1994 and 2010 and, following Brexit, the United Kingdom continues to follow the main pillars of the directive albeit under its own laws and regulations.

The directive lists some 200 habitat types which are of European importance, of which 73 require priority attention with regard to their conservation (Priority habitats), as well as over 1000 species of plant and animal which require statutory protection. Only two Priority habitats are listed under the category "Open sea and tidal areas," viz. Posidonia beds and coastal lagoons.

References and further reading

Borja, A., Elliott, M., Basurko, O.C., Fernández Muerza, A., Micheli, F., Zimmermann, F., Knowlton, N., 2022. #OceanOptimism: balancing the narrative about the future of the ocean. Front. Mar. Sci. 9, 886027. https://doi.org/10.3389/fmars.2022.886027.

Elliott, M., Houde, E.D., Lamberth, S.J., Lonsdale, J.-A., Tweedley, J.R., 2022. Management of fishes and fisheries in estuaries. In: Whitfield, A.K., Able, K.W., Blaber, S.J.M., Elliott, M. (Eds.), Fish and Fisheries in Estuaries—A Global Perspective. John Wiley & Sons, Oxford, ISBN: 9781444336672, pp. 706–797 (Chapter 12).

European Commission, 1992. Habitats and Species Directive 92/43/EEC. http://eur-lex.europa.eu/LexUriServ/LexUriServ.do?uri=CONSLEG:1992L0043:20070101:EN:PDF.

Linked

Nature conservation designations, Nature conservation organizations; Governance of the coastal and marine environment

Governance and management: 108. Integrated management of catchment, transitional and coastal waters

The management and governance of a small coastal area cannot be separated from the wider estuaries, coastal and marine, and other transitional water habitats because of the connectivity and interactions between these habitats. Within the framework of environmental legislation, the coastal industry environmental manager will be required by the regulators to consider their industrial activities against a background of the receiving area and the adjacent habitats extending into estuaries and their catchments. In turn, the management of these habitats, areas, and ecosystems cannot be separated from the management of the catchment. As such, major pieces of legislation such as the EU Water Framework Directive (WFD) and the US Clean Waters Act (CWA) take a catchment-to-coast approach as does similar legislation in many other countries worldwide.

While the details later cover the WFD, it is of note that many of its concepts and approaches are similar to the CWA. The WFD covers all inland freshwaters, groundwater, transitional waters (including estuaries, lagoons, rias, fjords), and coastal waters to 1 nautical mile from national baselines (increased to 3 nautical miles in Scotland and for water chemical status), including those that are heavily modified or artificial. Waters are grouped into similar types according to their physical and chemical characteristics and are divided up into water bodies which represent the smallest operational management unit (Table 1).

The water bodies are then classified by the relevant competent authority (usually an environmental protection agency) according to biological, chemical, and physical criteria and within a prescribed timetable. In the WFD, the ecology of all waters is classified according to Biological Quality Elements—phytoplankton, macroalgae, other macrophytes such as seagrasses, benthic invertebrates and, in freshwater and transitional waters, fishes (Fig. 1). Results are presented on a five-point scale with pre-defined colors for mapping and reporting following monitoring (Fig. 2): bad, poor, moderate, good, and high status and on the basis of "one out all out" (OOAO, i.e., whichever criterion fails one quality measurement then that determines the classification). In cases where good or high status cannot be reached because of anthropogenic modifications (such as a barrier or shoreline defense) to the water body then the water body is designated as having a Good Ecological Potential; that is, if the cause of the modification was removed then the water body could reach Good Ecological Status (GEcS) (Borja and Elliott, 2007).

The key chemical contaminants, physical criteria, and particular biological quality elements are specifically listed in the WFD with some chemical standards nationally derived and others agreed, in this case, across European Member States. The first important objective of the directive is to work toward all surface waters and groundwaters achieving at least good status. Some older European directives were integrated

Table 1 EU Water Framework Directive (a) process, (b) classification and presentation of results.

(a)
1. Decide on water body (TraC)
2. Decide on reference conditions
3. Determine pressures
4. Sample Biological Quality Elements
5. Test for Good Ecological Status (GEcS) via a multimetric approach
6. Use one-out-all-out approach
7. If GEcS not met then carry out action
8. Take account of economic factors (Highly Modified Water Body)

(b)
- Establish monitoring systems to estimate biological element values;
- Consider ecological potential in heavily modified waters;
- Express results as Ecological Quality Ratios (EcoQR);
- EcoQR = (parameter value for test area)/(parameter value for reference conditions);
- EcoQR has values 0 (bad)–1 (high);
- Divide EcoQR for each surface water category into five numerical classes ranging from high to bad ecological status;
- Use intercalibration exercises to define the boundaries between the upper classes.

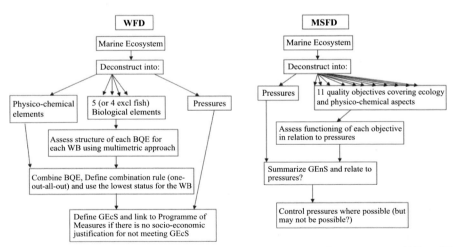

Fig. 1 Comparison of the status assessments in the WFD and the MSFD (see entry 109). *BQE*, biological quality elements; *GEcS*, Good Ecological Status; *GEnS*, Good Environmental Status; *WB*, water body.
From Borja, Á., Elliott, M., Carstensen, J., Heiskanen, A.-S., van de Bund, W., 2010. Marine management—towards an integrated implementation of the European Marine Strategy Framework and the Water Framework Directives. Mar. Pollut. Bull. 60, 2175–2186.

Fig. 2 Status classification colors for reporting on maps.

into the WFD, principally those relating to freshwater fish waters, shellfish growing and harvesting waters, groundwaters, and dangerous substances, and the original directives have been repealed. The EU Bathing Waters Directive remains and so areas where bathing is traditionally practiced will be treated separately within water bodies—any industry discharging sewage pathogens into designated bathing areas may therefore cause the areas to fail this directive on microbiological grounds.

A key aspect of the Water Framework Directive is the use of river basin districts to coordinate the approach over large areas, including the river system catchment. This requires countries to produce River Basin Management Plans and to adopt monitoring schemes and a Programme of Measures aimed at raising water bodies to acceptable standards (Good or High status). This also reinforces the need for cross-border integration where Member States share water bodies (termed an International River Basin District) such as in the case of the Solway in the United Kingdom, the Scheldt in Belgium and France, and the Rhine passing through several countries.

In essence, the determination of Good Ecological Status in the WFD using five biological quality elements and chemical elements is accompanied by monitoring and assessment with a Programme of Measures aimed at restoring any degraded areas (Fig. 2). This bears many similarities to the EU Marine Strategy Framework Directive in which the WFD Biological Quality and Chemical Elements are replaced by 11 Descriptors, where these are required to be in Good Environmental Status (GEnS), and where monitoring and Programme of Measures are also implemented by a country (see entry 109).

Wide stakeholder consultation is emphasized in implementing the WFD and the measurement tools used for classification are intercalibrated between Member States to ensure they are producing comparable results. Waters must not be allowed to deteriorate and all waters must be improved to at least good status, although there is a derogation clause allowing for consideration of excessive cost preventing good status being reached and the economic importance of the use of the water bodies. In many cases, the Environmental Protection Agency of a country acts as the competent authority for this legislation but it has to work closely with a range of other stakeholders.

In the case of the WFD, the progress of improvement is timetabled through 6-year cycles through stages of monitoring sites, assessment of results and classification, identifying pressures, drawing up improvement plans, and implementing improvement action (via the Programme of Measures). It is emphasized that transparency is important and there will be open access to all the data. The European Environment Agency is tasked with integrating and mapping the results to help the European Commission audit progress.

References and further reading

Borja, A., Elliott, M., 2007. What does 'Good Ecological Potential' mean, within the European Water Framework Directive? Mar. Pollut. Bull. 54, 1559–1564.

Borja, Á., Elliott, M., Carstensen, J., Heiskanen, A.-S., van de Bund, W., 2010. Marine management—towards an integrated implementation of the European Marine Strategy Framework and the Water Framework Directives. Mar. Pollut. Bull. 60, 2175–2186.

Borja, A., Elliott, M., Henriksen, P., Marbà, N., 2013. Transitional and coastal waters ecological status assessment: advances and challenges resulting from implementing the European Water Framework Directive. Hydrobiologia 704 (1), 213–229.

Hering, D., Borja, A., Carstensen, J., Carvalho, L., Elliott, M., Feld, C.K., Heiskanen, A.-S., Johnson, R.K., Moe, J., Pont, D., Solheim, A.L., van de Bund, W., 2010. The European Water Framework Directive at the age of 10: a critical review of the achievements with recommendations for the future. Sci. Total Environ. 408 (19), 4007–4019.

Perez-Dominguez, R., Maci, S., Courrat, A., Lepage, M., Borja, A., Uriarte, A., Neto, J.M., Cabral, H., Raykov, V.S., Franco, A., Alvarez, M.C., Elliott, M., 2012. Current developments on fish-based indices to assess ecological-quality status of estuaries and lagoons. Ecol. Indic. 23, 34–45.

Linked

Monitoring; Governance, sustainable management, integrated marine management

Governance and management: 109. Integrated marine management

As with freshwaters and transitional water bodies, there is a trend toward the integrated management of marine areas. This requires the activities, pressures, effects, and ecological components to be managed, not least within a system of maritime spatial planning. The European Union Marine Strategy Framework Directive (MSFD) and the Maritime Spatial Planning Directive (MSPD) create a coherent strategy for managing the features and activities in European marine areas; countries outside the EU such as Iceland and the United Kingdom have a corresponding system (e.g., the UK Marine Strategy) which reflects these two directives. Coastal industries therefore have to ensure they follow these procedures, especially in being the subject of Programme of Measures as a means of removing pressures and controlling activities such as discharges from coastal industries. As such, the implementation of these directives gives lessons for other coastal areas worldwide.

Marine governance has progressed from managing the environment sectorally, that is, by controlling each sector (fisheries, navigation, sea disposal, conservation, etc.) separately to adopting a holistic system in which all areas are managed in order to achieve the Ecosystem Approach. As a pre-eminent example of this, the European Marine Strategy Framework Directive (MSFD), adopted in July 2008, was proposed under the 6th Environment Action Plan to form a pillar of the European Commission Maritime Strategy (Fig. 1). It had the aim, firstly, to protect and preserve the marine environment, prevent its deterioration or, where practicable, restore marine ecosystems in areas where they have been adversely affected. Secondly, it aimed to prevent and reduce inputs in the marine environment, with a view to phasing out pollution in

Fig. 1 Recommendation of the way to develop a Marine Strategy (note that this sequence is then repeated at 6-year intervals) (https://ec.europa.eu/environment/marine/eu-coast-and-marine-policy/implementation/reports_en.htm).

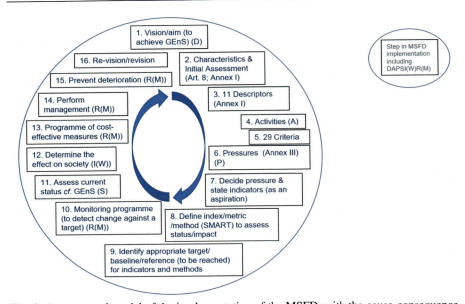

Fig. 2 A conceptual model of the implementation of the MSFD, with the cause-consequence-response model DAPSI(W)R(M) superimposed (see entry 93).
From Elliott, M., Borja, Á., McQuatters-Gollop, A., Mazik, K., Birchenough, S., Andersen, J. H., Painting, S., Peck, M., 2015. Force majeure: will climate change affect our ability to attain Good Environmental Status for marine biodiversity? Mar. Pollut. Bull. 95, 7–27. https://doi.org/10.1016/j.marpolbul.2015.03.015.

order to ensure that there are no significant impacts on or risks to marine biodiversity, marine ecosystems, human health, or legitimate uses of the sea.

The declared justification for the creation of this directive was to address gaps in previous disparate regulations and in the belief that European seas were deteriorating in quality through eutrophication, pollution, loss of biodiversity, and the detrimental effects of climate change. The overall strategy required by the directive has many similarities to that required by the Water Framework Directive in that it requires an initial assessment, development of a Good Environmental Status (GES) goal for each descriptor, establishment of targets, development of a monitoring program, and a Programme of Measures to be drawn up to achieve GES by 2020 (Fig. 2). GES for the MSFD is not to be confused with GES Good Ecological Status for the Water Framework Directive.

European seas are divided up into regional seas, all of which are also the subject of Regional Seas Conventions (RSC): the Baltic (HELCOM RSC), Mediterranean Sea (Barcelona Convention), the North-East Atlantic (OSPAR), and the Black Sea (Bucharest Convention). The aim for the MSFD was to work closely and be implemented with the RSC and so the RSCs have produced guidance and data relevant to the MSFD implementation. The RSC also produce Quality Status Reports showing the overall characteristics of the areas. As a further complication and area of overlap, the International Council for the Exploration of the Sea (ICES) also performs marine environmental characterization.

The sea area to be managed goes out to the limits of national jurisdiction (<200 nautical miles or the midline between countries) and where, for inshore waters, there is any overlap with the WFD (1 nm from the coast), then the MSFD only applies to aspects of GES that are not already addressed by the WFD. The MSFD aims to achieve Good Environmental Status based on a set of 11 descriptors against which are set targets and indicators to ensure these are achieved (originally by 2021 but this timetable has been delayed). The Qualitative Descriptors for determining GES are: (1) Biological diversity; (2) Non-indigenous species; (3) Commercial fish/shellfish; (4) Marine food webs; (5) Eutrophication; (6) Sea floor integrity; (7) Alteration of hydrographical conditions; (8) Contaminants; (9) Contaminants in seafood; (10) Marine litter; (11) Energy including noise.

The descriptors are linked and cover pressures and state changes to the system (as defined under the cause-consequence-response chain DAPSI(W)R(M)) (Figs. 2 and 3). They can be regarded as a hierarchy in that if the descriptors 1 and 4 are in good status then by definitions all the others could be regarded as being suitable to create a system in good status (shown by the linkages in Fig. 3). Fig. 4 describes the sequence of determining whether an area is in Good Environmental Status, GES (not to be confused with Good Ecological Status, GES in the WFD). Once less than Good Environmental Status has been determined by the monitoring program then actions under the Programme of Measures will be required to return an area to good status. Table 1 gives an example (for England) of the relevant competent authority and the available governance instruments required and available to effect that Programme of Measures.

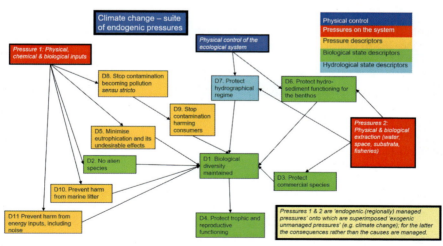

Fig. 3 The EU MSFD linkages between the 11 descriptors, whether they relate to state or pressures and their relationship to endogenic and exogenic pressures, including climate change. Modified from Borja, Á., Elliott, M., Carstensen, J., Heiskanen, A.-S., van de Bund W., 2010. Marine management—towards an integrated implementation of the European Marine Strategy Framework and the Water Framework Directives. Mar. Pollut. Bull. 60, 2175–2186.

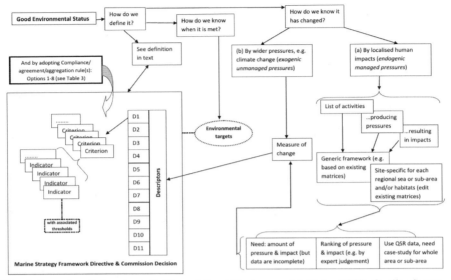

Fig. 4 Operationalizing the definition of Good Environmental Status (*QSR*, Quality Status Report).
From Borja, A., Elliott, M., Andersen, J.H., Cardoso, A.C., Carstensen, J., Ferreira, J.G., Heiskanen, A.-S., Marques, J.C., Neto, J., Teixeira, H., Uusitalo, L., Uyarra, M.C., Zampoukas, N., 2013. Good Environmental Status of marine ecosystems: what is it and how do we know when we have attained it? Mar. Pollut. Bull. 76, 16–27.

Table 1 The competent statutory body/bodies and the indicative relevant legal instruments to enable the GES to be determined: The context for England.

Descriptor	Competent statutory bodies	Indicative legal instruments
D1—Biological diversity (cetaceans, seals, birds, fish, pelagic habitats, and benthic habitats)	Natural England, Joint Nature Conservation Committee, IFCAs, Marine Management Organisation	Habitats Regulations, Marine and Coastal Access Act, NERC Act, Environment Act, Wildlife and Countryside Act (as amended by CROW), Conservation of Seals Act (as amended), Fisheries Act, COTES Regs.
D2—Non-indigenous species	Port Authorities	IMO Ballast Water Regulations, Invasive Alien Species Regs.
D3—Commercially exploited fish and shellfish	Marine Management Organisation, IFCA,	Fisheries Act, Fisheries Management Plans; Sea

Table 1 Continued

Descriptor	Competent statutory bodies	Indicative legal instruments
D4—Food webs (cetaceans, seals, birds, fish, and pelagic habitats)	selected Environment Agency, CEFAS, Sea Fish Industry Authority Natural England, Joint Nature Conservation Committee, Marine Management Organisation	Fisheries Act, Sea Fisheries (Wildlife Conservation) Act, Sea Fisheries (Shellfish) Act, Marine and Coastal Access Act Habitats Regulations, Marine and Coastal Access Act, NERC Act, Environment Act, Wildlife and Countryside Act (as amended by CROW), Conservation of Seals Act, Fisheries Act
D5—Eutrophication	Environment Agency, Marine Management Organisation	Diffuse and point source pollution controls, Water Environment Regs., Urban Waste Water Treatment Regs., Environment Act, Nitrate Pollution Prevention Regs., Bathing Waters Regs.
D6—Seafloor integrity (pelagic habitats and benthic habitats)	Natural England, Joint Nature Conservation Committee, Marine Management Organisation	Habitats Regulations, Fisheries Act, EIA, SEA
D7—Hydrographical conditions	CEFAS, BEIS, MoD, Crown Estate, Harbour Authorities	Marine and Coastal Access Act, EIA Regs., Planning Act, Habitat Regulations
D8—Contaminants	Environment Agency (for pipelines and catchment), MMO (for dredging and vessel disposal)	Water Framework Regs., Industrial Emissions Regs., Marine and Coastal Access Act, EIA legislation, Diffuse and point source pollution controls, Environmental Damage Regs.
D9—Contaminants in fish and other seafood for human consumption	Food Standards Agency, Environment Agency, CEFAS, Marine Management Organisation	EIA legislation, Water Environment Regs.

Continued

Table 1 Continued

Descriptor	Competent statutory bodies	Indicative legal instruments
D10—Litter	Local Authorities, Environment Agency, Marine Management Organisation	Environment Act, Merchant Shipping Regs.
D11—Introduction of energy, including underwater noise	Marine Management Organisation, BEIS	EIA legislation, Habitats Regs., Marine and Coastal Access Act

Whereas the MSFD is regarded as the quality assessment directive, its counterpoint the EU Maritime Spatial Planning Directive (MSPD) is regarded as the means of achieving an integrated planning for the seas and so is linked to the European Blue Economy strategy. The MSPD aim is to achieve "the sustainable growth of maritime and coastal economies and the sustainable use of marine and coastal resources." Maritime Spatial Planning (MSP) is about planning when and where human activities take place at sea—to ensure these are as efficient and sustainable as possible. This will then ensure a coordinated approach to MSP throughout Europe, enable the efficient and smooth application of MSP in cross-border marine areas, favor the development of maritime activities, and lead to the protection of the marine environment based on a common framework.

References and further reading

Borja, Á., Elliott, M., Carstensen, J., Heiskanen, A.-S., van de Bund, W., 2010. Marine management—towards an integrated implementation of the European Marine Strategy Framework and the Water Framework Directives. Mar. Pollut. Bull. 60, 2175–2186.

Borja, A., Elliott, M., Andersen, J.H., Cardoso, A.C., Carstensen, J., Ferreira, J.G., Heiskanen, A.-S., Marques, J.C., Neto, J., Teixeira, H., Uusitalo, L., Uyarra, M.C., Zampoukas, N., 2013. Good Environmental Status of marine ecosystems: what is it and how do we know when we have attained it? Mar. Pollut. Bull. 76, 16–27.

Cavallo, M., Elliott, M., Touza-Montero, J., Quintino, V., 2016. The ability of regional coordination and policy integration to produce coherent marine management: implementing the Marine Strategy Framework Directive in the North-East Atlantic. Mar. Policy 68, 108–116. https://doi.org/10.1016/j.marpol.2016.02.013.

Cavallo, M., Borja, A., Elliott, M., Quintino, V., Touza, J., 2019. Impediments to achieving integrated marine management across borders: the case of the EU Marine Strategy Framework Directive. Mar. Policy 103, 68–73. https://doi.org/10.1016/j.marpol.2019.02.033.

Elliott, M., Boyes, S.J., Barnard, S., Borja, Á., 2018. Using best expert judgement to harmonise marine environmental status assessment and maritime spatial planning. Mar. Pollut. Bull. 133, 367–377.

Linked

Governance; Nature conservation bodies; Marine nature conservation designations; Integrated management of catchment, transitional and coastal waters

Governance and management: 110. BAT, BATNEEC, best practice, and Integrated Pollution Prevention and Control (IPPC)

Environmental and sustainability managers of coastal industries will be charged with ensuring that their industry can reduce its environmental footprint in the most cost-effective and operationally effective manner. Best available technology (BAT) is a term applied within regulations on limiting pollutant discharges. Similar terms are best available techniques, best practicable means, or best practicable environmental option (BPEO). Interpretation of the terms has evolved over time since developing societal values and advancing techniques may change what is currently regarded as "reasonably achievable," "best practicable," and "best available." A literal understanding would identify with a "spare no expense" doctrine which prescribes the acquisition of the best state-of-the-art technology available, without regard for traditional cost-benefit analysis. In practical use, however, and because of commercial constraints especially in times of financial difficulties, the cost aspect is also taken into account.

Best practicable means was used for the first time in UK national primary legislation in section 5 of the Salmon Fishery Act 1861. The BAT concept was first time used in the 1992 OSPAR Convention for protection of the marine environment of North-East Atlantic for all types of industrial installations. In the United States, BAT or similar terminology is used in the Clean Air Act and Clean Water Act.

In the case of European Union directives, the term Best Available Techniques Not Entailing Excessive Costs (BATNEEC), sometimes referred to as best available technology, was introduced with the 1984 Air Framework Directive (AFD) and applies to air pollution emissions from large industrial installations. In 1996 the AFD was superseded by the Integrated Pollution Prevention and Control Directive (IPPC), 96/61/EC, which applies the framework concept of Best Available Techniques (BAT) to the integrated control of pollution to air, water, and soil. The concept is also part of the directive as modified in 2008 (2008/1/EC) and its successor directive (Industrial Emissions Directive) published in 2010. The IPPC framework is now taken to include emissions to all environments, air, land, and waters.

A series of reference documents have been issued by the European Joint Research Centre as part of the exchange of information carried out in the framework of Article 13(1) of the Industrial Emissions Directive (IED, 2010/75/EU). A series of Best Available Techniques (BAT) reference documents, the so-called BREFs, have been adopted under both the IPPC Directive (2008/1/EC) and the IED. As a consequence, relevant definitions are included in the IPPC directive:

- "best available techniques" shall mean the most effective and advanced stage in the development of activities and their methods of operation which indicate the practical suitability of particular techniques for providing in principle the basis for emission limit values designed

to prevent and, where that is not practicable, generally to reduce emissions and the impact on the environment as a whole.
- "techniques" shall include both the technology used and the way in which the installation is designed, built, maintained, operated, and decommissioned.
- "available" techniques shall mean those developed on a scale which allows implementation in the relevant industrial sector, under economically and technically viable conditions, taking into consideration the costs and advantages, whether or not the techniques are used or produced inside the Member State in question, as long as they are reasonably accessible to the operator.
- "best" shall mean most effective in achieving a high general level of protection of the environment as a whole.

In comparison to the EU IPPC Directive and the IED, in United States environmental law, the Clean Air Act requires that certain facilities employ Best Available Control Technology to control emissions. The Clean Water Act (CWA) requires national industrial wastewater discharge regulations (called "effluent guidelines"), which are based on BAT and several related standards. A further CWA provision for cooling water intake structures, for example, requires standards based upon "best technology available."

Best Practicable Environmental Option (BPEO)—this was first outlined as a concept in the United Kingdom by the Royal Commission on Environmental Pollution and subsequently used in Health and Safety assessments for industrial premises. A BPEO approach was described as meaning the reduction or modification of waste generation and the direction of what waste remained to be discharged to the environmental medium in which the least overall damage would be done. The definition developed was: "A BPEO is the outcome of a systematic consultative and decision-making procedure which emphasises the protection of the environment across land, air and water." The BPEO procedure establishes, for a given set of objectives, the option that provides the most benefit or least damage to the environment as a whole, at acceptable cost, in the long term as well as in the short term.

A best practice is a method or technique that has consistently shown results superior to those achieved with other means and that is used as a benchmark. In addition, a "best" practice can evolve to become better as improvements are discovered. Best practice is used to describe the process of developing and following a standard way of doing things that multiple organizations can use. Best practices are used to maintain quality as an alternative to mandatory legislated standards and can be based on self-assessment or benchmarking. Best practice is a feature of accredited management standards such as ISO 9000 and ISO 14001.

Integrated Pollution Prevention and Control

Of concern to coastal industries is IPPC which requires industrial activities with a high pollution potential to have an authorization which ensures that the industry must reduce or prevent any emissions causing pollution. The process covers all potentially polluting industries which must use all appropriate pollution-prevention measures, utilizing the best available techniques (which produce the least waste, use less hazardous substances, enable the substances generated to be recovered and recycled, etc.). This should prevent all large-scale pollution; prevent, recycle, or dispose of waste in the least polluting way possible; and use energy efficiently. It places a responsibility on the industry to ensure accident prevention and damage limitation and return sites to their original state following decommissioning.

The IPPC authorization (permit) will stipulate emission limit values for polluting substances; any soil, water, and air protection measures required; waste management measures; and measures to be taken in exceptional circumstances (leaks, malfunctions, temporary or permanent stoppages, etc.). The authorization will also stipulate the monitoring, recording, and reporting to be carried out. If carried out thoroughly, this should also minimize transboundary pollution.

The authorization will include a description of the installation and the nature and scale of its activities as well as its site conditions; the materials, substances, and energy used or generated; the sources of emissions from the installation; and the nature and quantities of foreseeable emissions into each medium, as well as their effects on the environment. Furthermore, the authorization gives the proposed technology and other techniques for preventing or reducing emissions from the installation, measures for the prevention and recovery of waste, measures planned to monitor emissions, and possible alternative solutions.

Finally, in common with the Environmental Impact Assessment guidelines and practices, IPPC should also allow wide consultation of stakeholders including the consideration of adjacent industries.

References and further reading

Bogan, C.E., English, M.J., 1994. Benchmarking for Best Practices: Winning Through Innovative Adaptation. McGraw-Hill, New York.
Council of the European Union, Council Directive 96/61/EC.
Sorrell, S., 2001. The Meaning of BATNEEC: Interpreting Excessive Costs in UK Industrial Pollution Regulation. http://www.sussex.ac.uk/Units/spru/publications/imprint/sewps/sewp61/sewp61.html.
IPPC Background—https://eur-lex.europa.eu/legal-content/EN/TXT/?uri=LEGISSUM:l28045.
US Clean Water Act of 1977.
US Clean Air Act of 1990.

Linked

Sustainable environmental management; Sustainable solutions

Governance and management: 111. Marine infrastructure environmental management

In addition to managing adverse environmental effects from emissions from the coastal industries, coastal industry managers will also be required to consider the management of infrastructure placements, especially where these impact on other coastal users. Procedures will vary with country but it is illustrative to consider the operation of harbors and their interference with adjacent industrial infrastructure as well as licensing to place any structure or infrastructure on the seabed. Given the extensive governance procedure and the long history of operations, the example of England will be given; however, other countries are expected to have similar practices.

The development and operation of harbors is controlled by the Harbours Act 1964, as amended. The Marine Management Organisation (MMO) has powers under the act to make various types of order which either create or amend legislation governing harbor authorities. Depending on who is applying and the reason for the application, the legislation provides for three different types of order: sections 14, 15, and 15A harbour revision orders, and section 16 harbour empowerment orders. The most common are harbour revision orders under section 14 and harbour empowerment orders under section 16.

Harbour revision orders (HROs) under section 14 are used to amend existing legislation. The authority engaged in improving, maintaining, or managing the harbor, or a person appearing to have a substantial interest or body representative of persons appearing to have such an interest, may apply for one of these orders. The order may be made to achieve all or any of the objects specified in Schedule 2 of the 1964 Act. These objects refer to the amendment or extension of existing powers.

In contrast, harbour empowerment orders (HEOs) under section 16 are to construct a new harbor or to gain powers to improve, maintain, or manage an existing harbor where no such powers already exist for a person to do so or to do so effectively. Those harbour orders that authorize a development are commonly called works orders. Where a harbour order is sought which would authorize a harbor authority to undertake works, a marine license may also be required. Marine licenses are required for permission for any material or structure being placed on the seabed, including the intertidal area, including long-sea outfalls and other discharge infrastructure.

Elsewhere in the United Kingdom, the Welsh Assembly Government has responsibility for marine licenses within the Welsh inshore region; however, the MMO has responsibility for harbour orders for all non-fishery harbors within Wales. In Scotland, a Harbour Empowerment or Revision Order is made as a Scottish Statutory Instrument under the 1964 Harbours Act by Scottish Ministers.

In the planning process, the MMO encourages anyone intending to apply for a harbour order for a project to consult them, other relevant authorities and organizations, and persons likely to be affected by the proposals, prior to making an application. A "project" in this context means the execution of construction works or other installations or schemes, as well as other interventions in the natural surroundings and landscape, including those involving the extraction of mineral resources. In licensing, the MMO will be stating whether or not the proposal will be subject to an environmental impact assessment, and as such, it details what the EIA should cover as well as indicating which other statutory consultees should be involved.

Certain projects will be of such a size and nature that they will require a statutory environmental impact assessment (EIA). Annex I to the EIA Directive (EC Directive 85/337/EEC2) lists categories of projects which always require an EIA. Annex II to the Directive lists categories of projects which may require an EIA if the project satisfies certain other criteria such as being of sufficient scale. Hence, a project will be treated as not falling within Annex II to the Directive unless: the area of the works comprised in the project exceeds 1 ha, any part of the works is to be carried out in a sensitive area, or the MMO determines that the project shall be treated as falling within that annex.

Following extensive stakeholder consultation (of both statutory and non-statutory consultees) and the application for construction, accompanied by an EIA, if there are contentious issues then the decision may be subject to a Public Inquiry with a legal hearing and written representations. Government-appointed inspectors will then oversee the Public Inquiry until a decision is reached; the decision then has legal force.

References and further reading

Harbour Orders—http://www.marinemanagement.org.uk/licensing/harbour.htm.
Marine Licensing—https://www.gov.uk/guidance/do-i-need-a-marine-licence.
Scottish Marine Licensing—https://www.gov.scot/collections/marine-licensing-and-consent/.
EIA Guidance—https://www.gov.uk/guidance/environmental-impact-assessment.

Linked

Governance to achieve marine environmental management; Environmental impact assessment; Habitats and Species directive; Marine licenses; Marine Strategy Framework directive; Sustainable solutions

Governance and Management: 112. Marine licenses and Maritime Spatial Planning

Coastal industries will often have a requirement to place structures in the sea or at least on the intertidal/shoreline area. Such structures may include pipelines, long-sea outfalls, barriers, breakwaters, erosion defenses, and cooling waters intakes and discharge culverts. Each of these is likely to both interfere with coastal physical and ecological structure and functioning and processes and occupy areas previously part of the marine environment. Hence the coastal industry will have to apply for permission to construct such a piece of infrastructure. As with licensing of discharges (see entry 101), the regulator does not have to show that the construction will have an adverse effect on the environment but the developer has to either show that it will not have an adverse effect or convince the planning authorities that the effects are negligible or acceptable and that there is no alternative.

Again, while each country will have its own processes and relevant legislation, the case of England is used here as an illustration. It is considered that the principles and processes of marine licenses will be the same in most countries. The marine licensing system in England comes under the Marine and Coastal Access Act 2009 (MCAA) which has been in force since 6 April 2011. This system consolidates and replaces some previous statutory controls, including:

- licenses under Part 2 of the Food and Environment Protection Act 1985
- consents under section 34 of the Coast Protection Act 1949
- consents under Paragraph 11 of Schedule 2 to the Telecommunications Act 1984
- licenses under the Environmental Impact Assessment and Natural Habitats (Extraction of Minerals by Marine Dredging) Regulations 2007.

The Marine Management Organisation (MMO), an Arms-length Body of the Department for Environment, Food and Rural Affairs (Defra), is responsible for most marine licensing in English inshore and offshore waters and for Welsh and Northern Ireland offshore waters. The Secretary of State is the licensing authority for oil and gas-related activities and administers marine licenses through the Department for Business, Energy and Industrial Strategy (BEIS).

What a marine license is for?

A marine license is required for many activities involving a deposit or removal of a substance or object below the mean high-water springs mark or in any tidal river to the extent of the tidal influence. This may be the construction of a port or wind farm, the construction of pipelines and culverts, the dredging of a channel, or the use of munitions. In some cases, a marine license will be required for activities outside UK waters, for example, where the activity takes place from a British vessel or where the vessel was loaded in UK waters. One marine license may cover a number of activities. A list of licensable activities and exemptions and further information on what may require a marine license is available on the MMO website.

In considering an application for a marine license, the MMO acts in accordance with government policy statements and guidance and with the principles of sustainable development, namely:

- achieving a sustainable marine economy
- ensuring a strong, healthy, and just society
- living within environmental limits
- promoting good governance
- using sound science responsibly.

The issuing of marine licenses will have repercussions for the marine spatial planning processes in which countries, especially those in Europe, are required to produce Maritime Spatial Plans. In the case of the United Kingdom, marine plans for many inshore and some offshore areas have already been published and other countries such as Belgium and Germany also have well-developed MSP processes. This in turn will provide a framework for marine licensing decisions. In addition, applications for a marine license will be subject to assessment under the prevailing nature conservation legislation and the effluent discharge and environmental quality legislation such as the Habitats and Species Directive and Regulations and the Water Framework Directive. As indicated before, they are also likely to be subject to an environmental impact assessment. In addition to a marine license, a project may require other consents either from the MMO or from other bodies, such as a harbour order or consent under Section 36 of the Electricity Act.

It is an offense to disturb or injure European protected species (EPS) under the Conservation of Habitats and Species Regulations, 2010. Seismic and other geophysical surveys could be associated with the disturbance or injury of marine EPS primarily though the emission of man-made sound. The MMO website provides more information on these surveys and the risk assessments needed to assess the likelihood of an offense being committed.

In Wales, licenses for inshore water under the MCAA are issued by Natural Resources Wales, while in Northern Ireland, the Department for Agriculture, Environment and Rural Affairs is the competent body. In Scotland, the primary legislation is the Marine (Scotland) Act 2010 with licenses issued by a dedicated Scottish Government Marine Scotland Licensing Operations team.

Marine Strategy and Maritime Spatial Planning

Marine infrastructure permissions also have to comply with legislation regarding marine and coastal environmental status assessments and protection and the sustainable planning of marine activities, again to ensure that ecological structure and functioning is not impaired. In European marine governance terms, this requires complying with the Marine Strategy framework Directive (MSFD) as the quality assessment directive, and the EU Maritime Spatial Planning Directive (MSPD) as the means of achieving an integrated planning for the seas.

The MSPD's aim is to achieve: "the sustainable growth of maritime and coastal economies and the sustainable use of marine and coastal resources." Maritime Spatial Planning (MSP) is about planning when and where human activities take place at sea—to ensure these are as efficient and sustainable as possible and hence it covers the positioning of marine infrastructure. It further aims to ensure a coordinated, effective, and efficient approach to MSP across different countries; to favor the development of maritime activities; and to lead to the protection of the marine environment based on a common framework.

In the case of the Marine Strategy (whether under European legislation such as the MSFD or nationally such as the UK Marine Strategy), permissions for marine infrastructure are given on the basis that the development does not adversely affect the environmental status (termed Good Environmental Status) and the 11 descriptors and their criteria, targets, and indicators upon which the determination of status is based. For example, those descriptors will be affected by placing a marine structure which creates litter, noise, and pollution; which interferes with hydrographic and sedimentary processes; which changes the biodiversity and prevents other exploitation such as fisheries.

References and further reading

Borja, A., Elliott, M., Andersen, J.H., Cardoso, A.C., Carstensen, J., Ferreira, J.G., Heiskanen, A.-S., Marques, J.C., Neto, J., Teixeira, H., Uusitalo, L., Uyarra, M.C., Zampoukas, N., 2013. Good Environmental Status of marine ecosystems: what is it and how do we know when we have attained it? Mar. Pollut. Bull. 76, 16–27.

Cavallo, M., Elliott, M., Touza-Montero, J., Quintino, V., 2016. The ability of regional coordination and policy integration to produce coherent marine management: implementing the Marine Strategy Framework Directive in the North-East Atlantic. Mar. Policy 68, 108–116. https://doi.org/10.1016/j.marpol.2016.02.013.

Elliott, M., Boyes, S.J., Barnard, S., Borja, Á., 2018. Using best expert judgement to harmonise marine environmental status assessment and maritime spatial planning. Mar. Pollut. Bull. 133, 367–377.

http://www.marinemanagement.org.uk/licensing/marine.htm.

http://www.defra.gov.uk/environment/marine/protect/licensing/.

Linked

Habitats and Species legislation; Marine infrastructure environmental management

Governance and Management: 113. Land-based infrastructural development—Planning process

In addition to placing structures in the sea or on the foreshore (intertidal area), coastal industries will require to build their plant on land but which still has an influence on the coastal environment. It is therefore important to consider the consenting and planning process for such infrastructure. While the precise regulations will differ with country, it is of value to give an example, in this case using UK planning processes.

Nationally significant infrastructural development projects (NSIPs) are usually large-scale developments such as new harbors, power generating stations (including wind farms), and electricity transmission lines, which require a type of consent known as "development consent" under procedures governed by the Planning Act, 2008 (as amended by the Localism Act, 2011). In England, government policy for infrastructural development is delivered through National Policy Statements (NPSs). The then Department of Energy and Climate Change published six NPSs for the energy sector (including one for nuclear power), in July 2011.

Prior to April 2012, NSIPs were dealt with by the Infrastructure Planning Commission (IPC). Under the Localism Act 2011, the IPC was abolished and its functions transferred to the Planning Inspectorate on April 1st 2012. The Planning Inspectorate now carries out certain functions related to national infrastructure planning on behalf of the Secretary of State. In England, the Planning Inspectorate examines applications for development consent from the energy, transport, waste, wastewater, and water sectors. In Wales, it examines applications for energy and harbor development, subject to detailed provisions in the act; other matters are for Welsh Ministers.

The planning process

Pre-application—Any developer wishing to construct a NSIP must first apply for development consent. The developer notifies the Planning Inspectorate that they intend to submit a future application. The relevant Secretary of State will appoint an "Examining Authority" to examine the application. The authority will be from the Planning Inspectorate and will be either a single inspector or a panel of three or more inspectors.

Before submitting an application, the developer is required to carry out extensive consultation on their proposals. Before formally consulting the local community, the developer must prepare a Statement of Community Consultation (SOCC), having first consulted relevant local authorities about what it should contain. The SOCC details the consultation the developer intends to undertake with the local community about the project. The developer is required to carry out the consultation as set out in the SOCC. While the first stage in the Environmental Impact Assessment process is screening, that is, the need to determine whether an EIA is required, this will always be the case for such large projects. The second stage in the EIA process will contain an environmental scoping report for wide consultation among stakeholders which eventually leads to describing all of the possible impacts of the project. This scoping will not yet have an allowance for appropriate design and mitigation but emphasizes the need to determine what aspects should be covered in the EIA. This then leads to environmental status reports and later, a draft environmental statement.

Acceptance—This stage begins when a developer submits a formal application for development consent to the Planning Inspectorate. The Secretary of State then has 28 days to consider whether or not the application meets the standards required to be formally accepted for examination. The Examining Authority will need to be satisfied that the consultation described before was carried out competently with a full audit trail available.

Pre-examination—If an application is formally accepted, upon notification, the applicant is required to publicize the fact that their application has been accepted and the arrangements for making representations about it. A period of at least 28 days is then provided in which representations by valid interested bodies can be made. Representations should relate specifically to the project and its impacts and must be registered with the Planning Inspectorate. All relevant representations will be published on the National Infrastructure portal. People who make valid registered representations become "interested parties" in the application and are invited to take part at relevant stages of the examination.

Once the period for making relevant representations is closed, the applicant must certify that they have complied with the statutory requirements for giving notice of the accepted application. The Secretary of State will then appoint the Examining Authority, who then has up to 21 days to review the application and all relevant representations and identify the principal issues for examination. The Examining Authority will then call all interested parties to a Preliminary Meeting to determine how the application will be examined. The authority will issue a procedural decision including the timetable for the various stages of the examination (including the periods allowed for submission of further written evidence and any hearings the authority has decided to hold).

Examination—The examination is a formal legal process, which must be completed within 6 months of the last day of the Preliminary Meeting. The examination includes the representations of all interested parties, evidence submitted, and answers provided. All interested parties are invited to provide further written evidence, known as a written representation. The Examining Authority is then likely to put written questions to the applicant and other interested parties to clarify points and seek further information. All interested parties have the opportunity to comment on the representations of others and to respond to any comments made on their representations. Deadlines for such dialogues will be included in the examination timetable. While the examination is mainly a written process, hearings may be held in particular circumstances. Information about hearings will be provided in the procedural decision following the Preliminary Meeting.

The Examining Authority's recommendation and the Secretary of State's decision—The authority must prepare a report and recommendation on the application to the relevant Secretary of State, within 3 months of the end of the examination.

The Secretary of State has a further 3 months to make the decision to grant or refuse development consent. All interested parties are notified and both the decision and the reasons for making it will be published on the National Infrastructural portal.

The Development Consent Order (DCO) is the legal order providing consent for the project and means that a range of other consents, such as planning permission and listed building consent, will not be required. A DCO can include provisions authorizing the compulsory acquisition of land or of interests in or rights over land which is the subject of an application.

Postdecision—Once a decision has been issued on behalf of the Secretary of State, there is a 6-week period in which the decision may be challenged in the High Court. This process of legal challenge is known as Judicial Review.

References and further reading

http://infrastructure.planningportal.gov.uk/application-process/the-process/.
http://infrastructure.planningportal.gov.uk/help/glossary-of-terms/.
http://www.planninghelp.org.uk/planning-explained/planning-for-major-infrastructure-projects.

Linked

Governance to achieve marine environmental management; Marine licenses and maritime spatial planning; Sustainable solutions

Index

Note: Page numbers followed by *f* indicate figures and *t* indicate tables.

A

Acoustic fish deterrents (AFDs), 153, 171–173
Activity-footprint, 110–112
Act of Parliament, 306–307
Actor maps, 8
Acute toxicity, 141
Air Framework Directive (AFD), 354–355
Alien and invasive species (AIS), 1–2
Analytical quality control/quality assurance (AQC/QA), 265–266
Antifouling procedures, 159
Argument mapping, 5–7, 5*f*

B

Barnacle fouling, 184–185, 185*f*
Bayesian Belief Network Modeling, 8–9
bCisive argument map, 6, 6*f*
Bedforms, 30
Before-after-control-impact—paired series (BACI-PS), 127
Best available technologies not entailing excessive costs (BATNEEC), 315, 354–356
Best available technology (BAT), 150–151, 171–172, 354–356
Best Practicable Environmental Option (BPEO), 315, 355
Bio-Acoustic Fish Fence (BAFF), 171–172
BioBullets, 201–202
Biocides, 210–212
Biodiversity Net Gain (BNG), 98
Biofouling, 210
  antifouling measures, 194–197
  BioBullets, 201–202
  chlorination control, 198–200
  colonization process, in sea/estuarine water, 181
  fouling organisms
    biology of, 192–193
    peak times of settlement, 190–191
  heat treatment, 203–204
  methods of control
    heat treatment/thermoshock, 194–195
    macrofouling, manual cleaning for, 194–197
    nonoxidizing biocides, 196–197, 196–197*t*
    oxidizing biocides, 196–197, 196*t*
    physical, 194
    sponge rubber balls, 195
  microbial and macrobial fouling, 182–187
  microbially influenced corrosion (MIC), 205–207
  planktonic organisms, settlement of, 188–189
Biological quality elements (BQE), 156
Bow-tie analysis, 316–317
Breaching regulations, 324–325
Bromate, 228
Bromine chemistry, for seawater, 216*f*, 217
Brundtland Report, 333

C

Carrying capacity, 33–35
Catch per unit effort (cpue), 86–87
Causal-loop diagrams, 8
Cause-consequence-response frameworks, 129, 290–293
Changing salinity, 261–262
Chemicals
  biocides, 210–212
  chlorination chemistry, 209
    free oxidant chemistry, 213–214
    terminology, 213–214
  complex industrial plants, 209
  continuous and pulse dosing, 223–225
  corrosion chemicals, 209
  corrosion control, oxygen scavengers, 235–239
  discharge permits, 209
  discharges, 229–231

Chemicals *(Continued)*
    priority hazardous substances, 229–230
    electro-chlorination plants (ECPs), 218–222
    microbial pathogens, chemical interactions, 232–234
    nonoxidizing residuals, 226–228
    process chemicals, 209
    toxicological assessments, 209
Chemical stressors, 51–53
Chlorination, 198–200
    chemicals, 209
    free oxidant chemistry, 213–214
    terminology, 213–214
Chlorine, 210–211, 211*f*
Chlorine by-products (CBPs), 226
Chlorine decay, 213
Chlorine demand, 213
Chlorine-produced oxidants (CPO), 214, 259
Chronic toxicity, 142
Clean Water Act (CWA), 355
Coastal degradation, 56–57
Coastal plant environmental monitoring, 272–274
Cohesive sediment, 29
Colonization, nonnative organisms, 49–50
Combined cycle gas turbine (CCGT) station, 151
Combined heat and power (CHP) schemes, 151
Compliance Classification Scheme (CCS), 324, 325*t*
Compliance monitoring, 275–276
Computer-aided argument mapping (CAAM), 5–6
Continuous dosing, 223–225
Convention for Biological Diversity (CBD), 326–328
Cooling water (CW)
    coastal power plants, 149
    direct cooling mechanisms, 149–152
    discharge guidelines, 156–157
    thermal power stations, 149
    water abstraction, 153–155
Cooling water system (CWS), 182–183, 185
Corrosion chemicals, 209
Corrosion control, oxygen scavengers, 235–239

D

"Damager debt principle", 3
DAPSI(W)R(M) framework, 291–292*f*, 293
Debris fouling, 186*f*
Development Consent Order (DCO), 362–364
Direct cooling mechanisms, 149–152
Direct toxicity assessment (DTA), 252, 323
Discharge plumes
    effluent discharge characteristics, 242
    impingement and entrainment (physical), 249–250
    plume characteristics and behavior, 243–246
    regulatory mixing zone, 251–253
    salinity tolerances of organisms, 261–262
    scouring of seabed, 247–248
    temperature thresholds, spawning times, 257–258
    thermal plume constituents (excluding heat), 259–260
    thermal tolerances of organisms, 254–256
        environmental stressor, tolerance levels to, 254
        temperature tolerances, 255–256
    wastewater discharges, 241
D:river-pressure-state-impact-response (DPSIR) framework, 290–293

E

Ecoengineering, 64–66, 313
Ecohydrology, 64–66
Ecologically/Biologically Sensitive Areas (EBSAs), 313
Ecosystem
    resilience, resistance and recovery, 67–70
    restoration, 58–63
Ecosystem Approach, 326, 327*t*
Ecotoxicology assessment, 141–144
Effects-footprint, 110–112
Effluent guidelines, 355
Electro-chlorination plants (ECPs), 198, 210, 218–222
English Environment Agency, 176–177
Entrained organisms (biota), 178–180
Environment Act 2012, 307
Environment Agency (EA), 324
Environmental assessment paradox, 130

Environmental Impact Assessment Directive (EIA) 85/337/EEC (updated to 2014/52/EU), 98–100
Environmental impact assessment (EIA), coastal industries, 3, 312, 358, 363
  activity-footprint, 110–112
  appropriate assessment, 139–140
  biological and ecosystem health, 106–109
  climate change mitigation, 98
  conservation features, 98–100
  decision-making process, 98
  ecotoxicology assessment, 141–144
  effects-footprint, 110–112
  environmental degradation, 98
  Environmental Statement (ES), 98
  eutrophication and organic wastes, 123–126
  hazards and risk, 113–117
  indicators, management, 129–135, 131–134*t*
  industrial plant and marine system, interactions of, 118–119
  marine processes and human impacts, 102–105
  objectives, 129–135
  pressures-footprint, 110–112
  public consultation, 98
  response-footprint, 110–112
  scoping, potential impacts, 98
  screening process, 98
  significant effect, determination of, 127–128
  standards, 129–135
  structure of, 98, 99*f*
  temporal and spatial physical scales, 136–138
  water and substratum quality considerations, 120–122
Environmental non-governmental organizations (eNGOs), 319–320
Environmental Performance Evaluation (EPE), 315
Environmental Protection Agency (EPA), 324
Environmental Quality Objectives/Standards (EQO/EQS) approach, 142, 321–322
Environmental Statement (ES), 98
Environmental stressor, 254
Epiflora, 38–39

Epiphytes, 38–39
Equivalent adult value (EAV), 167–170
Erosion corrosion, 236, 237*f*
Estuaries, 11–15
Estuarine degradation, 56–57
Estuarine quality paradox, 107
Eulerian method, 19
European Environment Information and Observation Network (EIONET), 33
European Marine Sites (EMS), 313
European protected species (EPS), 360
European Union Marine Strategy Framework Directive (MSFD), 347–349, 352
European Urban Waste-water Treatment Directive (UWWTD), 124
Eutrophication, 123–126
EU Water Framework Directive (WFD), 142, 343–345
Exogenic unmanaged pressure, 71
Extracellular polymeric substances (xPS), 182

F

Fisheries Act 2020, 307
Fisheries (management) terms, 93
  catch per unit effort (cpue), 86–87
  conservation species, 77
  fishing mortality (F), 84
  maximum sustainable yield (MSY), 88
  pawning-stock biomass (SSB/B), 80
  precautionary principle and approach, 78–79
  recruitment, 82–83
  reference points, 91
  total allowable catch (TAC), 90
Fish recovery and return (FRR) system, 51, 153, 173–175
"Fit-for-purpose" science, 3, 3*t*
Free chlorine/free available chlorine (FAC), 213
Free oxidant chemistry, 213–214
  nitrogenous compounds, 214–217
Free oxidants (FO), 198–200, 211

G

Galvanic corrosion, 236
Geoengineering, 313

Good Ecological Status (GEcS), 343–345, 344f
Good Environmental Status (GES), 348–349, 352
Good Laboratory Practice Monitoring Authority (GLPMA), 275–276
Governance and management
  administrative and regulatory aspects, 309–310
  amelioration, 329–331
  Best Available Techniques Not Entailing Excessive Costs (BATNEEC), 354–356
  Best available technology (BAT), 354–356
  best practice, 354–356
  breaching regulations, 324–325
  coastal and marine environment, 303–307
  compensation, 329–331
  Convention for Biological Diversity (CBD), 326–328
  discharge consent, permit, license and authorizations, 321–323
  ecological, socioecological and socioeconomic valuation, 334–340
  ecosystem approach, 302
  enhancement, 329–331
  environmental and operational managers, 315–317
  habitats and species legislation, EU Directive example, 341–342
  horizontal integration, 301
  Integrated Coastal Zone Management (ICZM), 308
  integrated marine management, 347–353
  Integrated Pollution Prevention and Control (IPPC), 355–356
  land-based infrastructural development, 362–364
    planning process, 362–364
  marine infrastructure environmental management, 357–358
  marine licenses and maritime spatial planning, 359–361
  mitigation, 329–331
  nature conservation bodies, 318–320
    environmental non-governmental organizations (eNGOs), 319–320
  nature conservation designations, 311–313
  UK local designations, examples of, 313–314
  sustainable solutions, 332–333
  UN Conference on Environment and Development (UNCED), 326–328
  vertical integration, 301

H

Habitat risk assessment, 139–140
Habitats and species legislation, EU Directive example, 341–342
Habitats regulations assessment (HRA), 100, 139–140, 312
Haloacetic acids, 227
Haloacetonitriles (HANs), 227
Halophenols, 227
Harbour empowerment orders (HEOs), 357
Harbour revision orders (HROs), 357
Heat treatment, 194–195, 203–204
Holoplankton, 40–41
"Horrendograms", 9
Humber Estuary, 11, 12f
Hydrodynamics, 19–20
Hypobromous acid (HOBr), 210, 216–217
Hypohalous acids, 217

I

Ichthyoplankton, 43–44
Imperative reasons of overriding public interest (IROPI), 139–140, 312, 318
Impingement and entrainment
  acoustic fish deterrent (AFD), 171–173
  antifouling procedures, 159
  biota, impingement of, 162–164
  cooling water system, 159
  entrained organisms (biota), 178–180
  equivalent adult value (EAV), 167–170
  fish recovery and return (FRR) system, 174–175
  impinged material, disposal of, 176–177
  planktonic organisms, 159
  seawater, 159
  source and receiving waters, 160–161
  stationary trawlers, 165
Infauna, 38–39
Infrastructure Planning Commission (IPC), 362

Index

Integrated Coastal Zone Management (ICZM), 308
Integrated marine management, 347–353
Integrated Pollution Prevention and Control (IPPC), 315–316, 355–356
Intermittently closed and open lakes and lagoons (ICOLLs), 16
Invasive alien species (IAS), 45–48
ISO 31000 risk management standard, 316–317
Issue maps, 8

K

KEMA biofouling monitor, 273f
k-epsilon models, 25

L

Lagoons, 16–18
Lagrangian method, 19
Long-term effects, 3–4
Lower incipient lethal temperature (LILT), 255
Low-velocity side-entry (LVSE), 153, 154f

M

Macrobial fouling, 182–187
Macrofauna, 36
Macroflora, 37
Macrofouling, 194–197
Management
  cause-consequence-response frameworks (DPSIR, DAPSI(W)R(M) approaches), 290–293
  ecosystem services, societal goods and benefits, 294–299
  endogenic managed pressures (EnMP), 286–289
  exogenic unmanaged pressures (ExUP), 286–289
  governance, 278
  marine, coastal and estuarine activities, 283–285
  PEST/PESTLE model, 277, 279
  socio-ecological system, 278
  sustainable development, 277
  sustainable environmental management, 279–282

Marine and Coastal Access Act 2009 (MCAA), 359
Marine degradation, 56–57
Marine license, 313–314
Marine Management Organisation (MMO), 357, 359–360
Marine Strategy and Maritime Spatial Planning, 360–361
Marine Strategy Framework Directive (MSFD), 229
Maritime Spatial Planning Directive (MSPD), 347–348, 352, 361
Maximum sustainable yield (MSY), 88
Mechanical stressors, 51–53
"Medium-term effects", 3–4
Megafauna, 36
Megaflora, 37
Meiofauna, 36
Meroplankton, 41
Metamorphosis, 183
Microbial fouling, 182–187
Microbial influenced corrosion (MIC), 205–207, 236
Microbial pathogens, chemical interactions, 232–234
Microbiological contaminants, 121
Microfauna, 36
Microflora, 36
Mind-maps, 8–9
Minimum biologically acceptable limit (MBAL), 91
Monitoring
  coastal plant environmental monitoring, 272–274
  compliance monitoring, 275–276
  survey, experimental and modeling approaches, 269–271
  types and definitions, management framework, 265–268
Mussel fouling, 183

N

Nationally significant infrastructural development projects (NSIPs), 362
National Policy Statements (NPSs), 362
National Pollutant Discharge Elimination System (NPDES), 252, 322–323, 322t

Nature conservation bodies, 318–320
Navier-Stokes equations, 25
"Need-to-know" aspects, 3
Nested modeling, 25
"Nice-to-know" aspects, 3
Nonindigenous species (NIS), 1–2, 45–48
Nonnative organisms, 49–50
Nonoxidizing biocides, 196–197, 196–197*t*
Nonoxidizing residuals, 226–228

O

"One out all out" (OOAO), 343–345
Operational Risk Assessment (OpRA) risk rating, 324
Organic wastes, 123–126
Oxidation-reduction/redox potential (ORP), 213
Oxidizing biocides, 196–197, 196*t*
Oxygen scavengers, corrosion control, 235–239

P

Pawning-stock biomass (SSB/B), 80
Pitting corrosion, 236
Planktonic bacteria, 182
Planktonic organisms, 159
  settlement of, 188–189
Planning Inspectorate, 362–363
Political, economic, social and technological (PEST) constraints, 277, 279
"Polluter-pays principle", 3
Pressures-footprint, 110–112
Pulse-Chlorination, 199
Pulse dosing, 223–225

Q

Quaternary ammonium compounds (QACs), 201, 211

R

Recruitment-overfishing, 83
Regionally Important Geological Sites (RIGS), 313
Regional Seas Conventions (RSC), 349
Response-footprint, 110–112
Rubber "Taprogge" balls, 184–185

S

Science, coastal industries
  argument mapping, 5–7, 5*f*
  baseline and reference conditions, 54–55
  carrying capacity of systems, 33–35
  chemical stressors, 51–53
  climate change and effects, 2, 71–73
  coastal degradation, causes and solutions, 56–57
  colonization, nonnative organisms, 49–50
  ecohydrology and ecoengineering, 64–66
  ecosystem resilience, resistance and recovery, 67–70
  ecosystem restoration, 58–63
  epiflora, 38–39
  epiphytes, 38–39
  estuaries, 11–15
  estuarine degradation, causes and solutions, 56–57
  geomorphological terms, 31–32
  hydrodynamics, 19–20
  ichthyoplankton, 43–44
  infauna, 38–39
  invasive alien species (IAS), 45–48
  lagoons, 16–18
  macrofauna, 36
  macroflora, 37
  marine degradation, causes and solutions, 56–57
  mechanical stressors, 51–53
  megafauna, 36
  megaflora, 37
  meiofauna, 36
  microfauna, 36
  microflora, 36
  nonindigenous species, 45–48
  nonnative species, 45–48
  phenology, 74–75
  physical terms
    currents and waves, 23–24
    numerical hydrodynamic modeling, 25–28
    sea level and tides, 21–22
  plankton, 40–42
  power generating stations, 1
  risk assessment and management process, 1
  role and adequacy, 3–4
  sediment, 29–30

Index 371

sessile and mobile epifauna, 38–39
systems analysis and conceptual models, 8–10
thermal stressors, 51–53
Self-monitoring, 276
Sessile and mobile epifauna, 38–39
Settled biofouling, 185
Short-term effects, 3–4
Site of Nature Conservation Interest (SNCI), 313
Sites of special scientific interest (SSSIs), 161
Socioecological system, 294–299
Sound projector array (SPA) system, 171–173
"So what?", 3–4
Special areas of conservation (SACs), 161, 341
Special protection areas (SPAs), 161, 341
Sponge rubber balls, 195
Statement of Community Consultation (SOCC), 363
Stationary trawlers, 165
Statutory Instruments, 306–307
Stoke's Law, 29
Strategic Environmental Assessment Directive (SEA) 2001/42/EC, 98–100
Stress corrosion cracking (SCC), 236
Stress-subsidy continuum, 254
Sustainable environmental management, 279–282
Swell waves, 23

T

Taprogge system, 195
Temperature thresholds, spawning times, 257–258
Temperature tolerances, 255–256

Thermal plume constituents (excluding heat), 259–260
Thermal power stations, 149
Thermal stressors, 51–53
Thermoshock, 194–195, 203
Total allowable catch (TAC), 78–79, 90
Total available chlorine (TAC), 213
Total economic value (TEV), 295–299
Total residual chlorine (TRC), 213
Total residual oxidants (TRO), 198–200, 211, 214, 259
Total system value (TSV), 295–299
Transitional and coastal (TraC) waters, 156
Trihalomethanes (THMs), 226
Turbidity maximum zone (TMZ), 11

U

UK Good Laboratory Practice (GLP), 275–276
Uniform Emission Standards (UES) approach, 142, 321–323
Upper incipient lethal temperature (UILT), 255
US Clean Waters Act (CWA), 343

W

Wastewater discharges, 241
Water abstraction, 153–155
Water Framework Directive (WFD), 229
Weathering process, 182
"What if?", 3–4
Wind waves, 23

Y

Yield-per-recruit curve, 84, 84*f*

Printed in the United States
by Baker & Taylor Publisher Services